# Reviewing

# The Living Environment: BIOLOGY

## SECOND EDITION

## *With Sample Examinations*

**Rick Hallman**

**Principal, Retired**

Benjamin N. Cardozo High School
Bayside, New York

AMSCO SCHOOL PUBLICATIONS, INC.,
a division of Perfection Learning®

The publisher wishes to acknowledge the helpful contributions of the following reviewers and consultants in the preparation of this book:

**Annick C. O'Reilly**
*Biology / Living Environment Teacher*
*Bainbridge-Guilford High School*
*Bainbridge, New York*

**Lauren Bolonda**
*Biology Teacher*
*Canandaigua High School*
*Canandaigua, New York*

**William Augrom**
*Biology Teacher*
*Perry High School*
*Perry, New York*

**Howard B. Fisher**
*Biology Teacher, Retired*
*Binghamton High School*
*Binghamton, New York*

Cover Photo: American Black Bear Cub (*Ursus americanus*), © John Conrad/CORBIS
Cover Design: Meghan J. Shupe

Composition: Brad Walrod/Kenoza Type
Text Design: Howard Petlack/A Good Thing, Inc.
Artwork: Hadel Studio

Please visit our websites at: ***www.amscopub.com*** and ***www.perfectionlearning.com***

When ordering this book, please specify:
*either* **1346501** *or* Reviewing The Living Environment: Biology, Second Edition

ISBN13: 978-1-56765-945-0



Printed in the United States of America

14 15 16 17 18 19 EBM 23 22 21 20 19 18

# CONTENTS

# STUDENT'S STUDY GUIDE

This guide explains how to use this review book to prepare for the New York State Regents High School Examination in Living Environment: Biology.

## I. WHAT IS THE REGENTS LIVING ENVIRONMENT EXAMINATION?

All students in New York State must pass one Regents examination in science in order to earn a high school diploma. The Regents Living Environment Examination can be used to meet this requirement. Students will usually take this examination after completing one year of high school biology.

The Regents Living Environment Examination is a three-hour test that consists of a variety of multiple-choice and open-ended questions that test comprehension of core topics (Parts A, B, and C) and mandated laboratory activities (Part D). Part A consists of multiple-choice questions. Parts B and C consist of graphing, short-answer, and short essay questions. Part D consists of multiple-choice and open-ended questions, which are based on three of the four required lab activities that are completed during the school year. Sample examinations are included at the back of this book. The format for the Living Environment Examination is as follows:

### THREE-HOUR TEST:

**Part A** Multiple-choice questions on key ideas and details of biology (approximately 35 questions, 35% of the exam grade)

**Part B** Multiple-choice, short-answer, and short essay questions based on key ideas and details of biology, laboratory skills, and experimental problems (approximately 30% of the exam grade)

**Part C** Short essay questions based on real-world problems and situations that require the application of knowledge of biology to the problem or situation (approximately 20% of the exam grade)

**Part D** Multiple-choice, short-answer, and short essay questions based on the required laboratory activities; completed as part of the three-hour test (approximately 15% of the exam grade)

## II. WHAT SHOULD I STUDY?

The following are the areas of study in which you will need to be prepared in order to be successful on the Regents Living Environment Examination.

### A. Key Ideas and Details in Biology

It will be necessary to know the key ideas and details in high school biology in order to answer the Regents examination questions. The main concepts are covered by the seven themes of this review book:

- Scientific and Laboratory Procedures
- Energy, Matter, and Organization
- Maintaining a Dynamic Equilibrium
- Reproduction, Growth, and Development
- Genetics and Molecular Biology
- Evolution: Change Over Time
- Interaction and Interdependence

The content in this review book and the questions in each chapter will help you study these ideas. Any words in **boldface** may appear on the examination without any explanation. You must know the meaning of these words. Other words in the review book are *italicized* for emphasis if they are important, but they do not need to be memorized. *Chapter 25: Human Evolution* is an advanced topic that will NOT be tested on the Regents examination. Also, the details in the review book on each of the human organ systems (for example: digestive, circulatory, reproductive) are details that are taught before high school. They also will not be specifically tested on the Regents Living Environment Examination; but knowledge of these systems is important as background information for concepts that may be tested.

### B. The Nature of Scientific Inquiry

What science is and how scientific research is conducted will also be tested. This topic is explained in more detail in the new *Chapter 1: Scientific Inquiry and Biology* and in Appendix A of this review book.

It is important to realize that science is both a body of knowledge and a way of learning about how the world works. Scientific explanations are developed using observations and experimental evidence to add to what has been learned previously. All scientific explanations may change if better ones are found. Good science involves asking questions, observing, experimenting, finding evidence, collecting and organizing data, drawing valid conclusions, and discussing results with other scientists.

In order to make useful observations to test proposed explanations, a research plan involving well-designed experiments must be prepared. Included in creating a research plan are the following:

- researching background information on the major concepts being investigated;

- ♦ developing proposals, which include hypotheses, to test the explanations; i.e., predict what should be observed under specific conditions if the explanation is true;
- ♦ designing experiments that include techniques to avoid errors or false conclusions; these techniques may include repeated trials, large sample size, objective data-collection techniques, and use of a control.

Observations made through research need to be analyzed to determine if they support or contradict the proposed explanations. Methods used to analyze data include the use of diagrams, tables, charts, and graphs as well as the use of statistical techniques. These statistical techniques should help determine the closeness of the predicted results in the hypothesis to the actual results, in order to reach a conclusion as to whether the explanation on which the prediction was based is supported.

It is important in science to make the results of an investigation public. It is assumed that different scientists will arrive at the same explanation if they are able to analyze similar evidence. Therefore, scientists tell each other how they did their experiments in order to allow others to repeat the investigations.

## C. Laboratory Skills

As a biology student, you need to be able to successfully conduct laboratory investigations. This topic is covered in more detail in the new *Chapter 2: Laboratory Techniques and Biology* and in Appendices B and C of this review book. Skills needed in the laboratory that may be tested on the Living Environment Regents Examination are listed below:

- ♦ Follows safety rules in the laboratory
- ♦ Selects and uses correct instruments
    - ✔ uses graduated cylinders to measure volume
    - ✔ uses metric ruler to measure length
    - ✔ uses thermometer to measure temperature
    - ✔ uses triple-beam or electronic balance to measure mass
- ♦ Uses a compound microscope/stereoscope effectively to see specimens clearly, using different magnifications
    - ✔ identifies and compares parts of a variety of cells
    - ✔ compares relative sizes of cells and organelles
    - ✔ prepares wet-mount slides and uses appropriate staining techniques
- ♦ Designs and uses dichotomous keys to identify specimens
- ♦ Makes observations of biological processes
- ♦ Dissects plant and/or animal specimens to expose and identify internal structures
- ♦ Follows directions to correctly use and interpret chemical indicators
- ♦ Uses chromatography and/or electrophoresis to separate molecules

- Designs and carries out a controlled, scientific experiment based on biological processes
- States an appropriate hypothesis
- Differentiates between independent and dependent variables
- Identifies the control group and/or controlled variables
- Collects, organizes, and analyzes data, using a computer and/or other laboratory equipment
- Organizes data through the use of data tables and graphs
- Analyzes results from observations/expressed data
- Formulates an appropriate conclusion or generalization from the results of an experiment
- Recognizes assumptions and limitations of the experiment

## III. HOW SHOULD I STUDY?

### A. Basic Strategies for Study

- Make a regular time and place for study at home, and plan a daily schedule. As a guideline, on a regular school night, 30 minutes *per subject* in high school should be spent on homework and study. Success will come—if you work hard!
- Prepare your own notes of important ideas and terms as you are reading. This helps to keep your mind concentrated on the subject while you are reading.
- Write your own questions about what you read, close the book, and then try to answer your own questions. After completing this task, open the book again, and review and correct your answers.
- Study when you are feeling alert and try to avoid distractions (such as the Internet, telephone, radio, and television).
- Take five-minute breaks every hour to help you stay alert and retain what you have studied.

### B. Test-Taking Strategies

*Multiple-Choice Questions*

For multiple-choice questions, **read** the statement or question carefully. If a diagram is included, study the diagram as well. Do not look at the answers until you have looked carefully at the questions first. Reread the question. It is helpful to underline key words as you read. Now look at each of the answers. Study all of the answers. Do not stop when you think you have found the right one! Look at all the answers. Be aware that you are looking for the word or expression that **best** completes the statement or answers the question. There will be only one answer that is the best choice.

As you study the answers for a multiple-choice question, eliminate the answers that you know are not correct. It is helpful to think about why answers are wrong as you eliminate them. This will assist you in selecting an answer. Refer back to the statement or question as often as necessary as you study the answers. After eliminating any answers you know are wrong, choose wisely, or in other words, guess from the remaining answers. **Be certain to make a selection. Do not leave a blank.** You will not be penalized for entering a wrong answer. Therefore, any attempt at an answer is better than none at all.

If you have additional time at the end of the test, and you are reviewing your answers for multiple-choice questions, be very careful before making any changes. After following the thinking process described above, the answer you choose the first time is more likely to be the correct one than another choice made later. In other words, be careful not to "second-guess" yourself.

### EXAMPLE:

The energy found in ATP molecules synthesized in animal cells comes directly from

**1** sunlight
**2** organic molecules
**3** minerals
**4** inorganic molecules

### ANALYSIS:

Although "energy" and "ATP" may seem to be the key terms in the question, you should realize that they describe the same thing: ATP *is* a form of chemical energy. There is something else important in the question—the word "animal." Now we know that the question is about energy and animals. At this point, we look at the answers. We recognize that Answer 1—"sunlight"—is wrong because sunlight provides plants with energy, not animals. Also, when we look back at the question, we realize from this wrong answer that another key word in the question is "directly." Animals get energy indirectly from sunlight when they eat plants (or other animals), not directly. Answer 2—"organic molecules"—describes substances from other living things, i.e., organic. This fits what we think would answer the question. We continue by looking at Answer 3—"minerals." Minerals are from the earth and do not contain stored energy. The same is true for Answer 4—"inorganic molecules." These answers help make us sure that the word "organic" in Answer 2 makes that answer the correct choice.

This kind of thinking process should be done for each multiple-choice question. While this requires time, it is the only method that will allow you to make use of what you have learned during the year and avoid making incorrect selections.

*Reading-Comprehension Questions*

The Living Environment test may have questions based on one or more reading passages. You will usually be required to prepare written answers to questions based on the reading passages. Prepare yourself for the reading passage by giving the passage you are about to read your **full attention**:

- make a decision that the reading passage **will** contain information that will interest you, even if the topic is something you have never paid any attention to before; comprehension of reading material is much higher when people are interested in what they are reading;
- stop worrying about other things in your life;

- ♦ stop looking around at other people in the room or at things happening outside the classroom;
- ♦ stop looking at your watch to check the time.

The following are suggested strategies for reading-comprehension questions. You may wish to use one or more of these strategies, based upon your personal preference:

- ♦ look over the questions before reading the passage; this may give you an idea of what to look for while you are reading; however once you start reading, do not stop for individual answers because this will break your concentration;
- ♦ as you read each paragraph, ask yourself what the paragraph was about; if a reading passage has more than one paragraph, this is because it has more than one important idea; each paragraph has one important idea;
- ♦ look carefully at the first sentence of the first paragraph; it usually states the topic for the reading passage;
- ♦ underline the sentence that states the main idea of the reading passage and of each paragraph;
- ♦ circle specific facts;
- ♦ reread the entire passage before working on the questions.

Work on one question for the reading passage at a time. Note the number of points indicated after the question for the answer. If the question is worth two or more points, be certain that you write a complete answer to earn these points. Underline the **key words** in the question and **include these words in your written answer**. This will make it much more likely that the teacher who reads your answer and marks your test will understand what you are trying to say. Also, write very neatly. If this is not possible, then print your answer. A teacher marking the test cannot give points for the answers that cannot be read!

*Data Interpretation and Graphing*

The Regents Living Environment Examination may include questions that require you to read about an experiment, study a data table showing the results of the experiment, graph the data, and answer one or more questions about the data. The data for a typical experiment may be presented in a Data Table with at least two columns. The first column is usually the independent variable. This is the condition or variable that the researcher set. The numbers in this column show each level at which the variable was set. The other columns are for the dependent variables. The values in these columns are the data that was measured or collected at each level of the independent variable. These values depend upon the independent variable.

As you read the passage about the experiment, study the data table. Use the reading passage to understand the data table. If you are required to construct a line graph for the data, follow the directions carefully. The title for a graph should refer to **both** the independent and the dependent variables. The most important part of a graph is the marking of an **appropriate scale**. The scale begins at the lower left, with the values increasing going to the right or going up. The scale for the independent variable is on the horizontal axis and the scale for the dependent variable is on the vertical axis. The scale must be prepared to include the lowest and the highest value for the variable. Each space on the graph paper must represent the same quantity. To plot the data points, refer to the data table. Each point on the line graph is located by two pieces of data from the data table, i.e., the value for each level of the independent variable with the corresponding value for the dependent variable.

A bar graph is used to compare data rather than to show relationships between variables. A bar graph has no scale along the horizontal axis. Rather, labels for each set of data are placed on the horizontal axis, with a scale for the data on the vertical axis.

## C. Final Test Preparation

During the review period at the end of the term, prior to taking the Regents examination, you should set aside time to take a complete practice exam all at once. Pretend that you are taking the real exam and do all the questions on the test. Record your answers according to the directions given in the practice test. Once you have taken the complete test, mark your answers. Note what topics were particularly difficult for you and return to the review book to study those topics further. Then take the other practice test provided in this review book. Finally, do not try to cram all the material the night before you take the actual test; it is important for you to be well rested when you take the Regents examination.

Remember: Hard work increases your abilities and will improve your test results.

# THEME I

## Scientific and Laboratory Procedures

# 1

# Scientific Inquiry and Biology

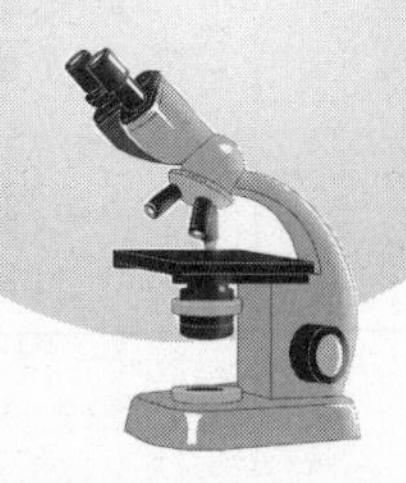

**Vocabulary**

| | | |
|---|---|---|
| biology | independent variable | scientific inquiry |
| dependent variable | observation | scientific method |
| experiment | science | theory |
| hypothesis | | |

### SCIENCE: LEARNING ABOUT THE WORLD

**Science** is a body of knowledge about our natural world. To make sense of all this knowledge, scientists have organized the information into major areas of understanding, such as chemistry, physics, biology, and earth science. **Biology** is the study of all living things, which includes humans. The study of biology also includes more specific areas, or *branches*, of study, such as anatomy, ecology, and genetics, among others. (See Table 1-1.)

Science is also a way of learning about the world. Everything that we now know in biology and every other area of science is a result of this way of learning. While it is important for us to understand how our bodies work and how all living things exist, it is also important for us to understand how we *learned* what we know and how we continue to expand our knowledge. In other words, we need to understand the nature of scientific *thinking*. This is called **scientific inquiry**, and it is where we begin our study of biology.

The thinking process known as science has been devised by people in order to learn about how the world works. Based on observations and evidence collected from experimentation, and using what is already known, scientific

**Table 1-1. The Main Branches of Biology**

| Branch | Specific Field of Study |
|---|---|
| Anatomy | Body structures of humans and animals |
| Botany | Structure and classification of plants |
| Biochemistry | Chemistry of life (internal) |
| Biotechnology | Genetic engineering and recombinant DNA technology |
| Cytology | Cell structure and function |
| Ecology | Interaction of organisms with their environment |
| Embryology | Early developmental stages in an organism's life |
| Genetics | How traits are expressed and passed from parent to offspring |
| Physiology | Internal functions of organisms |
| Marine biology | Living things in the ocean |
| Microbiology | Microscopic life-forms |
| Molecular biology | Molecular basis of inheritance and protein synthesis |
| Systematics/ Taxonomy | Classification and study of organisms |
| Zoology | Animals (behavior and classification) |

explanations are developed about the world. These explanations are always subject to change as new evidence and observations are presented. Scientific methods exist to constantly test and re-test what we know in relation to existing explanations. In this way, scientific knowledge advances toward a more complete understanding of the world around us. A **theory** is a general statement that is supported by many scientific observations and experiments; represents the most logical explanation of the evidence; and is generally accepted as fact (unless shown to be otherwise). Theories become stronger as more supporting evidence and experimental data are gathered.

Natural phenomena and events are understood by people on the basis of existing explanations. To develop scientific explanations, people use evidence that can be observed as well as information that is already known. Researchers think about these explanations by visualizing them; that is, they create mental pictures and develop mathematical models. It is also important to learn about the history of science and the particular individuals who have contributed to scientific understanding. While science provides knowledge about the world, it also challenges people to develop the values to use this knowledge ethically and effectively.

## SCIENCE BEGINS WITH QUESTIONS

To develop scientific ideas, a person needs to think, be curious, do library research, and discuss one's ideas with others, including experts. Scientific inquiry involves asking questions and locating information from a variety of sources. It also involves making wise judgments about how reliable and relevant the information is.

The questions that are studied often occur to us as we observe things around us. For example, you might have noticed that, over time, a plant near a window bends toward the light. (See Figure 1-1.) When you notice this, you are making an **observation**. Seeing many different plants do the same thing near many different windows is a discovery of a pattern or regularity. A scientific mind is very observant and also curious about why patterns exist. Science attempts to do much more than simply observe and describe. It seeks to find general explanations for why things are the way they are and why things behave the way they do. Thus, the question about plants might be posed as "What causes the tendency of plants to bend toward a light source?"

There are many questions in biology that

**Figure 1-1** You can observe that, over time, a plant near a window bends toward the sunlight.

science has studied and continues to study. Examples of these questions include:

- ♦ How is oxygen transported throughout the body by our blood?
- ♦ Why do birds migrate?
- ♦ Why are most plants green?
- ♦ Can skin tissue be grown in the laboratory and used to heal burn victims?
- ♦ How does DNA store genetic information?

For every biological question and problem studied by researchers, a similar approach is used to seek an answer. This approach, often called the *scientific method*, involves a way of conducting a scientific investigation, or **experiment**.

## THE SCIENTIFIC METHOD: STEPS IN AN INVESTIGATION

Experiments conducted by researchers demonstrate how scientific knowledge can be gained by using the **scientific method**, which is an organized approach to problem solving. The main steps in this procedure are:

- ♦ State the problem (in question form).
- ♦ Collect information about the problem.
- ♦ Form a hypothesis (a testable explanation).
- ♦ Design and conduct an experiment (use an experimental group with a variable; use a control group without the variable).
- ♦ Record observations and data.
- ♦ Check results; analyze data; redo experiment (if necessary).
- ♦ Draw your conclusions (accept or reject your hypothesis).
- ♦ Communicate your results.

### Making an Observation and Stating a Question

The work of a scientist always begins with asking a question. The question may result from making an observation or a series of observations. It may be a question that scientists have been working on for a long time, or it may be an entirely new question that no one has previously studied. In all cases, it is essential that the question be clearly defined and carefully stated. For example, "Why do plants like light?" is a poorly defined, vague question. "Why do plants bend toward the light?" is a well-defined, specific question. The next step in the scientific method will show why a scientific question must be clearly stated.

### Forming a Testable Hypothesis

Once a specific question has been asked, a scientist can think about possible answers to the question. A possible answer (actually an educated guess at this stage) is called a **hypothesis**. The scientist also begins to collect information that may help answer the question. The information may include results from other researchers' observations or experiments. Often, one question may lead to more than one hypothesis. For example, perhaps plants bend toward the light because of differences in temperature inside and outside the window. Perhaps the plant's roots push the stem toward the window. Or, perhaps structures inside the plant's stems are affected by the light, thus causing it to bend. Any one of these hypotheses might be the answer to the question. This is why the question must be very clear.

To conduct a scientific investigation, a researcher must have a testable hypothesis. There must be a possible answer that can be studied, tested, and shown to be true or false. Most often, once a hypothesis has been tested—and either supported or refuted—the result leads to another question, more hypotheses, and even more testing. This is the process of scientific experimentation.

### Designing a Scientific Experiment

A hypothesis is tested by conducting an experiment. The most common type of experiment in science is called a *controlled experiment*. It actually consists of at least two experiments carried out at the same time. In each one, all research conditions are kept the same except for one *factor* that is controlled by the researcher. This factor, called the **independent variable**, is the one difference (according to hypothesis) that might explain the observation and answer the question. The **dependent variable** is the change that occurs (such as bending toward the light) because of the independent variable. The setup that is exposed to the independent variable is called the *experimental group*; the setup that is not exposed to that factor is called the *control group*.

For example, to test our hypothesis that temperature differences cause the bending of plants toward light, we might design an experiment as follows: (a) Grow plants near a window in the winter, keeping the temperature inside the room as cold as the outside; and (b) grow plants near a window in the winter, making the temperature inside the room warmer than the outside. To test the hypothesis that structures in the stems cause the bending toward light, we might design this experiment: (a) Grow plants near the window with their stems covered by paper that blocks the light; and (b) grow plants near the window without any coverings on the stems.

## Conducting the Experiment

Once an experiment has been designed, the scientist will then carry out the experiment. To conduct an experiment, the researcher must consider the following:

- establish the number of organisms to be tested, called the *sample size* or *group size*; a larger group usually ensures a successful experiment, with more valid results, although the experiment may take more time and expense to complete;
- determine what measurements are to be made, what tools are to be used to make the measurements, and how often these data should be collected;
- maintain careful records of the data that have been collected and select the best methods to organize the data into charts and graphs; Figure 1-2 shows data organized in a table from an experiment designed to investigate the effect of temperature (the variable) on the action of an enzyme that breaks down starch into sugars; the same data also can be shown in the form of a *line graph*, as shown in Figure 1-3;
- finally, and perhaps most important, review and analyze the data to determine if the results of the experiment support or refute the hypothesis; and if it is possible to draw a conclusion from this experiment, determine what that conclusion could be.

| Temperature (°C) | Grams of starch hydrolyzed per minute |
|---|---|
| 0 | 0.0 |
| 10 | 0.2 |
| 20 | 0.4 |
| 30 | 0.8 |
| 40 | 1.0 |
| 50 | 0.3 |
| 60 | 0.2 |

**Figure 1-2** Data from an experiment about the effect of temperature on enzyme activity can be organized and displayed in a table.

## COMMUNICATING SCIENTIFIC RESULTS

No scientist should work alone. The process of scientific inquiry builds on previous knowledge and extends our understanding. For example, when scientists first investigated why plants bend toward light, they would have relied on information already known about the nature of light, the cellular structure of

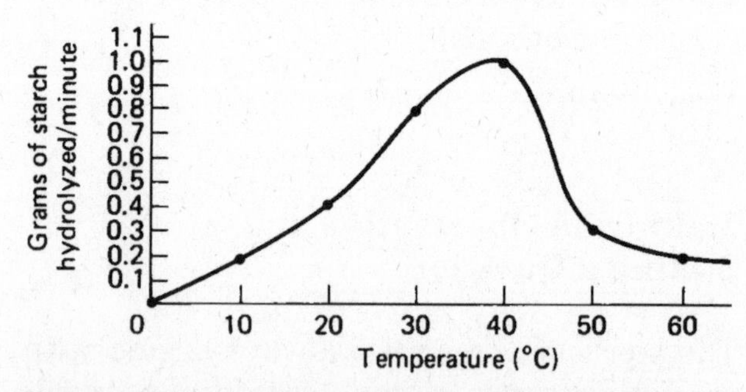


**Figure 1-3** Data from an experiment about the effect of temperature on enzyme activity can be organized in the form of a line graph, too.

plants, and how cells divide to make plants grow. Any new findings that scientists obtain from their experiments are shared with other scientists. For hundreds of years, this has been done through the publication of research reports. These reports are printed in scientific journals and sent around the world. Now, scientific findings are posted on the Internet as well, to be read by other researchers who work on similar topics. Experimental results are considered valid if other scientists can obtain the same results when they repeat the experiment under the same conditions.

It is through this community of scientific study that explanations about how the world works are constantly revised and expanded. As scientists study each other's work, new ideas develop. A conclusion that is based upon the results of an experiment is called an *inference*. In the plant investigation, the conclusion that light changes the growth of a plant stem is an example of an inference. In scientific inquiry, an inference always needs to be tested again (by conducting more experiments) to be confirmed. One way to do this is to make a new prediction from the inference. If the new prediction is tested and turns out to be true, it is strong evidence that the inference was valid. For example, based on the inference that light affects the growth of plant stems (causing them to bend), the prediction might be made that only plants with soft green stems can bend, whereas plants with hard wooden stems cannot. This would require a new experiment.

The study of science never ends because each new answer to a research problem leads to another question. This is *scientific inquiry*—the ongoing process of learning about the world around us.

## Chapter 1 Review

### *Part A—Multiple Choice*

**1.** Dr. Jones met a man who had been bitten by venomous snakes several times over many years, yet he was not sick. The doctor thinks that something produced in the man's blood protects him from harmful effects of the poison. This idea, or possible answer to a scientific question, is called a(n)

1 hypothesis
2 inference
3 observation
4 theory

**2.** A student placed slices of moist bread in a closed cupboard and noticed that mold grew faster on them than on slices of moist bread left out on a counter. The one difference in this experiment—the presence or absence of light—is called the

1 theory
2 hypothesis
3 dependent variable
4 independent variable

**3.** A logical explanation of natural phenomena that is supported by many scientific observations and experiments is called a(n)

1 hypothesis
2 inference
3 factor
4 theory

**4.** A student wants to test how much water is necessary to produce the most bread mold. She keeps one slice of bread dry while using varying amounts of water on other slices. The dry bread in this experiment is the

1 hypothesis
2 control
3 observation
4 theory

**5.** The number of organisms that are tested in an experiment is called the

1 variable size
2 controlled size
3 sample size
4 experimental factor

**6.** The scientific method is

1 a way of posing a research question only
2 used to organize data that is already known
3 an organized approach to problem solving
4 used by all scientists in an identical way

**7.** The best way to be sure that your experimental results are valid is to

1 ignore information from other sources
2 conduct your experiment one time only
3 use more than one variable in the experiment
4 test as large a sample size as possible

**8.** A researcher reviewing another scientist's experiment would most likely consider it valid if

1 the conclusions were not based on the observations
2 there was an experimental group but no control group
3 the sample size was extremely small
4 other researchers were able to duplicate the results

**9.** An experimental setup is shown in the diagram below. Which hypothesis would most likely be tested using this setup?

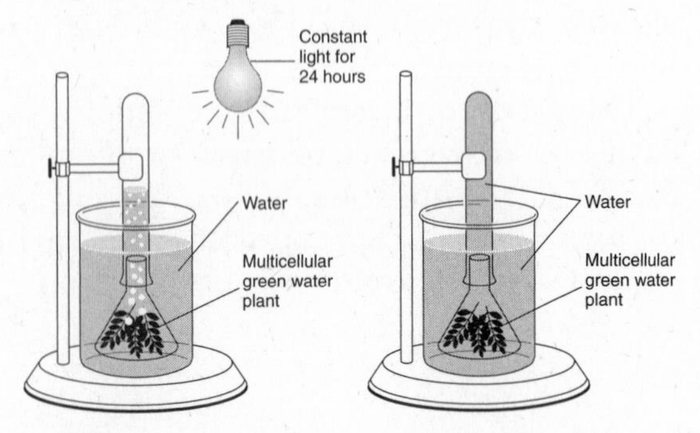


1 Water plants release a gas in the presence of light.
2 Roots of plants absorb minerals in the absence of light.
3 Green plants need light for cell division.
4 Plants grow best in the absence of light.

**10.** Which statement best describes a scientific theory?

1 It is a collection of data designed to provide support for a prediction.
2 It is an educated guess that can be tested by experimentation.
3 It is a scientific fact that no longer requires any evidence to support it.
4 It is a general statement that is supported by many scientific observations.

**11.** An experiment was carried out to determine which mouthwash was most effective against bacteria commonly found in the mouth. Four paper discs (A through D) were each dipped into a different brand of mouthwash. The discs were then placed onto the surface of a culture plate that contained food, moisture, and bacteria commonly found in the mouth. The following diagram shows the growth of bacteria on the plate after 24 hours. Which change in procedure would have improved the experiment?

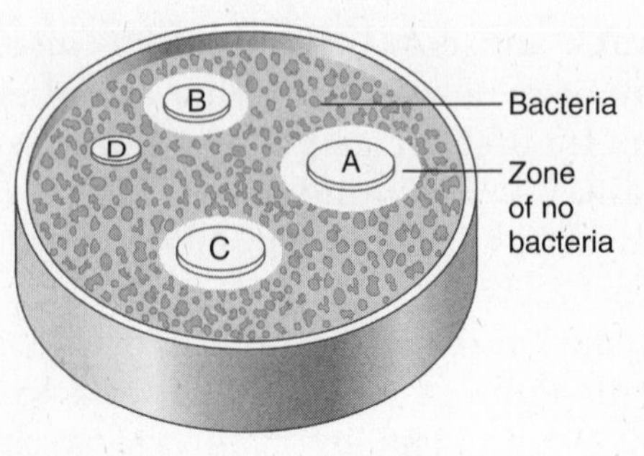


1 using a smaller plate with less food and moisture
2 using bacteria from many habitats other than the mouth
3 using the same size paper discs for each mouthwash
4 using the same type of mouthwash on each disc

**12.** Which source would provide the most reliable information for use in a research project investigating the effects of antibiotics on disease-causing bacteria?

1 the local news section of a newspaper from 1993
2 a news program on national television about antigens produced by various plants
3 a current professional science journal article on the control of pathogens
4 an article in a weekly news magazine about reproduction in pathogens

**13.** Researchers performing a well-designed experiment should base their conclusions on

1 the original hypothesis of the experiment
2 data from repeated trials of the experiment
3 a small sample size to ensure valid results of the experiment
4 results predicted before performing the experiment

**14.** A scientist is planning to carry out an experiment on the effect of heat on the function of a certain enzyme. Which would be the *least* appropriate first step to take?

1 doing research in a library or on the Internet
2 having discussions with other scientists
3 completing a data table of expected results
4 using what is already known about the enzyme

**15.** Which statement best describes the term *theory* as used when describing a scientific theory?

1 A theory is never revised even when new scientific evidence is presented.
2 A theory is an assumption made by scientists and implies a lack of certainty.
3 A theory is a scientific explanation that is supported by a variety of experimental data.
4 A theory is a hypothesis that has been supported by one experiment only.

**16.** The analysis of data gathered during a particular experiment is necessary to

1 formulate a hypothesis for that experiment
2 develop a research plan for that experiment
3 design a control group for that experiment
4 draw a valid conclusion for that experiment

**17.** When starch is digested, it is broken down into smaller units of sugar. The graph below represents data obtained from an experiment on starch digestion. Which statement best describes point *A* and point *B* on the graph?

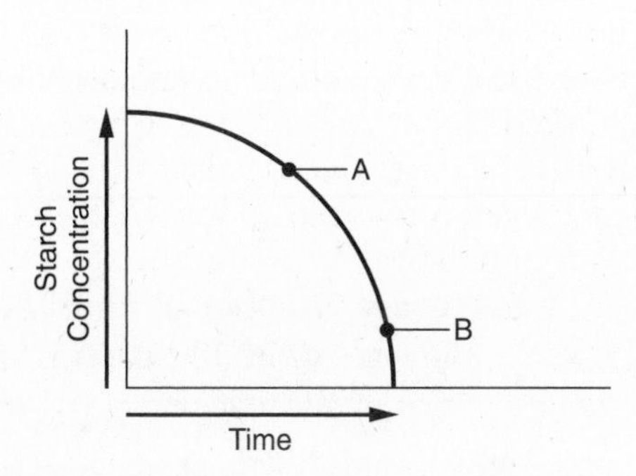


1 The concentration of sugars is greater at point *A* than it is at point *B*.
2 The concentration of sugars is greater at point *B* than it is at point *A*.
3 The starch concentration is the same at point *A* as it is at point *B*.
4 The starch concentration is greater at point *B* than it is at point *A*.

**18.** A great deal of information can now be obtained about the future health of people by examining the genetic makeup of their cells. There are concerns that this information could be used to deny an individual health insurance or employment. These concerns best illustrate that

1 scientific explanations depend upon evidence collected from a single source
2 scientific inquiry involves the collection of information from a large number of sources
3 acquiring too much knowledge in human genetics will discourage future research in that area
4 while science provides knowledge, values are essential to making ethical decisions when using this knowledge

**19.** A biologist reported success in breeding a tiger with a lion, producing healthy fertile offspring. Other biologists will accept this report as fact only if

1 research shows that other animals can be crossbred
2 the offspring are given a new scientific name
3 the biologist included a control in the experiment
4 they can repeat the experiment and get the same results

**20.** Scientists could consider any testable hypothesis valuable because it might

1 explain the results of their other experiments
2 lead to further research even if not supported by the evidence
3 require no further investigation if not supported by the data
4 explain their conclusion even if it is refuted by the experiment

### *Part B—Analysis and Open Ended*

**21.** Describe the basic steps of the scientific method in the sequence in which they should be used.

**22.** Define and distinguish between the terms *hypothesis* and *theory*.

**23.** Why is the development of a research plan necessary before testing a hypothesis?

**24.** Why is a clear statement of a hypothesis very important to the experiment that will follow?

**25.** Describe three ways in which a scientist can collect and organize data.

**26.** Why is it so difficult for the results of one experiment to become a theory?

**27.** Compounds containing phosphorus that are dumped into the environment can upset local ecosystems because phosphorus acts as a fertilizer. The following graph shows measurements of phosphorus concentrations at two sites taken during the month of June, from 1991 to 1997.

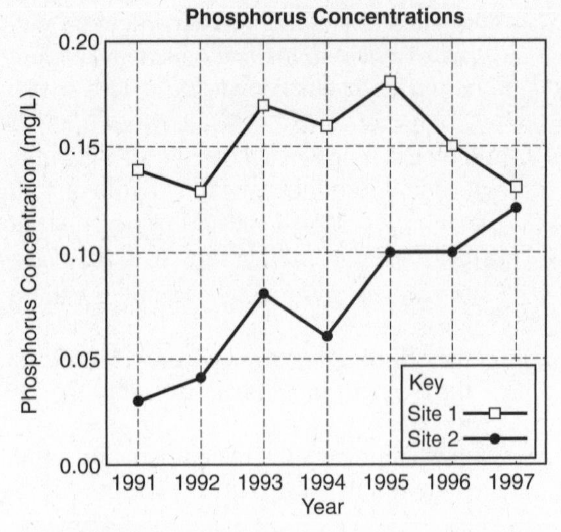


Which statement represents a valid conclusion based on information in the graph?

1 There was a decrease in the amount of phosphorus dumped at Site 2 from 1991 to 1997.
2 Pollution controls may have been put into effect at Site 1 starting in 1995.
3 There was most likely no vegetation present near Site 2 from 1993 to 1994.
4 There was a greater variation in phosphorus concentration at Site 1 than at Site 2.

**28.** The concentration of salt in water affects the hatching of brine shrimp eggs. Brine shrimp eggs will develop and hatch at room temperature in glass containers of salt solution. Describe a controlled experiment having three separate experimental groups that could be used to determine the best concentration of salt solution in which to hatch brine shrimp eggs. Your answer must include:

- a description of how the control group and each of the three experimental groups will be different;
- *two* conditions that must be kept constant in the control group and the experimental groups;
- what data that should be collected, and how often; and
- *one* sample of experimental results that would indicate the best concentration of salt solution in which to hatch brine shrimp eggs.

**29.** A student hypothesized that lettuce seeds would not sprout (germinate) unless they were placed in the dark. The student planted 10 lettuce seeds in soil that was kept in the dark and planted another 10 lettuce seeds in soil that was exposed to light. The data collected are shown in the following table.

| Experimental Conditions | Number of Seeds Germinated |
|---|---|
| Seeds in soil kept in the dark | 9 |
| Seeds in soil exposed to light | 8 |

One way to improve the reliability of these results would be to

1 conclude that darkness is necessary for lettuce seed germination
2 conclude that light is necessary for lettuce seed germination
3 revise the hypothesis
4 repeat the experiment

**30.** To determine which colors of light are best used by plants for photosynthesis, three types of underwater plants of similar mass were subjected to the same intensity of light of different colors for the same period of time. All other environmental conditions were kept the same. After 15 minutes of exposure to light, a video camera was used to record the number of gas bubbles each plant gave off (due to photosynthesis) in a 30-second time period. Each type of plant was tested six times. The average of the data for each plant type is shown in the table below.

**Average Number of Bubbles Given Off in 30 Seconds**

| Plant Type | Red Light | Yellow Light | Green Light | Blue Light |
|---|---|---|---|---|
| *Elodea* | 35 | 11 | 5 | 47 |
| *Potamogeton* | 48 | 8 | 2 | 63 |
| *Utricularia* | 28 | 9 | 6 | 39 |

Which statement is a valid conclusion based on the data?

1 Each plant carried on photosynthesis best in a different color of light.
2 Red light is better for carrying on photosynthesis than blue light is.
3 Each plant carried on photosynthesis best in red light and in blue light.
4 Water probably filters out the red light and the green light.

**31.** A certain plant that usually grows in slightly basic soil produces flowers with white petals. Sometimes the same plant species produces flowers with red petals. A company that sells the plant wants to know if soil pH affects the color of the petals in this plant. Design a controlled experiment to determine if soil pH affects petal color. In your experimental design be sure to:

- state the hypothesis to be tested in the experiment;
- state *one* way the control group will be treated differently from the experimental group;
- identify *two* factors that must be kept the same in both the control and the experimental group;
- state *one* piece of evidence that would support your hypothesis.

**32.** A researcher conducts an experiment on corn seedlings and uses the results to make predictions and conclusions about all plants. Explain the researcher's error in using the results in this way.

**33.** Why is it important for a scientist to include all the steps that he or she used to conduct an experiment when writing the research report for other scientists to read?

### *Part C—Reading Comprehension*

*Base your answers to questions 34 to 36 on the information below and on your knowledge of biology. Use one or more complete sentences to answer each question.*

The main purpose of science is to look at events, occurrences, and patterns in nature and develop explanations for them. These explanations can always be changed as new observations are made and new evidence is found. A possible explanation of a natural event or pattern is called a hypothesis. Charles Darwin, in his own words, showed why he was a true scientist:

"From my early youth I have had the strongest desire to understand or explain whatever I observed—that is, to group all facts under some general laws. I have steadily endeavored to keep my mind free, so as to give up any hypothesis, however much beloved (and I cannot resist forming one on every subject), as soon as facts are shown to be opposed to it. I followed a golden rule that whenever a published fact, a new observation or thought came across me, which was opposed to my general results, to make a memorandum of it without fail and at once. During some part of the day I wrote my Journal, and took much pains in describing carefully and vividly all that I had seen; and this was good practice . . . and this habit of mind was continued during the five years of the voyage. I feel sure that it was this training which has enabled me to do whatever I have done in science."

On the voyage of the *Beagle*, Darwin saw seeds that had washed ashore on a small island near Australia. He wondered whether seeds could travel long distances in the ocean and still be able to grow. Back in England, he enthusiastically filled his home with bottles of seawater. Darwin soaked many different kinds of seeds in the salty water, for short and long periods of time. He used a variety of crop seeds—such as cabbage seeds, lettuce seeds, and celery seeds—23 kinds in all. He then tried to grow them in soil. Darwin's experiments showed exactly what the scientific method is: State the problem; collect information; form a hypothesis; perform an experiment; record observations and data; draw a conclusion; and share your results.

**34.** When does a scientist find it necessary to change a hypothesis?

**35.** Why did Charles Darwin keep a journal?

**36.** Describe how Darwin followed each of the steps of the scientific method in his experiment with seeds and salt water.

# 2
# Laboratory Techniques and Biology

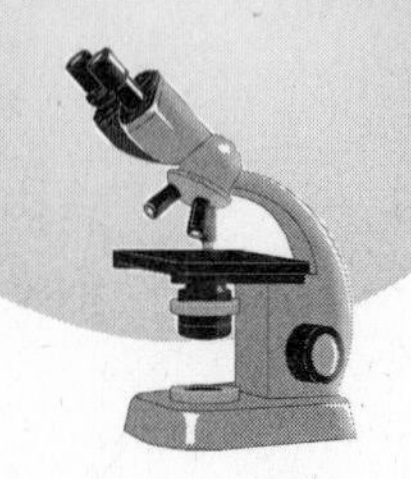

**Vocabulary**

| | | |
|---|---|---|
| centrifuge | microscope | technology |
| compound light microscope | statistical analysis | wet-mount slide |

## THE BIOLOGY LABORATORY: TOOLS AND DISCOVERIES

Scientific investigations are based on observations and measurements. Biologists also use a wide variety of tools and procedures to increase the range and accuracy of their natural senses. For example, we use eyeglasses in everyday life to aid our sense of sight. Similarly, a biologist uses microscopes, thermometers, weighing scales, and many other instruments to make detailed and precise observations and measurements. Careful attention and proper safety procedures are also essential in the laboratory at all times. Therefore, safety precautions related to laboratory tools will be examined in this chapter.

### Early Discoveries in Biology

The invention of the **microscope** had an enormous influence on the development of biology as a science. Anton van Leeuwenhoek, a Dutch lens maker who lived more than 300 years ago, is considered the first person to have built a simple microscope. With his invention, the first single-lens microscope, van Leeuwenhoek was able to magnify the image of tiny objects more than 270 times their normal size. This allowed him to see things that had never been seen before, such as bacteria, protozoa, yeast cells, and blood cells. (See Figure 2-1.) In 1665, Robert Hooke, an English physicist, made a compound light microscope

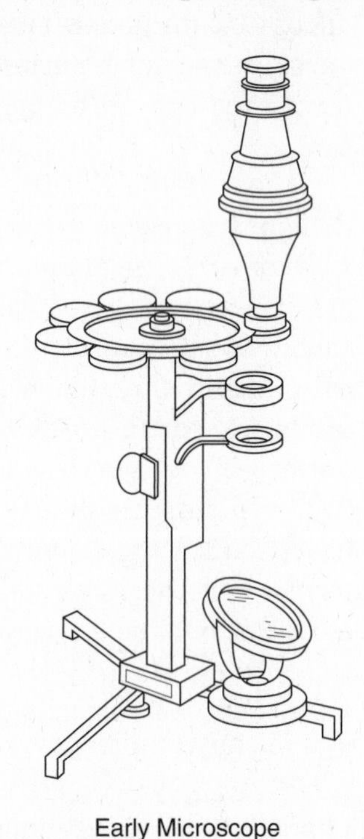

Early Microscope

**Figure 2-1** An early microscope, used to observe tiny organisms and cells.

using two lenses. With this device, he observed the inside structure of a thin piece of cork and called the spaces he saw "cells." In the 1800s, scientists in other countries used microscopes to learn much more about cells. Robert Brown in Scotland discovered the central structure in plant cells: the nucleus. Two German researchers—Matthias Schleiden studying plants and Theodor Schwann studying animals—concluded that the tissues of all living things are composed of cells. This is the basis for the *cell theory*, which states that all living things are made of cells (which come from other cells) and that the cell is the basic unit of structure and function of all living things.

The biological studies continued. In the mid-1800s, an Austrian monk, Gregor Mendel, investigated the simple garden pea plant to develop his fundamental principles of heredity. In 1859, English naturalist Charles Darwin published his theory of evolution by natural selection, based on years of observations and detailed plant and animal studies. In the early 1900s, Thomas Hunt Morgan studied the common fruit fly to learn about chromosomes and genes. Then, in 1953, scientists James Watson and Francis Crick, using precise measurements and molecular models, discovered the double-helix structure of DNA, allowing us to understand how this most important molecule serves as the basis for all life on Earth.

### Science and Technology

Science and technology affect the lives of people all over the world. As previously noted, *science* is a body of knowledge, discovered through the process of seeking answers to questions about our natural world. **Technology** is the process of using that scientific knowledge and other resources to develop new products and processes. These products and processes help solve problems and meet the needs of both the individual and society. While the emphasis in science is on gaining knowledge of the natural world, the emphasis in technology is on finding practical ways to apply that knowledge to solve problems. The fields of science and technology frequently help to advance each other. Scientific discoveries often lead to the development of even better technological devices, such as high-powered microscopes. These technologies may, in turn, lead to new discoveries or to a better understanding of scientific principles.

## MICROSCOPES AND MEASUREMENTS

### Using a Microscope

The **compound light microscope** uses two lens types, the *objective lens* and the *ocular lens*, to magnify the image of a specimen. The magnified image seen through the objective lenses is further enlarged by the ocular lens, or *eyepiece*. (See Figure 2-2.) When viewing a specimen, one should begin with the low-power objective, first using the coarse-adjustment knob to focus and then the fine-adjustment knob. Further study of specimen details can then be done by using the high-power objective and the fine-adjustment knob only. The area seen while looking through a microscope is called the *field of view*. The actual diameter of the field being viewed decreases as you switch from low power to high power. For example, if the magnification is 100X under low power and 400X under high power, then the field diameter

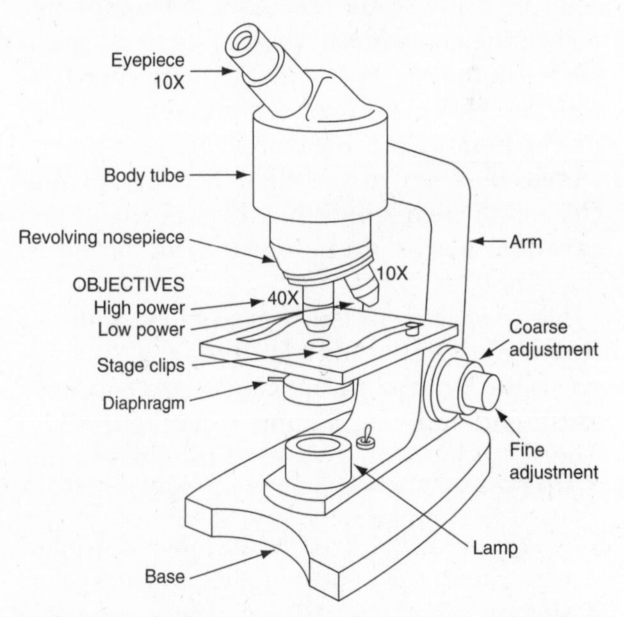


**Figure 2-2** Diagram of a compound light microscope, which uses two types of lenses.

**Table 2-1. Parts of the Compound Light Microscope and Their Functions**

| Part | Function |
|---|---|
| Base | Supports the microscope |
| Arm | Used to carry microscope; attaches to the base, stage, and body tube |
| Body tube | Holds the objective lens and eyepiece |
| Stage | Platform on which the glass slide with the specimen is placed (over the hole in the stage through which light passes) |
| Clips | Hold the slide in position on the stage |
| Nosepiece | Holds the objective lenses; rotates so that the different objective lenses can be moved in line with the specimen and eyepiece |
| Coarse adjustment | Larger knob used for rough-focusing with the low-power objective |
| Fine adjustment | Smaller knob used for focusing with the high-power objective and for final-focusing with the low-power objective |
| Lamp or mirror | Directs light to the specimen (on the stage) |
| Diaphragm | Controls the amount of light reaching the specimen |
| Objective lenses | Lenses (high-power and low-power) mounted on the nosepiece |
| Ocular lens | Lens at the top of the body tube; commonly called the eyepiece |

under high power will be one-fourth of what it is under low power. In this example, if four cells in a row could be observed under low power, then only one of those cells could be observed under high power, but with more detail. The field also appears dimmer under high power than under low power, because the objective is closer to the slide. But by opening the diaphragm (underneath the stage), you can allow more light in to reach the specimen. (See Table 2-1.)

Many specimens, such as tissue samples or tiny organisms, need to be studied with a microscope by first preparing a **wet-mount slide** and applying a stain. (See Figure 2-3.) The following steps are used to make a wet-mount slide:

1. Use a medicine dropper to place a drop of water onto the center of a clean slide.
2. Use the medicine dropper to place the specimen into the water on the slide.
3. Gently lower a coverslip (placing one edge down first, at a 45° angle) over the drop of water to avoid trapping air bubbles.
4. Add a drop of a stain such as iodine or methylene blue at one edge of the coverslip (to see more detail in the specimen). Draw the stain (dye) out with a piece of paper toweling.

Many cell parts and structures are not visible with a compound light microscope. University and professional laboratories use *electron microscopes* to magnify images more than 250,000 times. In place of light beams and optical lenses, the electron microscope has an electron beam and electromagnetic lenses. Beams of electrons passing through the specimen are focused on a television screen for viewing. This is called a *transmission electron microscope* (TEM). Another piece of equipment, just as advanced and expensive, is the *scanning electron microscope* (SEM), in which an electron beam is passed back and forth over the surface of a specimen, revealing very fine details about the surface structure of the object.

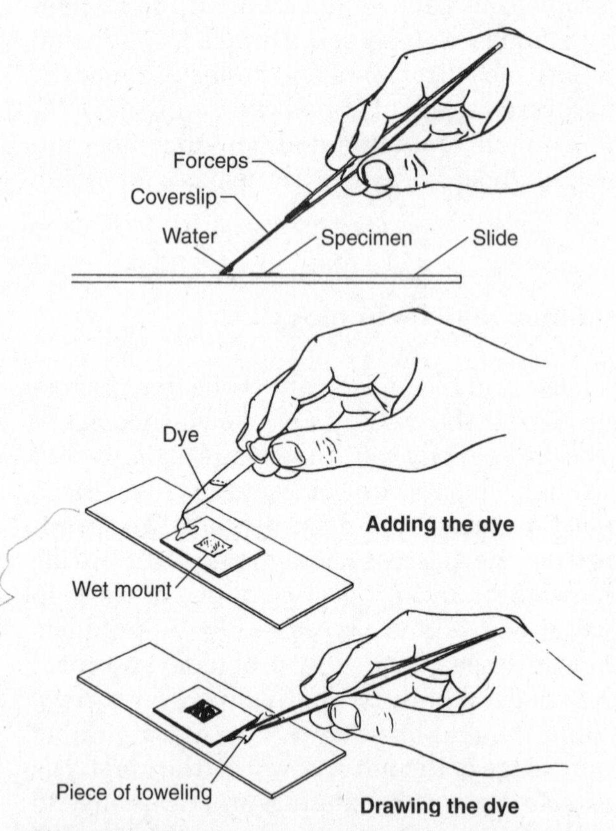


**Figure 2-3** The wet-mount technique is used to prepare a biological specimen for viewing under a microscope.

## Taking Measurements

As part of their scientific investigations, researchers take a variety of measurements. One or more of the following factors may be measured while conducting an experiment: length, volume, temperature, and mass. The tools for measuring these factors are described below.

- *Length:* The *metric ruler* is used to determine the length of a specimen. A one-meter ruler, or meterstick, is divided into 100 centimeters, with each centimeter further divided into 10 millimeters. One meter equals 1000 millimeters. (See Figure 2-4.)
- *Volume:* A clear numbered column called a *graduated cylinder* is used to measure the volume of a liquid in liters and milliliters. One liter equals 1000 milliliters. (See Figure 2-5.) The surface of the liquid is compared to the measurement scale on the cylinder to determine the volume. To be most accurate, the measurement should be read at the bottom, or *meniscus*, of the curved surface of the liquid. (See Figure 2-6.)

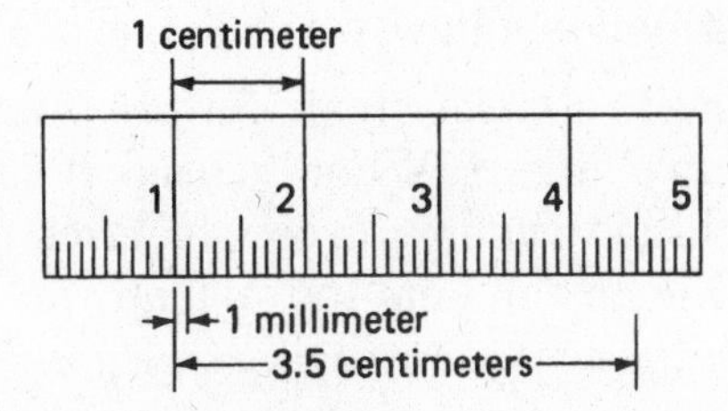


**Figure 2-4** A centimeter ruler.

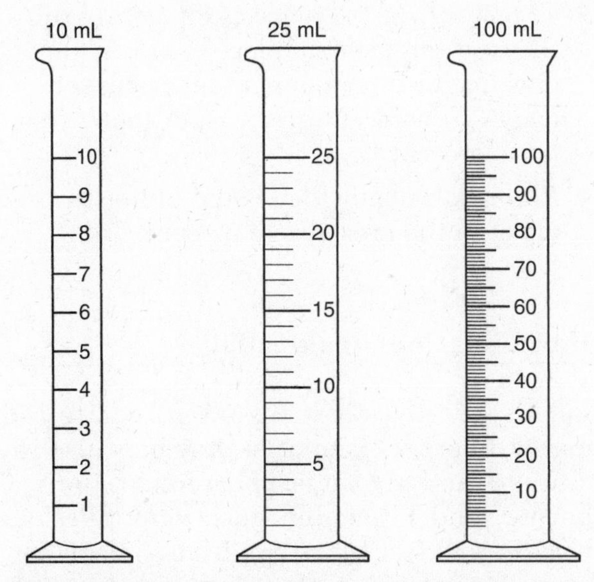


**Figure 2-5** Graduated cylinders are used to measure the volume of a liquid; measurements are in milliliters (mL).

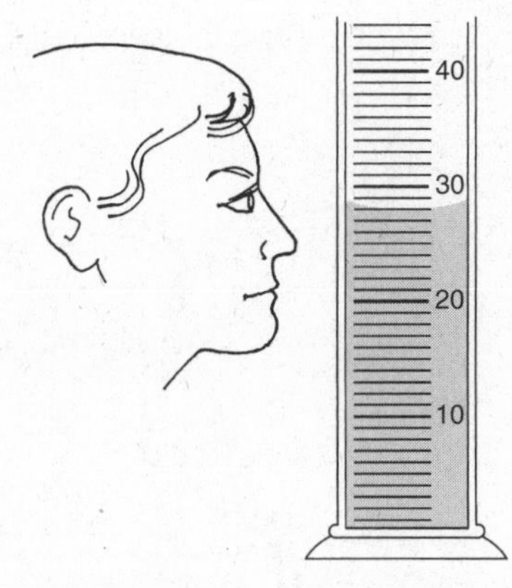


**Figure 2-6** Always read the volume of a liquid at the bottom of the meniscus (not at the top of the curve).

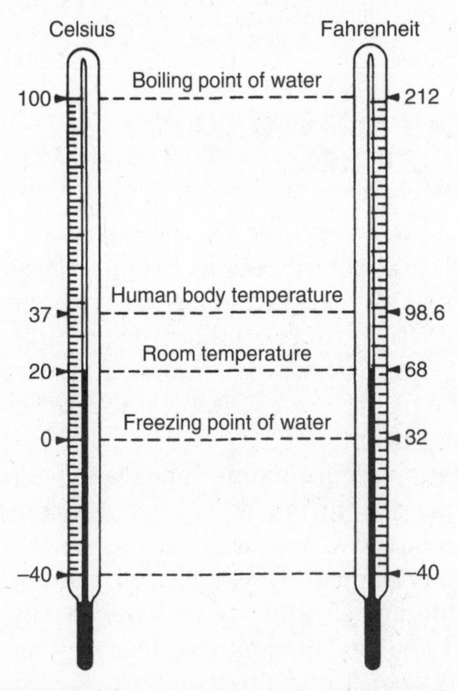


**Figure 2-7** Some important readings on the Celsius and Fahrenheit thermometers are shown for comparison.

- *Temperature:* The temperature of a substance can be measured by using a *Celsius thermometer*. Zero degrees Celsius (0°C) is the freezing point of freshwater; 100 degrees Celsius (100°C) is the boiling point of freshwater. (See Figure 2-7.)
- *Mass:* A *triple-beam balance* or *electronic balance* is used to measure the mass of an object in kilograms and grams. One kilogram equals 1000 grams. (See Figure 2-8.)

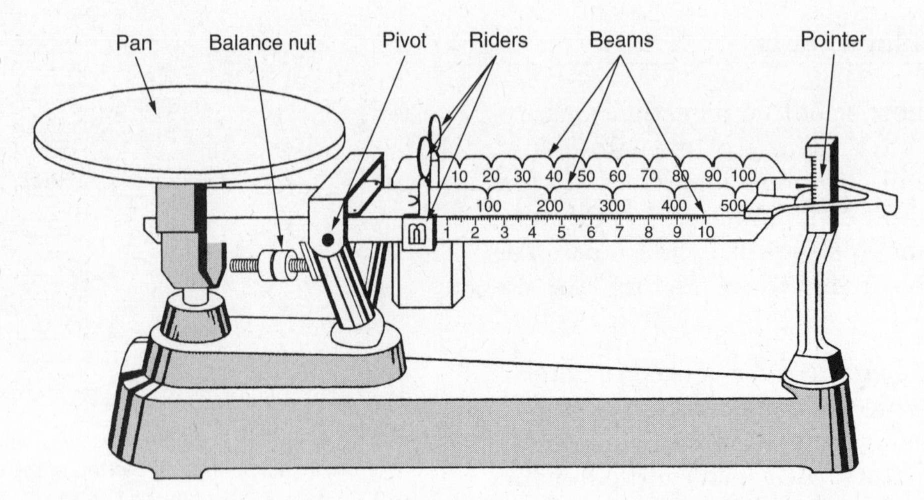


**Figure 2-8** The triple-beam balance is used to measure mass; measurements are in grams.

## CHEMICALS AND SAFETY PROCEDURES

One of the most important things to do *before* you go to work in a lab is: Prepare. Study the assigned investigation before you come to class. When in the laboratory, maintain a clean, open work area that is free of everything except those materials needed for the investigation.

If any lab equipment appears to be broken or in poor condition, do not use it. Report it to your teacher. If any accident or injury occurs, report it immediately; and if any damage occurs to your clothing or personal belongings, report that to your teacher, too.

### Chemical Indicators and Reagents

Chemical *stains* (such as iodine and methylene blue) are used in the preparation of microscope slides to make tiny features more visible. Chemical *indicators* are used to test for the presence of specific substances or to determine chemical characteristics of a material. The following list describes typical chemical tests used in a biology laboratory.

- *Litmus paper* is an indicator used to determine whether a solution is acidic or basic. Blue litmus turns red in an acid; red litmus turns blue in a base.
- *pH paper* has an indicator soaked into the paper that changes to one of many colors based upon the actual pH of the solution. The color of the pH paper is matched against a color chart to determine the pH of the solution tested.
- *Bromthymol blue* is an indicator that turns a solution yellow in the presence of carbon dioxide. If the carbon dioxide is removed from the solution containing the bromthymol yellow, it turns back to blue.
- *Benedict's solution* tests for the presence of simple sugars. When heated, it turns from blue to yellow, green, or brick red, depending on the amount of sugar present.
- *Lugol's solution*, or *iodine solution*, tests for the presence of starch. In its presence, it turns from dark red to blue-black.
- To test for the presence of fat, gently rub the food sample on a piece of brown paper toweling or brown paper (supermarket) bag. A translucent grease spot shows that fat is present.
- *Biuret solution*, which is light blue, turns violet in the presence of protein.

### Handling Chemicals Safely

Laboratory investigations often require the use of different chemicals; they may also require the use of heat, such as from a Bunsen burner, and breakable glassware, such as beakers and test tubes. Spilled chemicals can cause burns; and heated chemicals can give off toxic vapors. Because of these risks, it is very important to keep alert when working in a lab and to follow the safety procedures outlined below.

- ♦ Do not handle chemicals or equipment unless you are told by your teacher to do so.
- ♦ Never eat, drink, chew gum, or apply makeup in the laboratory.
- ♦ Read the label on every container before you use it. Do not use chemicals from containers that are not clearly labeled.
- ♦ Wear protective clothing and keep chemical stains off counters and other materials. Do not touch stains with your fingers; use medicine droppers to transfer stains from container to slide.
- ♦ Tie back long hair and remove all dangling jewelry. Roll up long sleeves and do not wear loose-fitting sleeves.
- ♦ Never taste chemicals or inhale the vapors from a chemical, since they can be toxic and dangerous; gently wave hand over open container to smell nontoxic chemicals.
- ♦ When you heat a liquid in a test tube, be certain that the opening of the test tube is pointed away from you and away from anyone else nearby. (See Figure 2-9.)
- ♦ Never pour excess reagents back into stock bottles. (Place the excess reagents in a designated waste container.) Never exchange the stoppers between different bottles.
- ♦ Wear safety goggles and a laboratory apron, especially when heating substances in the laboratory.

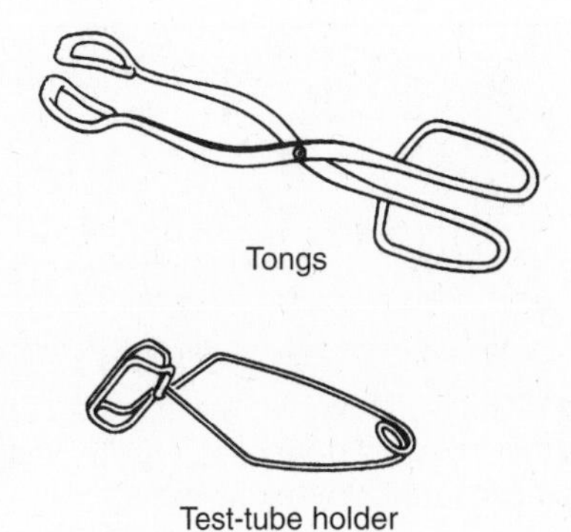


**Figure 2-10** Lab equipment needed to handle hot objects.

- ♦ Always handle hot test tubes with a test-tube holder and hot beakers with tongs or an oven mitt. (See Figure 2-10.)
- ♦ Keep all flammable substances, such as alcohol, far away from an open flame.
- ♦ Know the locations of the fire extinguisher, fire blanket, first aid kit, safety shower, eyewash station, fire exit, and fire alarm.
- ♦ Clear counters carefully after use and then wash your hands thoroughly.
- ♦ Inform instructor immediately of any problems, concerns, or accidents.

## OTHER LABORATORY TECHNIQUES AND PROCEDURES

*Chromatography* is a method used for separating and analyzing mixtures of complex chemical substances. The mixture to be separated is placed on a material such as chromatography or filter paper. A solvent is added,

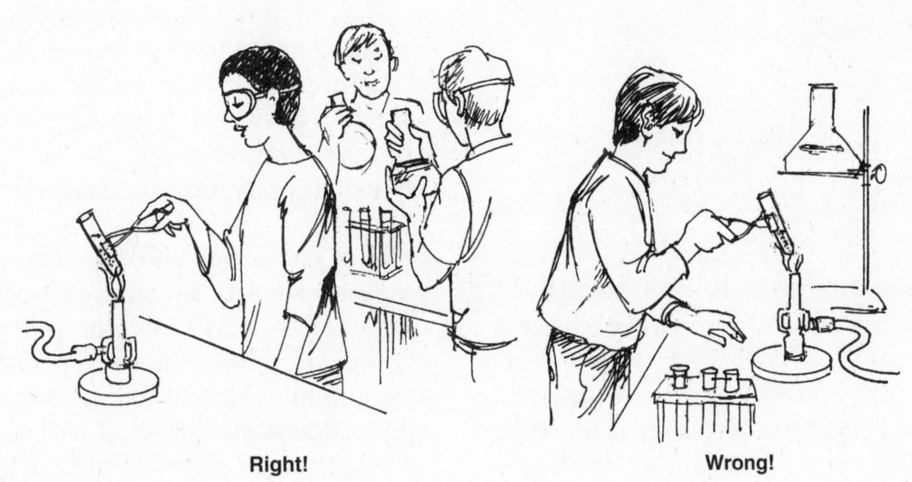


**Figure 2-9** Always wear safety goggles and use caution when heating chemicals in a test tube.

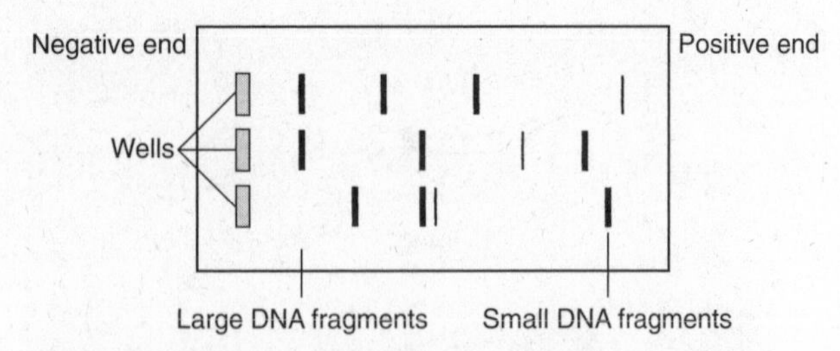


**Figure 2-11** Gel electrophoresis is used to separate molecules that have different charges; the smaller DNA fragments move farthest from the wells.

which begins to move through the material. Different substances in the mixture will move along with the solvent at different rates, causing them to be separated and allowing them to be studied.

*Electrophoresis* is a technique used to separate molecules that have different electrical charges. An electric current is run through a material, usually a gel, in which the mixture has been placed. Different substances move at different rates in the electrical field. The resulting pattern of bands shows the different substances that were originally mixed together. Gel electrophoresis is the process used, for example, to analyze one's DNA and compare it to the DNA of another person. (See Figure 2-11.)

*Centrifugation* is the method used to separate materials of different densities from one another. The original liquid mixture is placed in a test tube, which is spun around in a **centrifuge**. The heaviest particles settle to the bottom; the least-dense material forms a layer on the top. The *ultracentrifuge*, the most powerful of these machines, can spin at rates of 100,000 revolutions per minute, allowing the lightest-weight particles in cells to be separated from one another.

## SCIENTIFIC INVESTIGATIONS: PUTTING IT ALL TOGETHER

Data collected and observations made during scientific research need to be analyzed to see what they mean. The data may be organized and represented in a variety of ways. For example, diagrams, tables, charts, graphs, and equations can represent data. When the data are interpreted, the result may be the statement of a new or revised hypothesis. Another result may be the conclusion that a general understanding or explanation of a natural phenomenon is, in fact, supported by the data.

The mathematical processes of **statistical analysis** are used to determine if the results obtained are valid or if they might have been simply due to chance. Statistics also allows a researcher to find out the degree to which the predicted results (based on the hypothesis) match the actual results. Based on this "matching," the scientist can conclude whether the proposed explanation is, or is not, supported by the data. Data and observations are *valid* if they measure what they are supposed to be measuring. They are *reliable* if repeated trials give the same results. Statistical procedures help indicate the reliability of a set of observations. *Accuracy* is a general term that includes both the validity and reliability of a set of observations.

The analysis of the data, followed by public discussion, can lead to a revision of the original explanation, the development of new hypotheses, and the design of new research plans. When claims are made based on the collected evidence, the reliability of the claims should be questioned if the design of the experiment was faulty. For example, if there were small sample sizes, incomplete or misleading use of data, or a lack of controlled conditions, the results would not be reliable. Also, great care should be taken not to confuse facts with opinions.

When all the research and data analysis are concluded, a written report is prepared for the public to study. This report includes a literature review, the research, the results, and suggestions for further research. One purpose of making the results public is to allow the research to be repeated by someone else. Science assumes that, through the collection of similar evidence, different researchers will come to the same conclusions and explanations of natural phenomena. *Peer review*—the study of research reports by fellow scientists—is important as a check on the quality of the research. It also results, at times, in the suggestion of alternative explanations for the same observations.

# Chapter 2 Review

## *Part A—Multiple Choice*

**1.** A scientist wants to study the internal structure of a chloroplast (part of a plant cell) in great detail. The best instrument for this detailed examination would be a(n)

1 compound microscope
2 simple light microscope
3 electron microscope
4 ultracentrifuge

**2.** Blood plasma can be separated from red blood cells because they have different densities. To separate these parts of blood tissue, a biologist would use a(n)

1 microdissection instrument
2 centrifuge or ultracentrifuge
3 compound light microscope
4 electron microscope

**3.** To view cells under the high power of a compound light microscope, a student places a slide of the cells on the stage and moves the stage clips to secure the slide. She then moves the high-power objective into place and focuses on the slide with the coarse adjustment. Two steps in this procedure are incorrect. For this procedure to be correct, she should have focused under

1 high power first, using the fine adjustment, then low power using the fine adjustment
2 low power using the coarse and fine adjustments, then high power using only the fine adjustment
3 low power using the coarse and fine adjustments, then high power using only the coarse adjustment
4 low power using only the fine adjustment, then high power using the fine adjustment

**4.** A student used iodine solution to test a material, which turned very dark blue after the iodine was added. This proved the presence of

1 proteins
2 simple sugars
3 oxygen
4 starch

**5.** When a student adds bromthymol blue to a solution containing carbon dioxide, the solution should

1 shows no color change
2 turn yellow
3 turn blue-black
4 turn red-orange

**6.** An unknown solution may contain glucose. Which reagent could be used to test for the presence of this simple sugar?

1 litmus paper
2 bromthymol blue
3 Lugol's solution
4 Benedict's solution

**7.** Which piece of equipment should be used to transfer a one-celled organism onto a microscope slide?

1 pair of scissors
2 dissecting needles
3 pipette
4 test tube

**8.** While a student is heating a liquid in a test tube, the mouth of the tube should always be

1 corked with a rubber stopper
2 pointed toward the student
3 pointed straight upward
4 pointed away from everybody

**9.** During the observation of a microscope slide, a sharp crack is heard as the objective presses against the slide and breaks it. Which of the following actions might have caused this result?

1 changing the diaphragm opening under high power
2 using the fine adjustment to focus under low power
3 using the coarse adjustment to focus under high power
4 switching back to low power after focusing under high power

**10.** You are observing a wet-mount slide of a paramecium culture under the low power of a compound microscope. After a time you notice that the range of movement of the organisms is shrinking toward the center of the mount. What should you do to restore the activity of the organisms?

1 Add a drop of iodine stain.
2 Switch to high power and refocus.
3 Use a brighter light source.
4 Add a drop of water to the edge of the coverslip.

**11.** An acid-base indicator that changes to either red or blue only, and is easy to use, is called

1 methylene blue
2 bromthymol blue
3 Lugol's solution
4 litmus paper

**12.** When a person exhales into a solution of bromthymol blue, the solution changes color. It changes to what color and why?

1 dark blue due to presence of oxygen
2 yellow due to presence of oxygen
3 dark blue due to presence of carbon dioxide
4 yellow due to presence of carbon dioxide

**13.** When you heat a test tube that contains liquid, which of the following is a dangerous action?

1 wearing safety goggles
2 using a metal test-tube holder
3 wearing a laboratory apron
4 looking into the tube opening to observe a reaction

**14.** In the laboratory, which of the following actions is *not* recommended?

1 wearing loose-fitting clothing and dangling jewelry
2 having your setup checked by the teacher before starting a procedure
3 reading the instructions through to the end before starting an investigation
4 being familiar with the location and use of the emergency equipment

**15.** Safety goggles must be worn in the laboratory, most especially when

1 testing pH levels
2 heating solids
3 heating liquids
4 cleaning up after an experiment

**16.** A student performed the following experiment: a dry piece of white bread was tested with iodine solution; the bread turned blue-black. Another piece of white bread was chewed and then tested with Benedict's solution; the mixture turned red. The student therefore concluded that when a piece of bread is chewed, starch is changed to sugar. One error in the student's procedure was that he did *not*

1 test the chewed piece of bread for starch
2 test the dry piece of bread for sugar
3 consider the age of the piece of bread
4 use Biuret solution to test for starch

**17.** An unknown solution was tested with Benedict's solution and the liquid remained blue. The most reasonable conclusion from this evidence is that the solution contained no

1 sugars
2 protein
3 starch
4 fats

**18.** To measure and pour 10 milliliters (mL) of Benedict's solution into a test tube, which is the best procedure to ensure accuracy of the measurement?

1 Use a 10-mL graduated cylinder.
2 Use a 100-mL graduated cylinder.
3 Weigh out 10 mL on a metric balance.
4 Fill a 1-mL medicine dropper 10 times.

**19.** A slide of human blood cells was observed under the low-power objective of a compound light microscope that had clean lenses. When the microscope was switched to high power, the image was dark and fuzzy. Which two parts of the microscope should be used to correct this situation?

1 nosepiece and coarse adjustment
2 diaphragm and ocular lens
3 objective and fine adjustment
4 diaphragm and fine adjustment

**20.** Which structure is best seen by using a compound light microscope?

1 a cell's nucleus
2 a paramecium
3 a DNA sequence
4 a mitochondrion

### *Part B—Analysis and Open Ended*

**21.** The diagrams below show four different one-celled organisms (shaded) in the field of view of the same microscope using different magnifications. Which illustration shows the largest one-celled organism?

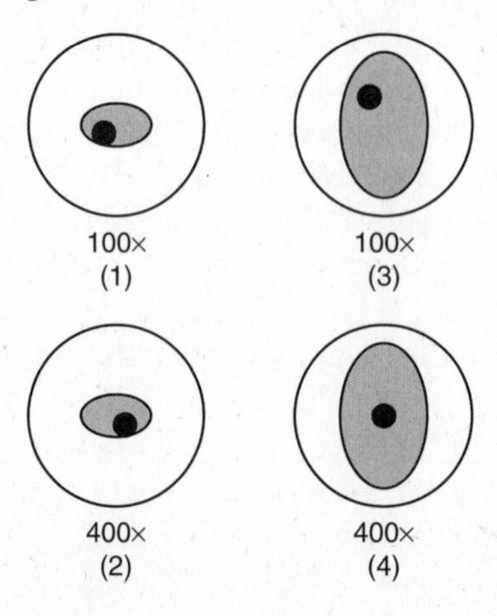

1 diagram 1
2 diagram 2
3 diagram 3
4 diagram 4

**22.** A peppered moth and part of a metric ruler are represented in the diagram below.

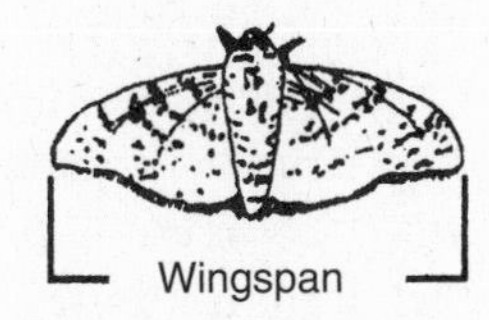


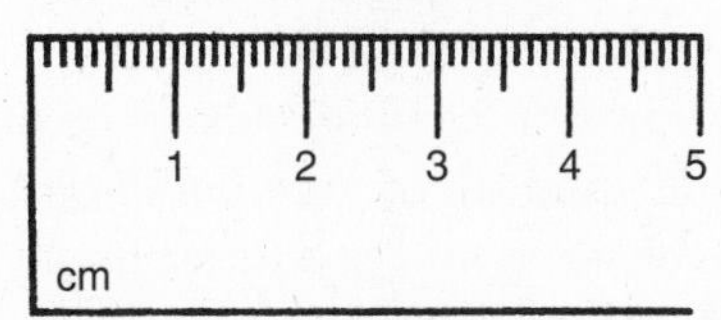


Which row in the chart below best represents the ratio of body length to wingspan of the peppered moth?

| Row | Body Length:Wingspan |
|---|---|
| (1) | 1:1 |
| (2) | 2:1 |
| (3) | 1:2 |
| (4) | 2:2 |

1 row 1
2 row 2
3 row 3
4 row 4

**23.** How much water should be removed from the graduated cylinder shown below in order to leave exactly 5 milliliters of water in the cylinder?

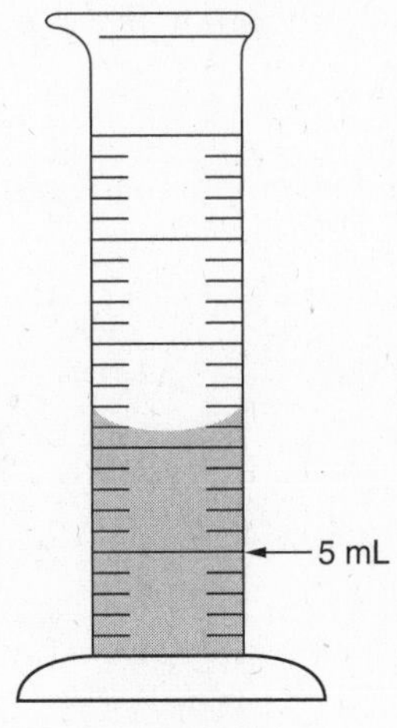


1 6 mL
2 7 mL
3 11 mL
4 12 mL

**24.** The following diagram shows how a coverslip should be lowered onto some single-celled organisms during the preparation of a wet-mount slide. Why is this the preferred procedure?

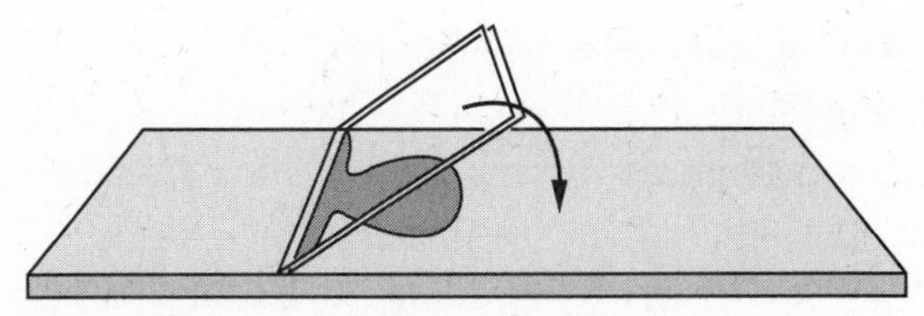

1 The coverslip will prevent the slide from breaking.
2 The organisms will be more evenly distributed.
3 The possibility of breaking the coverslip is reduced.
4 The possibility of trapping air bubbles is reduced.

**25.** Identify a piece of laboratory equipment that normally would be used to measure 5 milliliters of glucose solution for an experiment. Explain how the reading should be done to get an accurate measurement.

**26.** Why do scientists take measurements when recording data?

**27.** Explain why scientists use tools and instruments during experiments.

**28.** Describe the procedure for preparing a wet-mount slide.

**29.** Explain the similarities and differences between the transmission electron microscope (TEM) and the scanning electron microscope (SEM).

**30.** Identify the metric units used for each of the following: length, volume of a liquid, temperature, and mass.

**31.** Suppose that a new biology laboratory room has been opened in your school. Prepare a poster listing the safety rules and procedures that should be followed by students working in the lab.

**32.** A person is determined to follow a very low-fat or fat-free diet. He has a piece of cheddar cheese and a slice of apple. Explain how he could test these two food items for the presence of fat. (If possible, conduct this test yourself at home and describe your results.)

## *Part C—Reading Comprehension*

*Base your answers to questions 33 to 35 on the information below and on your knowledge of biology. Use one or more complete sentences to answer each question.*

Light microscopes are used by scientists to view extremely small objects, such as cells. However, due to the physical properties of light, objects below a certain size cannot be seen in a sharp, focused image—no matter how well made and powerful the light microscope is. Bacteria with a diameter of 0.5 micrometer are about the smallest living things that can be observed with a high-quality light microscope. (A micrometer is 1/1,000,000 of a meter.) During the twentieth century, scientists learned to use a beam of electrons (similar to what is used in a television set), instead of a beam of light, to produce images of very small objects. Electron microscopes have now been used to observe and produce photographs of the detailed, internal structures of plant and animal cells, as well as of the smallest living things, such as bacteria and viruses.

Modern electron microscopes are at least 1000 times more powerful than light microscopes, allowing clear observations of objects that are as small as 0.5 nanometer. (A nanometer is 1/1,000,000,000 of a meter.) Two types of electron microscopes are the transmission electron microscope (TEM) and the scanning electron microscope (SEM). In a TEM, a beam of electrons is passed through an object to show the details within it. In an SEM, a beam of electrons is passed over an object, producing a detailed, three-dimensional image of its surface. Electrons normally bounce off the gas molecules that are present in the air. So there must be a vacuum where the specimen is placed inside an electron microscope. In addition, specimens are usually treated with stains that interact with the electron beams in order to produce the images. As a result, cells cannot be viewed with an electron microscope while they are still alive.

**33.** How have scientists been able to overcome the limitations of light microscopes?

**34.** In two or three sentences, describe the similarities and differences between light microscopes and electron microscopes.

**35.** If scientists were to invent a new kind of microscope, how might they combine the best of a light microscope with the best of an electron microscope?

# THEME II

## Energy, Matter, and Organization

# 3
# From Atoms to Cells

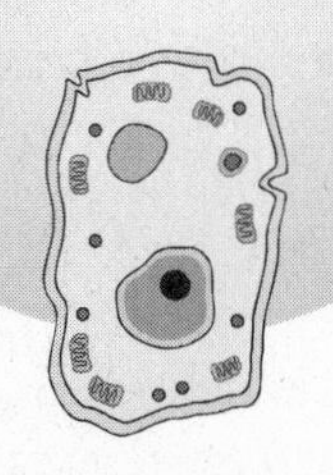

### Vocabulary

| | | |
|---|---|---|
| amino acids | glucose | organisms |
| atoms | lipids | organ systems |
| cells | metabolism | proteins |
| cell theory | molecules | starch |
| compounds | multicellular | system |
| covalent bond | organ | tissues |
| energy | organic compounds | unicellular |

## LEVELS OF ORGANIZATION

The study of how living things are put together begins with *atoms*. The next level of organization above atoms is *molecules*, then the families of organic compounds, and then organelles and **cells**. Beyond that, the organization of living things continues. Plants and animals, which are composed of enormous numbers of cells, have groups of similar cells called **tissues** that work together to do specific functions. For example, the nervous tissue in your brain consists of billions of nerve cells functioning together.

A group of tissues, in turn, works together as an **organ**. The brain is an organ made up of different types of tissue, including nervous, blood, and connective tissue. Organs that work together make up **organ systems**. The nervous system, for example, includes the brain, spinal cord, and sensory organs. Finally, different organ systems work together

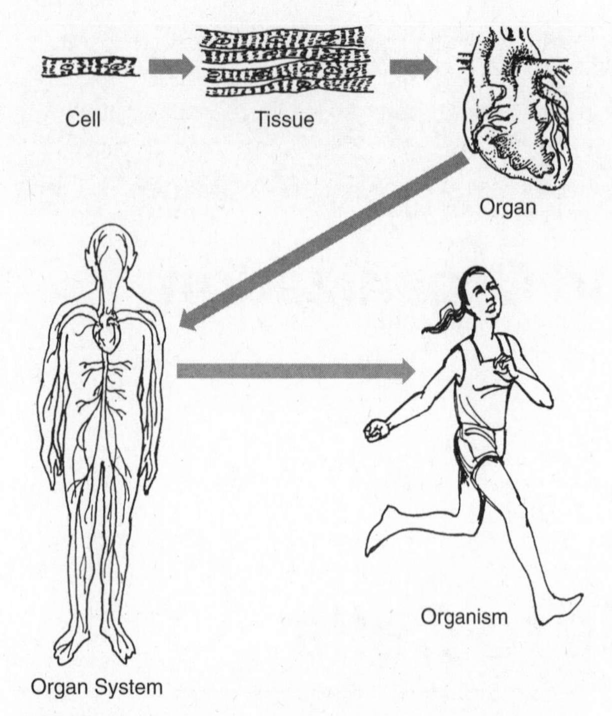


**Figure 3-1** The levels of organization in a living organism start with the cell.

as a functioning whole **organism**. (See Figure 3-1.)

## ENERGY, MATTER, AND ORGANIZATION

Everything in the world tends to get more disorganized as time passes. Physicists recognize that natural events tend to increase disorder in the universe; or, put more simply, the universe is running down. It becomes more disordered over time.

Living things are highly organized. How do they keep themselves carefully arranged and functioning properly in a universe where things are constantly running down?

To maintain the state of organization necessary for life to exist, all living things require **energy**. Energy exists in different forms, such as heat, motion, light, and electricity. And organisms always need a continuous input of energy to stay organized and remain alive.

## MATTER, ATOMS, AND LIFE

All matter is made up of **atoms**, particles far too small to be seen with the unaided eye or even through an ordinary microscope. Each atom has an extremely dense nucleus in its center. Distributed in the mostly empty space around the nucleus are *electrons*. Electrons have energy. The orbitals in which electrons move make up shells, or energy levels. The electron shells closest to the nucleus have the least energy; the electron shells farthest from the nucleus have the most energy. It is possible for electrons to move from one shell to another. A ball rolling from the top of a hill loses energy as it moves lower. So, too, does an electron as it moves from an outer, higher-energy level to an inner, lower-energy level. The movement of electrons between energy levels in atoms is what produces all changes, or *transformations*, of energy.

An organism is actually a very complex **system** for transforming energy. Through natural selection, evolution has produced species of organisms that are efficient energy transformers. Where does the energy come from that keeps organisms as different as grass, ants, and elephants alive? Sunlight is the main source of energy for most life on Earth. Plants trap this light energy; then plants, animals, and other organisms use it to live, grow, and reproduce.

## ATOMS BOND TOGETHER

There are more than 112 different types of atoms. Of these basic types, called *elements*, 92 are found naturally on Earth. Carbon, hydrogen, nitrogen, and oxygen are the four most important elements for organisms. Also vital for living things are phosphorus, sulfur, and a few other elements. (See Figure 3-2.)

Atoms of most elements combine with other

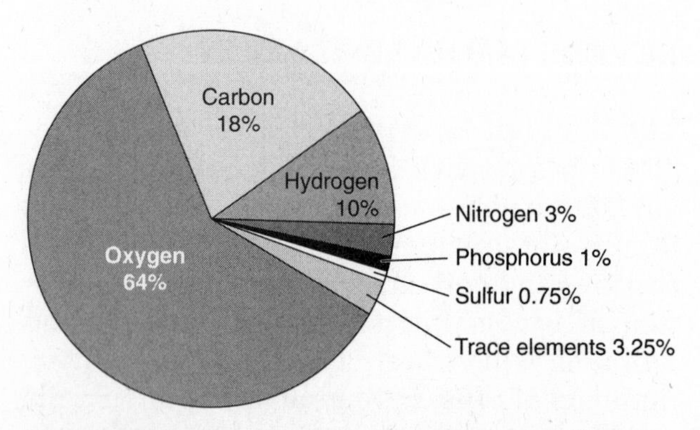


**Figure 3-2** The graph shows the percentages of the most important elements for organisms' bodies.

atoms to form larger structures called **molecules**. The atoms in a molecule are kept together by a kind of partnership called a *chemical bond*. One type of chemical bond is the **covalent bond**. In such a bond, atoms share electrons with each other, thus making each atom more stable.

Elements combine to form thousands of different **compounds**, but only a small number of compounds are necessary for living things. Many of the compounds found in organisms are called **organic compounds**. All organic compounds contain the elements carbon and hydrogen. Carbon atoms are of special importance because organic compounds usually have a skeleton of carbon atoms bonded to each other. Also, each carbon atom can form bonds with up to four other atoms, including other carbon atoms.

## ORGANIC COMPOUNDS

Organic compounds carry out many different, complex tasks that keep us and all other organisms alive. These tasks include capturing and transforming energy, building new structures, storing materials, repairing structures, and keeping all chemical activities in the body working properly. In organisms, chemical reactions are always putting things together or taking things apart. All together, these chemical activities in an organism are called its *metabolic activities*, or **metabolism**.

Organic molecules are often very large. A large molecule, such as the blood protein hemoglobin, is called a *polymer*. (*Note:* The term *poly* means "many.") Polymers are formed by the linking together of smaller molecules, or subunits, called *monomers*. (*Note:* The term *mono* means "one.") In polymers, these subunits are held together by covalent bonds.

## THE FAMILIES OF ORGANIC COMPOUNDS

Even though organisms contain many different molecules, there are relatively few different types of subunits that make up these molecules. The subunits located in the organic compounds found in living organisms, from bacteria to whales, are almost identical. However, the way these monomers are put together creates an enormous variety of polymers. Some polymers have more than 100 subunits; others have thousands of subunits. As a result, there are many different kinds of polymers.

Organic compounds are grouped into four major families: carbohydrates, lipids, proteins, and nucleic acids. These are the families of compounds found in all living things.

## CARBOHYDRATES

*Carbohydrates* include the simple sugars as well as polymers, which are made up of sugar subunits. Carbohydrates are formed from the elements hydrogen, oxygen, and carbon. In organisms, the main functions of carbohydrates are energy storage and providing strong building materials for certain types of cells. (See Figure 3-3.)

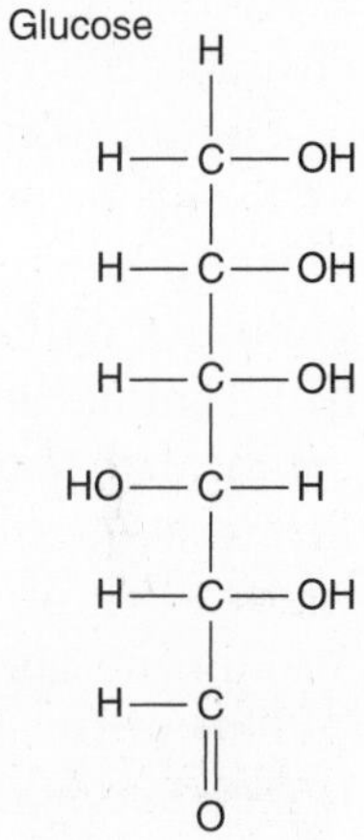

*Polysaccharides* are composed of many sugar subunits (called *monosaccharides*) joined together. Energy is stored in these large molecules. Our muscles must contain stored energy to allow them to work at a moment's notice. In humans, energy is stored in the polysaccharide *glycogen* in muscles and in the liver. It is made up of many **glucose** (simple sugar) subunits.

Glucose

H
|
H—C—OH
|
H—C—OH
|
H—C—OH
|
HO—C—H
|
H—C—OH
|
H—C
||
O

**Figure 3-3** Glucose is a simple sugar. Like most other organic compounds, it has a skeleton of carbon atoms to which other atoms form bonds.

Plants store energy in the form of **starch**, a type of polysaccharide, which is contained in such foods as corn and rice. Cellulose, another important polysaccharide found in plants, helps build up tough structures such as wood.

## LIPIDS

Oils and fats make up the second major family of organic compounds, the **lipids**. One type of lipid makes up the basic structure of the cell membrane. However, the main purpose of lipids is energy storage, which they do more efficiently than carbohydrates. Energy stored in lipids is for long-term use. During physical activity, the carbohydrate glycogen gets used up quickly. However, lipids, or fat deposits, do not disappear quickly.

## PROTEINS

Hundreds of thousands of different **proteins** exist; they determine a wide variety of functions and traits in organisms. Proteins are used to build materials, transport other substances, send signals, provide defense, and control chemical and metabolic activities. No two people other than identical twins have exactly the same proteins (and thus the same characteristics).

The building plan for proteins is the same as for other organic compounds. Proteins are large polymers that are made up of smaller subunits called **amino acids.** There are 20 different types of amino acid molecules. By combining these amino acids in a row, in different sequences, many types of proteins are made. Proteins can easily have more than 100 amino acid subunits. (See Figure 3-4.)

The order in which its amino acids are linked determines the characteristics of a protein molecule. Every different sequence produces a different protein. However, the sequence does not behave like a string of letters; the protein chains twist, turn, and bend into specific three-dimensional shapes. The shape of a protein molecule is called its *conformation*. Every protein molecule has a very specific conformation, and that is what determines its function.

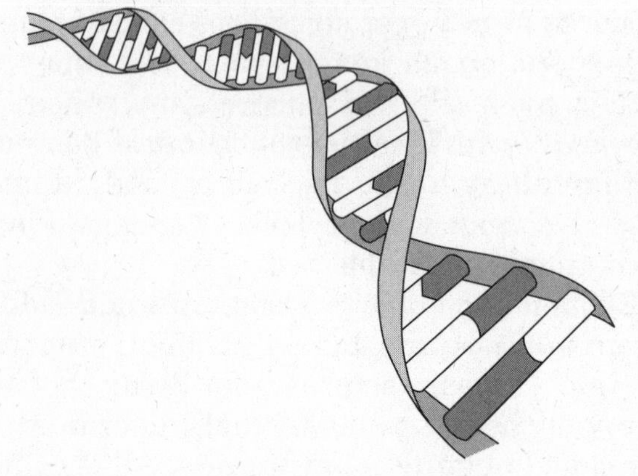

**Figure 3-5** DNA contains the instructions for building the protein molecules that make up an organism.

## NUCLEIC ACIDS

*Nucleic acids* consist of *deoxyribonucleic acid (DNA)* and *ribonucleic acid (RNA).* DNA and RNA are responsible for storing the genetic information that contains the directions for building every molecule that makes up an organism. (See Figure 3-5.)

The pattern for building nucleic acids is similar to that of the other organic molecules. Individual subunits, called *nucleotides*, are combined in a linear sequence to build the polymers DNA and RNA. DNA molecules usually contain thousands of nucleotides linked together in a specific sequence. The sequence in the DNA, which gets copied into RNA, is used as a building plan to construct a protein molecule. Specific portions of the DNA molecule are used to make specific proteins. Because proteins make us who we are—and because our DNA makes our proteins—it is really our DNA that makes us who we are.

## CELLS ARE US

To study life is to learn about all the levels of life's organization, sometimes one at a time, sometimes all together. At each level there are characteristics that were not present on the previous level. While we may not be able to define life, we can say what living things do.

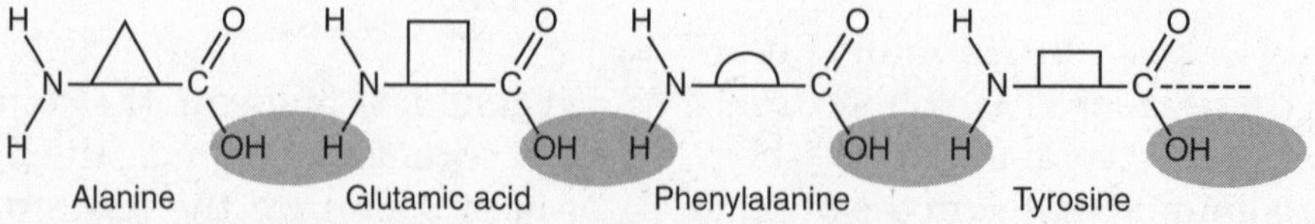

Alanine Glutamic acid Phenylalanine Tyrosine

**Figure 3-4** Amino acids combine in linear sequences to form a great variety of proteins.

Properties of life include order and use of energy; reproduction, growth, and development; digestion, nutrition, excretion, and respiration; movement, response, and coordination; and immunity. Tiny structures found within single-celled, or **unicellular**, organisms also function much like the tissues and organ systems found in many-celled, or **multicellular**, organisms. Called *organelles*, these cell parts perform life processes for unicellular organisms, just as tissues and organs do for multicellular life-forms.

The idea that organisms are made up of cells is one of the central ideas of modern biology. Referred to as the **cell theory**, its main points are as follows:

- All organisms are made up of one or more cells.
- The cell is the basic unit of structure and function of all living things.
- All cells arise from previously existing cells. (*Note:* The one exception would be the first cell that ever arose.)

## Chapter 3 Review

### *Part A—Multiple Choice*

**1.** As time passes, the universe becomes

1 more organized
2 less organized
3 simpler
4 smaller

**2.** To maintain organization, living things need

1 time
2 patience
3 money
4 energy

**3.** The densest region of an atom is the

1 space around its nucleus
2 nucleus at its center
3 inner electron shells
4 outer electron shells

**4.** Which event is most like an electron moving from an outer shell to an inner shell?

1 a ball rolling down a hill
2 a fish swimming upstream
3 a soccer ball rolling across a field
4 a pole-vaulter rising into the air

**5.** All transformations of energy involve

1 a change in height
2 the use of sunlight
3 the organization of atoms
4 the movement of electrons

**6.** How do molecules differ from atoms?

1 Atoms occur naturally, whereas molecules do not.
2 Atoms are individual particles, whereas molecules are combinations of atoms.
3 There many different types of atoms, but only a few different types of molecules.
4 Atoms are found in living things, whereas molecules are found in nonliving things.

**7.** The two elements found in every organic compound are

1 nitrogen and oxygen
2 oxygen and hydrogen
3 carbon and hydrogen
4 carbon and oxygen

**8.** In a cell, all organelles work together to carry out

1 diffusion
2 information storage
3 active transport
4 metabolic processes

**9.** Which statement concerning simple sugars and amino acids is correct?

1 They are both wastes resulting from protein synthesis.
2 They are both needed for the synthesis of larger molecules.
3 They are both building blocks of starch.
4 They are both stored as fat molecules in the liver.

*Refer to the diagram below to answer question 10.*

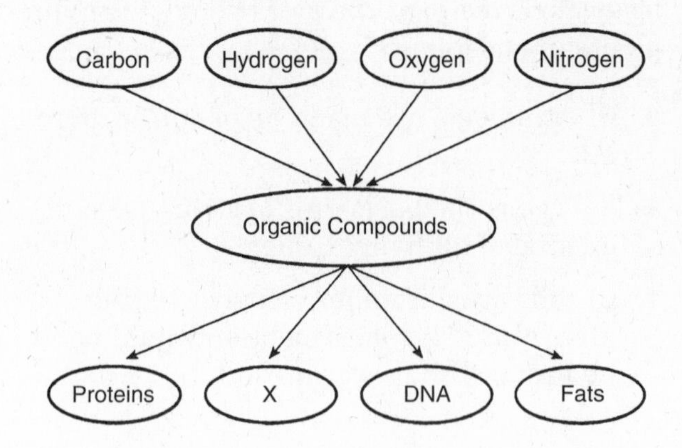


**10.** What substance could the letter *X* represent?

1 carbohydrates
2 carbon dioxide
3 ozone
4 water

**11.** Which family of organic compounds is used mainly to store energy and to build certain materials in cells?

1 lipids
2 carbohydrates
3 proteins
4 nucleic acids

**12.** An iodine test of a tomato plant leaf revealed that starch was present at 5:00 P.M. on a sunny afternoon in July. When a similar leaf from the same tomato plant was tested with iodine at 6:00 A.M. the next morning, the test indicated that less starch was present. This reduction in starch content most likely occurred because starch was

1 changed directly into proteins
2 transported downward toward the roots through tubes
3 transported out of the leaves through the guard cells
4 changed into simple sugars

**13.** The subunits that make up proteins are

1 amino acids
2 single atoms
3 fats and lipids
4 nucleic acids

*Refer to the diagram below, which provides some information about proteins, to answer question 14.*

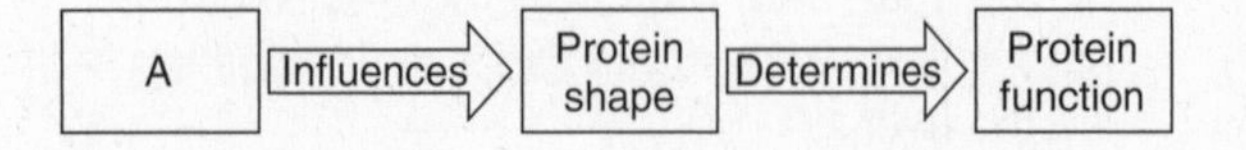


**14.** Which phrase does the letter A most likely represent?

1 sequence of amino acids
2 sequence of starch molecules
3 sequence of simple sugars
4 sequence of ATP molecules

**15.** The subunits of DNA are called

1 amino acids
2 nucleotides
3 polysaccharides
4 cell units

**16.** How is RNA related to proteins?

1 Proteins are made up of RNA molecules.
2 RNA determines which proteins are made.
3 RNA is copied into DNA to build a protein.
4 DNA is copied into RNA to build a protein.

**17.** DNA molecules are important because they store

1 fats for energy
2 genetic information
3 carbohydrates
4 polysaccharides

**18.** Which of the following is *not* an idea of the cell theory?

1 Organisms are made up of one or more cells.
2 Cells bond together much like atoms do.
3 The cell is the basic unit of structure in all living things.
4 All cells arise from previously existing cells.

**19.** Which sequence represents the correct order of levels of organization found in a complex organism?

1 cells→organelles→organs→organ systems→tissues
2 organelles→cells→tissues→organs→organ systems
3 tissues→organs→organ systems→organelles→cells
4 organs→organ systems→cells→tissues→organelles

**20.** In terms of levels of organization, a biosphere is most like

1 a cell
2 a tissue
3 an organ
4 an organism

*Refer to the diagrams of the organisms shown below to answer question 21.*

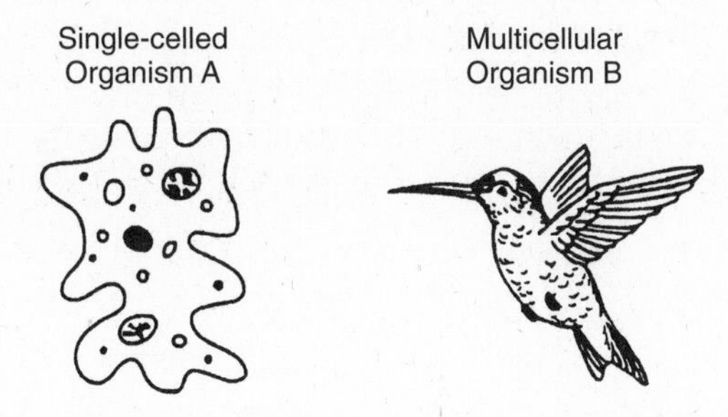


**21.** Which statement concerning organism A and organism B is correct?

1 Organism A contains tissues and organs, while Organism B lacks these structures.
2 Organism A and Organism B have structures that help them maintain homeostasis.
3 Organism A and Organism B have the same organs to perform their life functions.
4 Organism A lacks structures that maintain homeostasis, while Organism B has them.

**22.** Every single-celled organism is able to survive because it carries out

1 metabolic activities
2 heterotrophic nutrition
3 autotrophic nutrition
4 sexual reproduction

### *Part B—Analysis and Open Ended*

**23.** How do living things maintain the high level of organization that they need to survive?

**24.** Why is carbon particularly important for the existence of life on Earth?

**25.** List *four* important functions of organic compounds in living things.

*Refer to the diagram at right to answer questions 26 and 27. (Molecule is shown as a straight chain.)*

Glucose

H
|
H—C—OH
|
H—C—OH
|
H—C—OH
|
HO—C—H
|
H—C—OH
|
H—C
||
O

**26.** Based on the elements in glucose, and the way the atoms are attached, you could determine that glucose is an example of

1 a carbon molecule
2 a hydrogen molecule
3 an organic compound
4 an inorganic compound

**27.** When many of these glucose subunits join together, they make up a

1 protein molecule
2 polysaccharide
3 lipid molecule
4 DNA molecule

**28.** Explain why athletes need to eat lots of complex carbohydrates during training.

**29.** The diagram below illustrates a reaction in which

1 several amino acids join to form a protein molecule
2 inorganic compounds form an organic compound
3 simple sugars join to form a larger sugar molecule
4 polysaccharides are broken down into simple sugars

**30.** Identify three important characteristics of proteins. Your answer should include the following:

- what the subunits are that make up proteins
- four main functions of proteins in living things
- what determines structure and function of a protein

Glucose $C_6H_{12}O_6$ + Fructose $C_6H_{12}O_6$ → Sucrose $C_{12}H_{22}O_{11}$ + Water $H_2O$

**31.** Hemoglobin and hair are both proteins, yet they have different structures. Explain.

**32.** In what way do the particular proteins in our bodies depend on our DNA?

**33.** In terms of levels of organization, what is the difference between a *tissue* and an *organ*; that is, how do their structures and functions differ? Give an example of each.

**34.** List the main levels of organization of living things, from atoms to organism.

### *Part C—Reading Comprehension*

*Base your answers to questions 35 through 37 on the information below and on your knowledge of biology. Use one or more complete sentences to answer each question.*

Millions of cells die naturally each day in a person. Scientists have now discovered that these dying cells send out a chemical signal to attract other cells that specialize in disposing of cellular corpses.

Over the past few years, biologists have begun to understand how macrophages and other cells recognize dying cells. For example, a cell about to die sprouts what scientists refer to as eat-me signals, which tell a macrophage to consume the cell before it falls apart and triggers inflammation

But what if there is no macrophage close at hand to a dying cell? No problem, say Sebastian Wesselborg of the University of Tübingen in Germany and his colleagues. In the June 13 *Cell,* they report that dying cells from monkeys, mice, and people secrete a molecule called lysophosphatidylcholine. Previous research showed that the chemical attracts macrophages and other immune cells that may be some distance away. This lure ensures that dying cells are removed efficiently, Wesselborg's group concludes.

**35.** State the purpose of the chemical signal that is sent out by dying cells.

**36.** Explain why it is important for dying cells to be properly removed.

**37.** Explain the role played by the chemical lysophosphatidylcholine in the removal of dying cells.

# 4

# Chemical Activity in the Cell

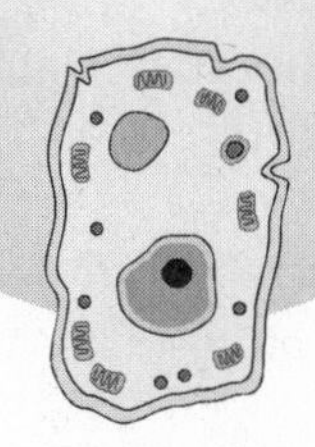

## Vocabulary

| | | |
|---|---|---|
| active transport | diffusion | organelles |
| carbon dioxide | enzymes | osmosis |
| catalysts | Golgi complex | passive transport |
| cell membrane | lysosomes | pH |
| cell wall | mitochondria | ribosomes |
| chloroplasts | nucleus | vacuoles |
| cytoplasm | | |

## INTRODUCTION TO THE CELL MEMBRANE

The inside of a single-celled organism is very much alive. However, the physical environment outside the cell is the opposite—a nonliving place where many changes occur. What stands between a cell and the potentially hostile environment that surrounds it? An ultrathin, extremely important layer separates the living world inside a cell from the nonliving world outside. This is the **cell membrane**, or *plasma membrane*.

The cell membrane performs two primary, yet very different, functions: it separates the cell from its environment and it enables communication and movement of materials between the cell and its environment. Without a cell membrane, there would be no cell. Protein molecules, which float within lipids in the membrane, enable much of the movement of materials across the cell membrane. These protein molecules often extend from one side of the membrane to the other. In multicellular organisms, some cell membrane proteins may also function as *receptor molecules* to which chemical signals may attach as one cell communicates with another cell. (See Figure 4-1.)

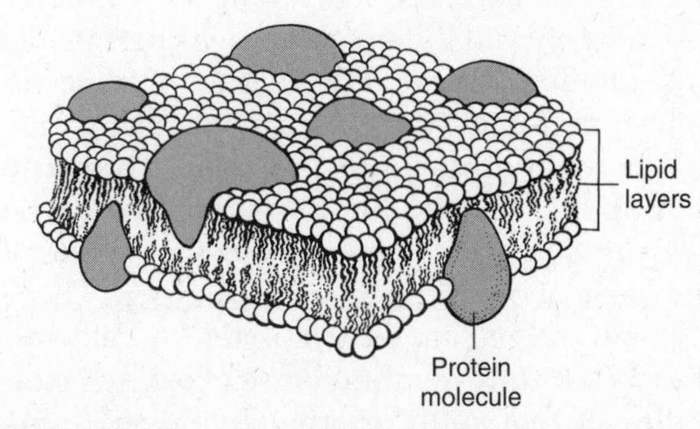


**Figure 4-1** The cell membrane acts as a barrier, separating the inside of the cell from the outside. Protein molecules, floating within lipids, enable the movement of materials across the cell membrane.

## TRANSPORT ACROSS THE CELL MEMBRANE

For a cell to remain alive, it must have a very special collection of chemicals inside it. These chemicals may be quite different from the chemicals located in the outside environment. Some substances that are abundant outside the cell are not found inside the cell. Other substances that are scarce outside the cell are present in larger quantities inside the cell. The cell membrane creates and maintains this special environment inside the cell. How does it do this?

The cell membrane allows some substances —that is, molecules—to pass through but keeps other substances out. This ability to determine which molecules can pass through is called *selective permeability*. The cell membrane is selectively permeable—it determines which molecules move through it and whether the molecules go into or out of the cell. It also makes possible the rapid transport of some molecules across it, while other molecules pass through slowly.

## PASSIVE TRANSPORT

Typically, there is an overall or net movement of molecules from an area of high concentration—a place where molecules are crowded together—to an area of low concentration. This kind of movement is called **diffusion**. Molecules are constantly in motion and they naturally move from where they are more concentrated to where they are less concentrated. This movement happens automatically with a cell if its membrane is permeable to the molecules and if there is a difference in concentration of the molecules on either side of the membrane. This is called **passive transport**, because no energy is used by the cell and no work is done. For example, one of the basic needs of most cells is oxygen. There are few oxygen molecules inside a cell, but there is usually an abundance of oxygen molecules in the water or other liquid that surrounds the cell. Thus, oxygen molecules diffuse across the cell membrane into the cell by passive transport. (See Figure 4-2.)

The diffusion of water molecules across a cell membrane—so important for living cells —is given a special name, **osmosis**. When plant cells are put in a strong salt solution, the abundant freshwater inside the plant cells automatically moves out of the cells, by osmosis, to where there is more salt and relatively fewer water molecules. Plant cell membranes can even be seen pulling away from the cell walls as the cells lose water; this process is called *plasmolysis*.

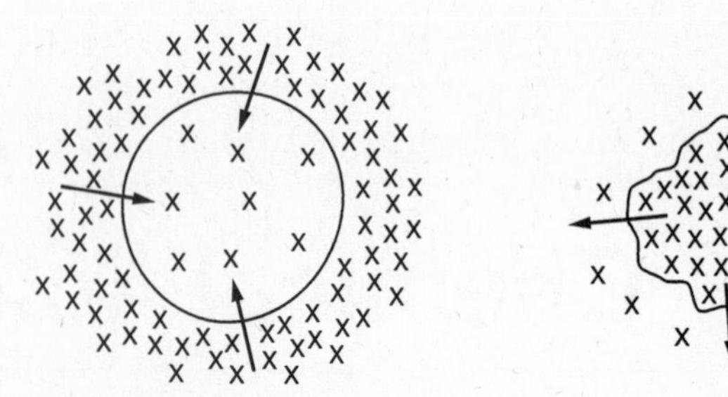

**Figure 4-2** Diffusion is the movement of molecules from an area of higher concentration to an area of lower concentration.

The reverse happens when limp celery stems are put in freshwater. The celery stems are limp because their cells have too little water in them. When the celery is put in the water, osmosis occurs and water molecules move into the cells. The cells expand, the cell membranes push against the cell walls, and the cells—and thus the celery stems—become firm again; this effect is known as *turgor pressure*.

The same effects can also be observed with animal cells. For example, red blood cells placed in a high salt concentration become shriveled as water leaves the cells. Red blood cells surrounded by freshwater swell and burst as water enters them by osmosis; this bursting process in cells is called *cytolysis*.

## ACTIVE TRANSPORT

The movement of a substance against the concentration gradient is called **active transport**. When substances are moved from an area of lower concentration to an area of higher concentration, energy is used and work is done. This kind of transport of materials across the cell membrane is one of the most important activities of cells. Other than using energy from your food to keep you warm, the most important use of energy in your body is to help pump substances across the membranes of your cells by active transport—a process that goes on all the time. Cells get the energy for active transport from ATP molecules.

## ENERGY TRANSFORMATIONS INSIDE THE CELL

The energy used in the chemical reactions that take place inside a cell is associated with the electrons of atoms. The greater an electron's distance from its nucleus, the more stored, or potential, energy it has. When some atoms join to form a new chemical compound, the electrons shared between them form covalent bonds. Each type of covalent bond has a specific amount of stored energy. Whenever covalent bonds are formed or broken, the amount of stored energy changes. Chemical reactions are mainly energy transformations in which the energy stored in chemical bonds is transferred to other newly formed chemical bonds or is released as heat or light. (See Figure 4-3.)

For example, the glucose in the food you eat is used by your cells after it is digested. In a chemical reaction that requires oxygen, the high-energy chemical bonds in the glucose molecules are broken. When glucose molecules are broken apart, **carbon dioxide** and water are formed. The energy levels of the chemical bonds in the carbon dioxide and water are lower than the energy levels of the chemical bonds in the glucose. Thus, energy has been released. ATP is the substance in which cells store this released energy until it is needed.

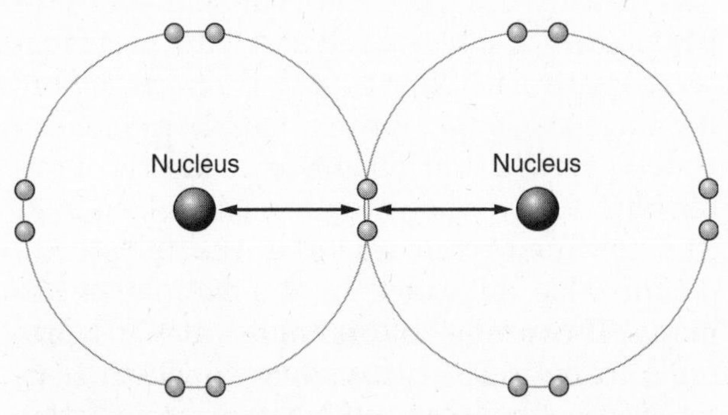


**Figure 4-3** The energy in a covalent bond depends on how far the shared electrons are from the nuclei of the atoms. When covalent bonds are formed or broken, the amount of stored energy changes.

## ENZYMES: THE CELL'S MIRACLE WORKERS

For chemical reactions in the cell to take place, they must usually occur in a series of small steps rather than in a single large burst of activity (such as occurs in a heated test tube). The steps must also be very precise. They must occur in the correct order, one after the other. One problem for the cell is that the reactants must get changed into exactly the right products and not into something else. Also, the reaction must occur at a relatively low temperature so that the cell is not harmed.

These problems are solved by substances called **enzymes**. Because a particular enzyme is needed for each type of chemical reaction, cells have thousands of different kinds of enzymes. The correct enzyme can enable a reaction to occur 10 times a second that otherwise might occur only once every 100 years. And the same enzyme molecule does its job over and over again without itself being changed or used up.

Substances that are responsible for greatly changing the rate at which chemical reactions occur, without being changed themselves, are called **catalysts**. Enzymes are organic catalysts, because they are either proteins or nucleic acids. Enzymes are very specific in their work because they are usually proteins that have a particular shape. The shape of an enzyme molecule includes a spot on its surface that is like a pocket. This "pocket" is exactly the right size and shape for a particular substance. If two different substances are involved in the chemical reaction, there will be a precise spot in the enzyme's pocket for each substance. This place on the enzyme molecule is called the *active site*. The fit of a substrate in an active site is so precise that it is often referred to as the lock-and-key model of enzyme action. (See Figure 4-4.)

An enzyme does its work by joining with the substances in this close fit. This is a temporary

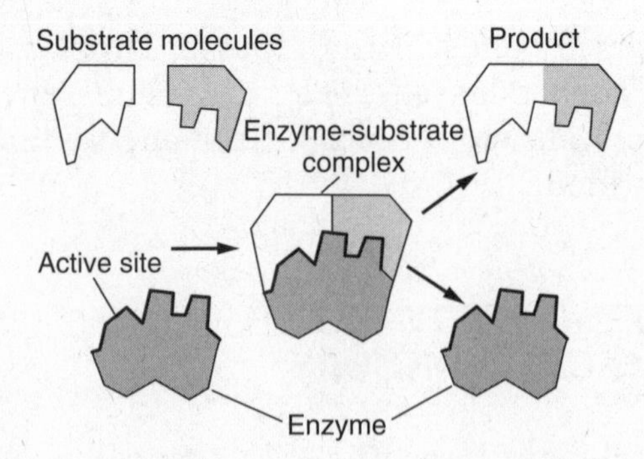


Figure 4-4 Enzymes speed up the rate of chemical reactions in the body by temporarily joining, at their active site, with other molecules.

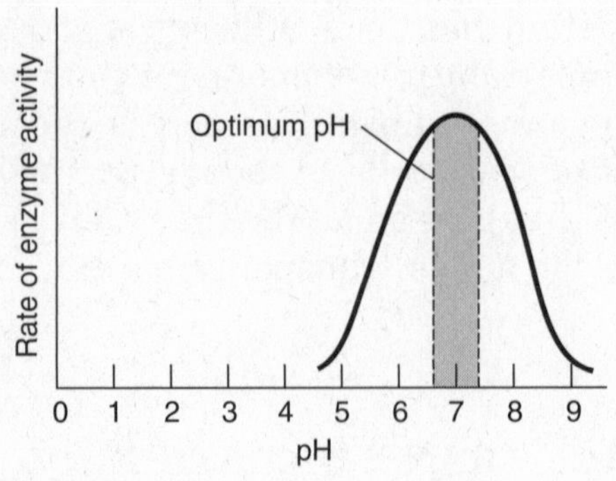


Figure 4-6 Most enzymes function best at about pH 7, which is neutral.

association only. The chemical reaction occurs while the enzyme and substance are fitted together. The substance changes in a specific way, but the enzyme does not change. The product of the reaction is released from the active site and moves to the next step in the process. The unchanged enzyme, with its open active site, gets used again to catalyze the same reaction on another molecule of the same substance. In fact, the enzymes are recycled.

## HELPING ENZYMES DO THEIR JOBS

For an enzyme to work properly, the protein molecule must maintain its correct shape. Two conditions in the cell, temperature and **pH**, are very important for maintaining the shape of an enzyme molecule. The main reason animals need to maintain a constant body temperature is to allow cellular enzymes to function properly. (That is why having a high fever can be so dangerous.) (See Figure 4-5.) The pH scale is used to measure how acidic or basic a solution is. Slight changes in pH quickly change an enzyme's shape and its ability to affect a substance. Most enzymes function best at about pH 7, which is neutral. Our cells must maintain that pH to survive. (See Figure 4-6.)

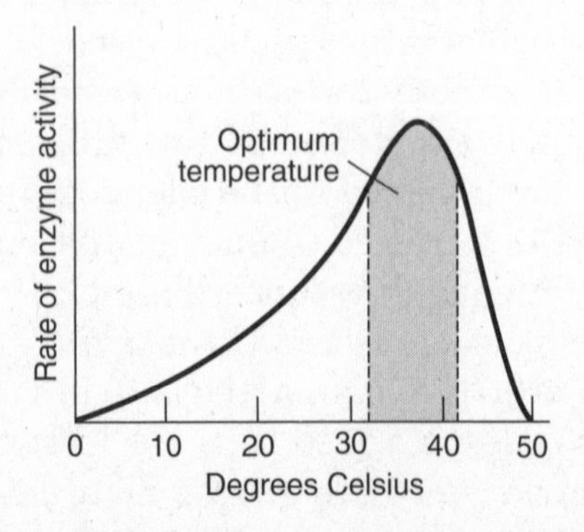


Figure 4-5 A typical enzyme in the body works best at temperatures between 35°C and 40°C.

## A TOUR OF THE CELL: ORGANELLES

Some living things are made up of a single cell, but most have many cells that work together on behalf of the organism. Yet, almost everything an organism does to stay alive is accomplished by each individual cell: getting food, using food for energy, transporting substances, growing, reproducing, and eliminating wastes. Each of these activities involves a large number of chemical reactions. Organization is needed for all of these reactions to take place under the precise control of so many enzymes. In many cells, these reactions take place in special internal structures called **organelles**.

In a typical *eukaryotic cell*, membrane-bound organelles are dispersed throughout the **cytoplasm**, which fills the cell and transports materials within it. (Note that *prokaryotic cells* do not have these membrane-bound organelles.) Cytoplasm is a thin gel, made up mostly of water, with many other chemicals dissolved in it. The cell membrane encloses the cell's cytoplasm. Also dispersed in the cytoplasm are many **ribosomes**, the organelles at which proteins are built. The **lysosomes**, which are scattered throughout the cell and contain digestive enzymes, are the structures involved in breaking down food. The complicated-looking **mitochondria**, shaped like tiny kidney beans, are

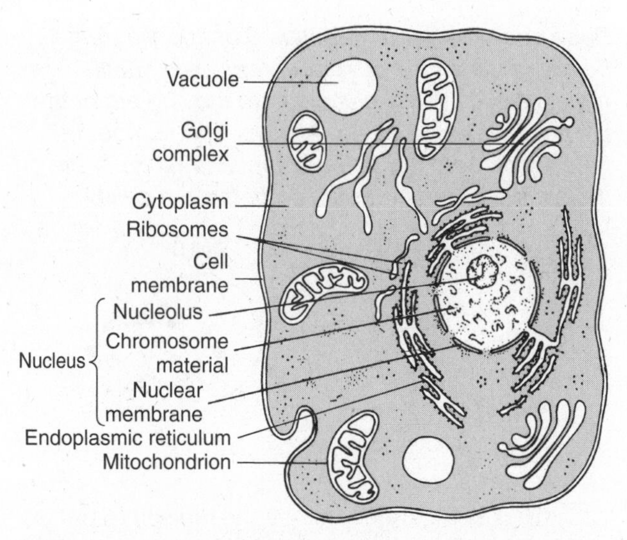


Figure 4-7 The organelles of a typical animal cell.

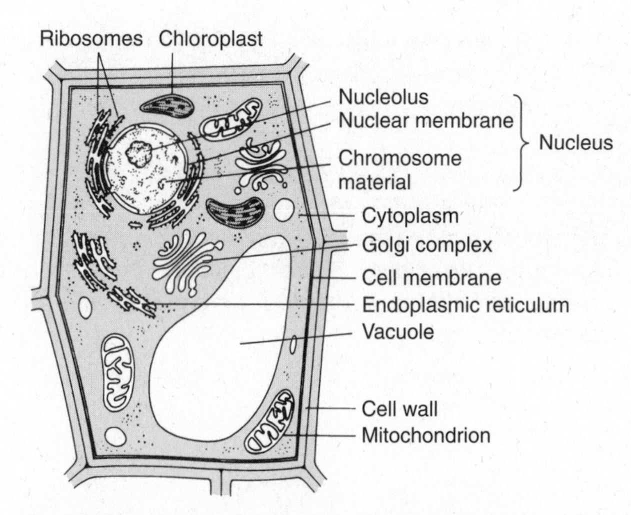


**Figure 4-8** The organelles of a typical plant cell.

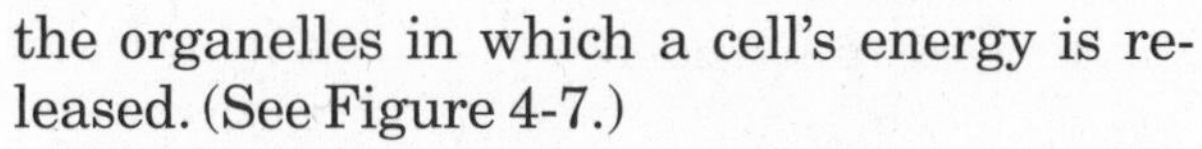

the organelles in which a cell's energy is released. (See Figure 4-7.)

The largest structure in a cell is generally the **nucleus**, which is responsible for information storage. The nucleus often fills the entire central portion of the cell. In addition, other specialized structures may be present in a cell to handle other functions, such as: the **Golgi complex**, which packages and distributes many materials; **vacuoles**, which store materials such as wastes; and, in plants, **chloroplasts**, the organelles in which plants convert energy from the sun into food; and the **cell wall**, a rigid layer that surrounds the entire plant cell. (See Figure 4-8.)

## Chapter 4 Review

### *Part A—Multiple Choice*

**1.** Which letter in the diagram indicates the cell structure that directly controls the movement of molecules into and out of the cell?

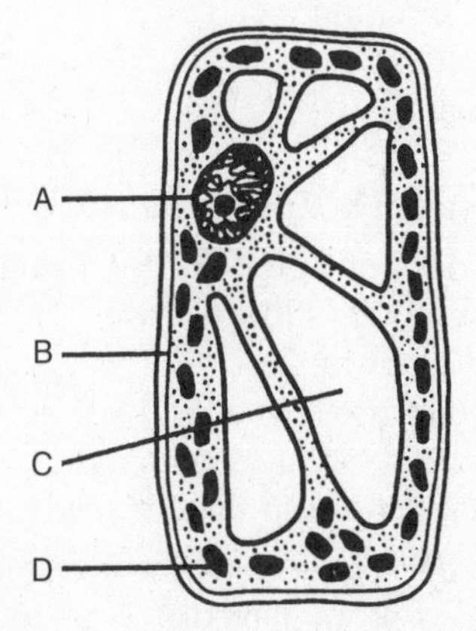


1 letter A
2 letter B
3 letter C
4 letter D

**2.** Which statement about the functioning of the cell membrane of all organisms is *not* correct?

1 The cell membrane forms a boundary that separates the cell's contents from the outside environment.
2 The cell membrane forms a barrier that keeps all substances that might harm the cell from entering it.
3 The cell membrane is capable of receiving and recognizing chemical signals.
4 The cell membrane controls the movement of molecules into and out of the cell.

**3.** What happens during diffusion?

1 Molecules move automatically from an area of higher concentration to an area of lower concentration.
2 Molecules are pumped from an area of lower concentration to an area of higher concentration.

3 An enzyme joins with a particular molecule.
4 A catalyst speeds up the rate of a chemical reaction.

*Base your answer to question 4 on the following diagram, which represents a cell in water. Formulas of molecules that can move freely across the cell membrane are shown. Some molecules are located inside the cell and others are in the water outside the cell.*

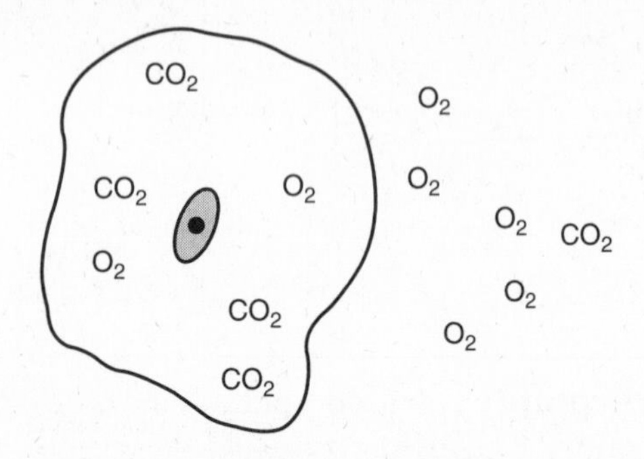


**4.** Based on the distribution of these molecules, what would most likely happen to them after a period of time has passed?
1 The concentration of $O_2$ will increase inside the cell.
2 The concentration of $O_2$ will remain the same outside the cell.
3 The concentration of $CO_2$ will remain the same inside the cell.
4 The concentration of $CO_2$ will decrease outside the cell.

**5.** A plant cell shrinks when placed in salt water due to the osmosis of
1 water molecules out of the cell
2 water molecules into the cell
3 salt into the cell
4 salt out of the cell

**6.** Placing limp celery in water will make the celery stalk firm again due to
1 diffusion
2 osmosis
3 active transport
4 a catalyst

**7.** A high concentration gradient means that the concentration of a substance is
1 low on both sides of the cell membrane
2 high on both sides of the cell membrane
3 about the same on both sides of the cell membrane
4 high on one side of the cell membrane and low on the other side

*Base your answer to question 8 on the diagram below, in which the dark dots represent small molecules. These molecules are moving out of the cells, as indicated by the arrows. The number of dots represents the relative concentrations of the molecules inside and outside of the two cells.*

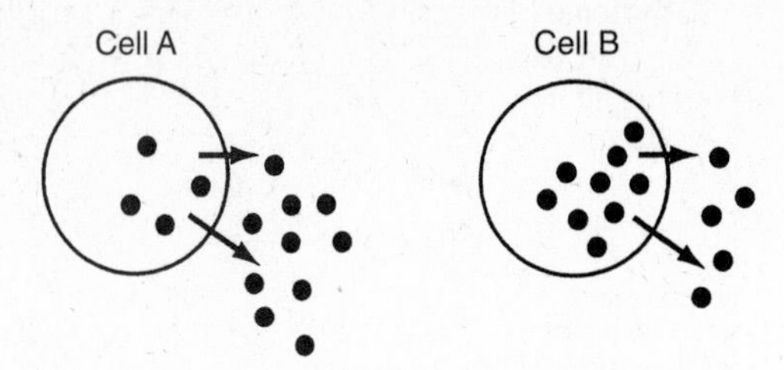


**8.** ATP is being used to move the molecules out of
1 cell A only
2 cell B only
3 both cell A and cell B
4 neither cell A nor cell B

*Refer to the set of diagrams below, which shows the movement of a large molecule across a cell membrane, to answer question 9.*

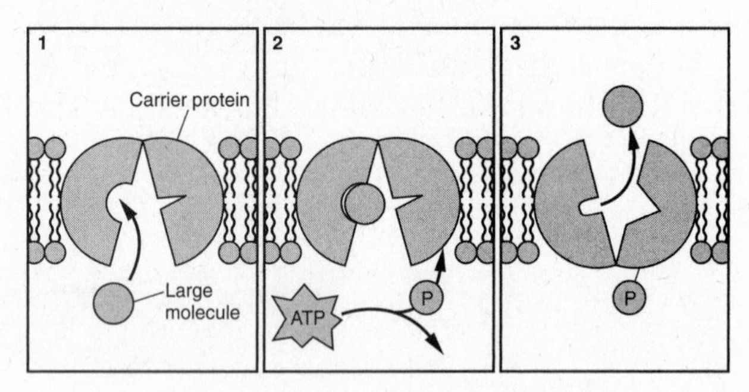


**9.** Which process is best represented by this set of diagrams?
1 active transport
2 protein building
3 diffusion
4 gene transfer

**10.** When covalent bonds are formed or broken,
1 the amount of stored energy changes
2 energy is always lost
3 energy is produced
4 new atoms are formed

**11.** Chemical reactions in the cell take place
1 over extremely long periods of time
2 in a series of small steps
3 all at once in a single burst
4 during short intervals of time

**12.** How do chemical reactions occur at the relatively low temperature found within cells?

1 Some energy is destroyed before it heats up the cell.
2 Some energy is stored temporarily in ATP molecules.
3 Enzymes are used to slow the rate of the reactions.
4 Enzymes are used to speed the rate of the reactions.

**13.** To carry out its chemical reactions, each cell contains

1 one specific type of enzyme
2 fewer than 10 different enzymes
3 thousands of different enzymes
4 thousands of copies of the same enzyme

**14.** The equation below represents a chemical reaction that occurs in humans. What data should be collected to support the hypothesis that enzyme C works best in an environment that is slightly basic?

$$\text{Substance X} + \text{Substance Y} \xrightarrow{\text{ENZYME C}} \text{Substance W}$$

1 the amino acid sequence of enzyme C
2 the shapes of substances X and Y after the reaction occurs
3 the amount of substance W produced in 5 minutes at various pH levels
4 the temperature before and after the reaction occurs

*Base your answer to question 15 on the diagrams below, which show an enzyme and four different molecules.*

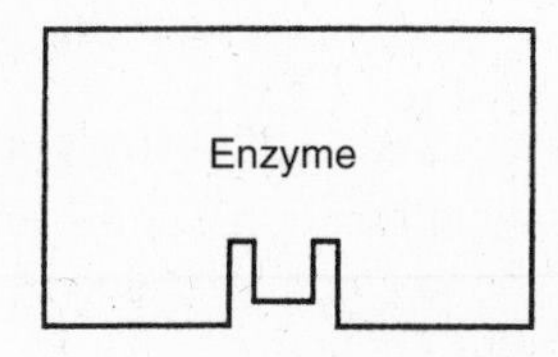


Molecules:

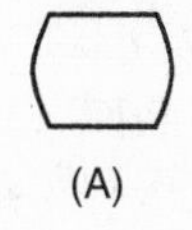
(A)

(B)

(C)
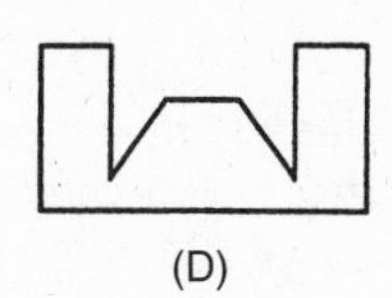
(D)

**15.** The enzyme would most likely affect reactions that involve

1 molecule A only
2 molecule C only
3 molecules B and D
4 molecules A and C

**16.** Which best describes the interaction between an enzyme and another substance?

1 a temporary association in which the substance changes
2 a temporary association in which the enzyme changes
3 the final product of a series of chemical reactions
4 the pocket into which the enzyme and the substance fit

**17.** Two conditions that must be maintained in a cell in order for enzymes to work properly are

1 pH and oxygen content
2 surface area and temperature
3 temperature and pH
4 volume and pressure

**18.** While viewing a slide of rapidly moving sperm cells, a student concludes that these cells require a large amount of energy to maintain their activity. The organelles that most directly provide this energy are known as

1 vacuoles
2 chloroplasts
3 ribosomes
4 mitochondria

**19.** The organelle that stores wastes for the cell is the

1 vacuole
2 chloroplast
3 ribosome
4 Golgi complex

**20.** Chloroplasts are important because they

1 are necessary to release stored energy
2 store wastes in both plants and animals
3 use energy from the sun to make food
4 are in the nuclei of both plant and animal cells

### *Part B—Analysis and Open Ended*

*Refer to the following diagram to answer questions 21 and 22.*

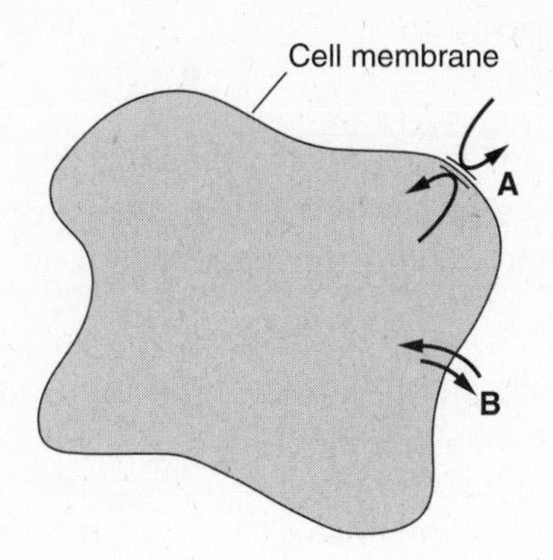

**21.** Which part of the diagram shows the cell membrane acting to allow materials into and out of the cell?

1 A only
2 B only
3 both A and B
4 neither A nor B

**22.** Which part of the diagram shows the cell membrane acting to separate the inside of the cell from the outside environment?

1 A only
2 B only
3 both A and B
4 neither A nor B

**23.** Discuss the meaning of "selective permeability" for a cell membrane. Your answer should explain the following:

- why the cell membrane is said to be selectively permeable
- why this characteristic is important to the health of a cell

*Base your answers to questions 24 and 25 on the diagram below, which represents a unicellular organism in a watery environment. The small triangles represent molecules of a specific substance.*

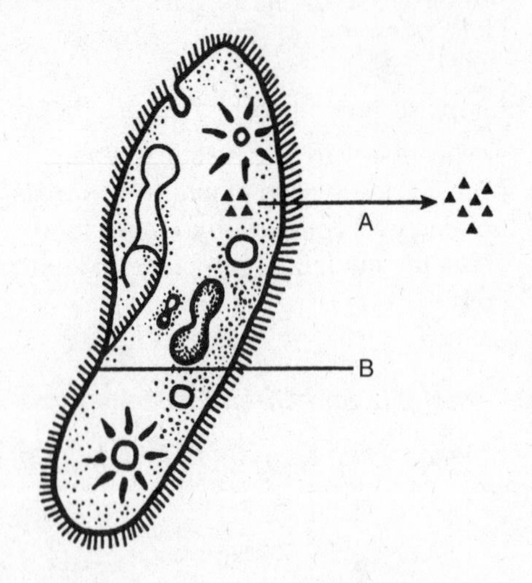


**24.** What kind of transport is represented by arrow *A*? State two ways in which active transport is different from diffusion (passive transport).

**25.** In the cells of multicellular organisms, structure *B* often contains special proteins known as *receptor molecules*. What specific function do these protein molecules carry out for the cell?

*Refer to the diagrams below to answer questions 26 and 27.*

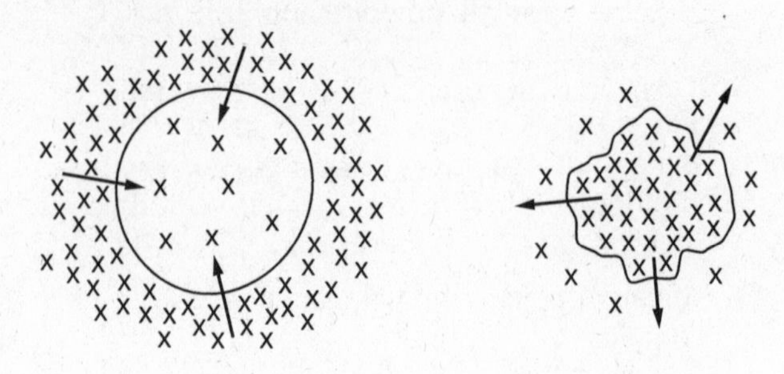

**26.** The diagrams represent the movement of molecules from an area of

1 low concentration to an area of high concentration
2 high concentration to an area of low concentration
3 low concentration to an area of equal concentration
4 high concentration to an area of equal concentration

**27.** The diagrams could be used to illustrate all of the following types of transport *except*

1 diffusion
2 osmosis
3 active
4 passive

**28.** Use the following terms to construct a concept map that shows how a cell regulates its internal environment: *active transport; diffusion; uses energy; does not use energy; cell membrane; osmosis.*

**29.** Identify the *two* main things that can happen (during chemical reactions) to the energy stored in chemical bonds. For what important activity do cells use this energy?

*Base your answers to questions 30 and 31 on the diagram below, which represents an enzyme and four types of molecules present within a solution in a flask.*

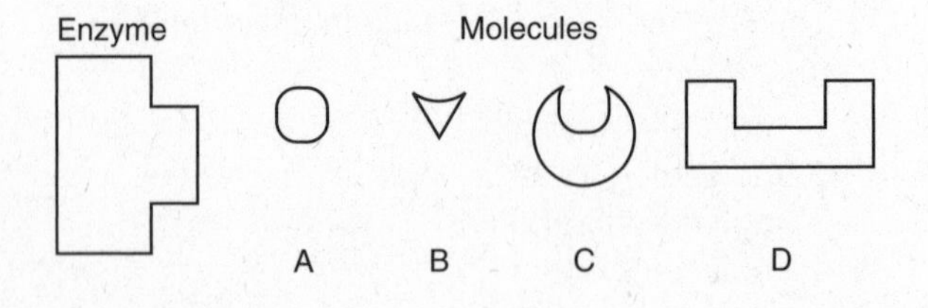


**30.** Which molecule would most likely react with the enzyme? Why?

**31.** What would happen to the rate of reaction if the temperature of the solution in the flask were increased gradually from 10°C to 30°C?

**32.** Enzyme molecules are affected by changes in conditions within organisms. Explain how a prolonged, high body temperature could be fatal to a human. Your answer must include:

- the role of enzymes in the human body
- the effect of a high body temperature on enzyme activity

**33.** Explain why it is important for the cells of our bodies to maintain a neutral pH.

**34.** What, specifically, would happen to a cell if its mitochondria were removed? Explain why.

**35.** Just like complex organisms, cells are able to survive by coordinating various activities. Complex organisms have a variety of systems, and cells have a variety of organelles. Describe the roles of two organelles. Be sure to include:

- the names of two organelles and the function of each
- an explanation of how these two organelles work together
- the name of an organelle and the name of a system in the human body that have similar functions

## *Part C—Reading Comprehension*

*Base your answers to questions 36 and 37 on the information below and on your knowledge of biology. Use one or more complete sentences to answer each question.*

Bilirubin, the bile pigment that yellows the skin of babies born with jaundice, is generally considered a toxic molecule. According to a new study, however, bilirubin may actually protect cells from dangerous oxygen-containing molecules called free radicals.

Bilirubin forms during the breakdown of hemoglobin, the oxygen-carrying protein in blood cells, and can build up to high concentrations in the blood. Several lines of evidence indicate that bilirubin is toxic, but why then is there a specific enzyme that converts the seemingly harmless molecule known as biliverdin into bilirubin?

Scientists puzzled by this question have unearthed data suggesting that bilirubin, when present at the right concentration, is helpful instead of harmful. A research team headed by Solomon H. Snyder of Johns Hopkins University School of Medicine in Baltimore reports that bilirubin protects brain cells growing in lab dishes from the damage typically caused by hydrogen peroxide, a free radical.

The scientists compared normal cells with ones in which the bilirubin-making enzyme was inhibited. The normal cells were able to survive a dose of hydrogen peroxide 10,000 times greater than the lethal dose for the bilirubin-deprived cells. The investigators report their findings in an upcoming *Proceedings of the National Academy of Sciences*.

Snyder and his colleagues also garnered evidence for a mechanism by which bilirubin, which is altered when it defuses free radicals, is recycled back into its original form. This reuse amplifies its protective powers. A protective role for bilirubin may explain previous findings that have linked low blood concentrations of the molecule to cancer, heart attacks, and other diseases, the scientists note.

**36.** Describe how bilirubin is produced in the body.

**37.** Explain why scientists have been confused about bilirubin.

# 5
# Photosynthesis and Respiration

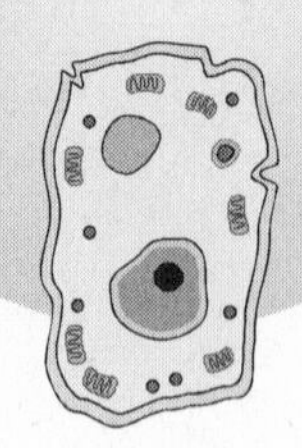

**Vocabulary**

| | | |
|---|---|---|
| ATP (adenosine triphosphate) | heterotrophic | organic |
| autotrophic | inorganic | photosynthesis |
| cellular respiration | | |

## FOOD: MATTER AND ENERGY

An apple is a type of food. It contains complex organic compounds; its atoms are held together as molecules by chemical bonds that are rich in stored energy. When you eat an apple, you get both the matter and the energy you need to build your body and to stay alive.

The apple tree that produced the fruit represents the group of organisms that are **autotrophic**, meaning "self-feeding." Like other plants, the apple tree makes its own food, taking in the **inorganic** substances carbon dioxide ($CO_2$) and water ($H_2O$) and changing them into **organic** compounds, such as sugars and starches. Humans represent the other group of organisms, which are **heterotrophic**, meaning "other-feeding." Since they cannot make their own food, humans and all other animals must get their complex organic compounds by eating other organisms. (See Figure 5-1.)

**Figure 5-1** The grass, like all other plants, is autotrophic because it makes its own food. The cows, like all other animals, are heterotrophic because they have to eat other organisms in order to survive.

For the apple tree to combine the inorganic raw materials of $CO_2$ and $H_2O$ into organic compounds such as sugar and starch, it needs a source of energy. The rays of sunlight, as they fall on the leaves of the apple tree, provide that energy. The process of making this food, by using light as the source of energy, is called **photosynthesis**. All green plants are photosynthetic autotrophs. Without plants to capture the energy of sunlight and convert it into the chemical forms that are edible, most animals would have no constant source of food and could not exist.

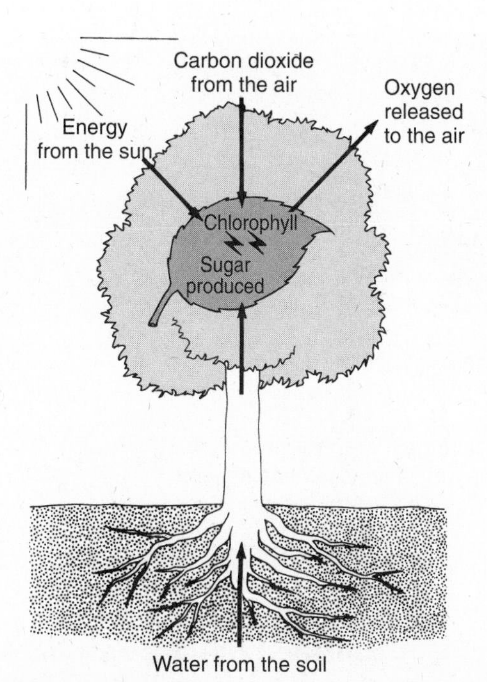


**Figure 5-2** The diagram illustrates the basic process of photosynthesis—plants take in inorganic substances from the environment and produce organic substances such as glucose (a sugar).

## PHOTOSYNTHESIS

Plants (and algae) are able to make their own energy-rich carbon compounds. In particular, they make the simple sugar glucose, whose chemical formula is $C_6H_{12}O_6$. Plants get the carbon for these glucose molecules from inorganic $CO_2$ in the air. In addition, plants release oxygen ($O_2$) to the air. (See Figure 5-2.) Scientists discovered that photosynthesis requires the green pigment *chlorophyll*. The chemical reactions of photosynthesis occur within the chlorophyll-containing organelles called *chloroplasts*, found within plant leaves and stems. Some scientists consider this process of photosynthesis the single most important chemical reaction that occurs on Earth. This all-important reaction can be summarized by the following chemical equation:

$$6\,CO_2 + 6\,H_2O \xrightarrow[\text{CHLOROPHYLL AND ENZYMES}]{\text{LIGHT ENERGY}} C_6H_{12}O_6 + 6\,O_2$$

CARBON DIOXIDE + WATER → GLUCOSE + OXYGEN

## LEAVES: PHOTOSYNTHETIC FACTORIES

The structures present inside the leaf are well organized. Such organization allows cells that contain chlorophyll to get maximum exposure to light. At the same time, the leaf controls the amount of water lost to the air. It also makes possible the movement of $CO_2$ and $O_2$ into and out of the leaf. (See Figure 5-3.)

## THE RATE OF PHOTOSYNTHESIS

As with any chemical reaction, the reactions of photosynthesis can occur at different rates. The factors that affect the rate at which photosynthesis occurs are temperature, light intensity, $CO_2$ concentration, availability of water, and the presence of certain minerals. (See Figure 5-4 on page 40.)

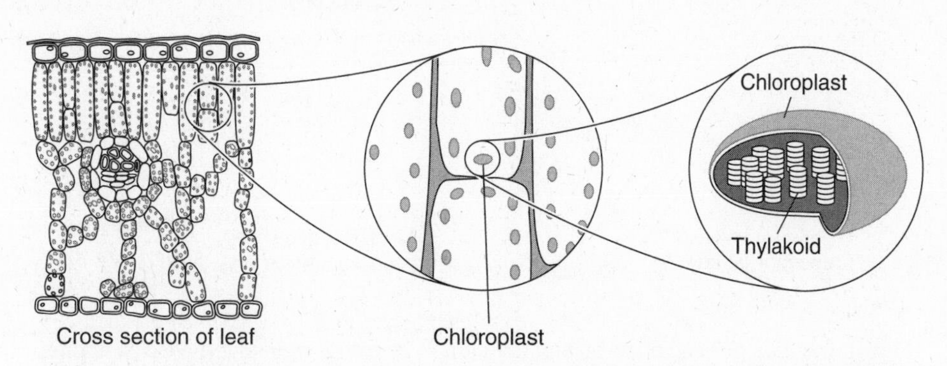


**Figure 5-3** Cells inside a leaf contain chloroplasts, which capture the sunlight that is used for photosynthesis. The structure of the leaf allows these cells to get maximum exposure to the light.

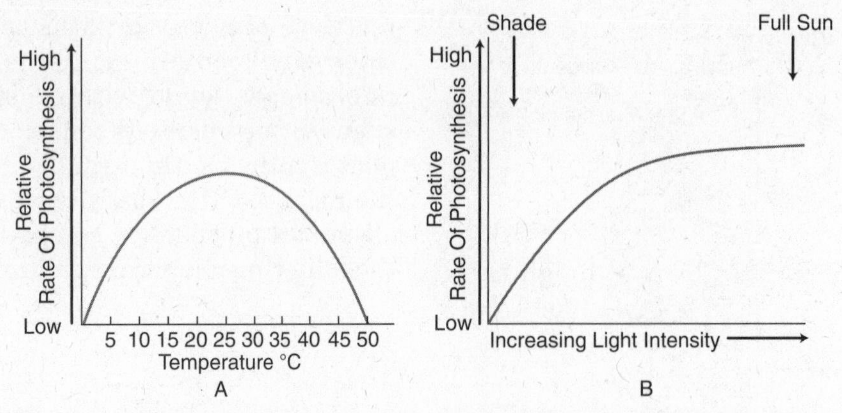


**Figure 5-4** Factors that influence the rate of photosynthesis: graph A shows the effect of different temperatures, while graph B shows the effect of increasing light intensity. Other factors, such as the availability of $CO_2$, water, and minerals, also affect the rate of photosynthesis.

## CELLULAR RESPIRATION: RELEASING THE STORED ENERGY IN FOOD

Consider the relationship between the sunlight that falls on the leaves of an apple tree and the chemical process of photosynthesis. During photosynthesis, the light energy of the sun is converted into the stored chemical energy of glucose in the apple. After you eat the apple, your cells are ready to use that stored chemical energy. How does this happen?

The release of energy cannot occur all at once. Too much heat would be released inside your cells. Instead, the release of energy occurs in a series of enzyme-controlled small steps. The energy stored in glucose is converted into a usable form, the energy source of all cells, **adenosine triphosphate**, or **ATP**. This process is known as **cellular respiration**. (See Figure 5-5.) Cellular respiration is basically the opposite process of photosynthesis, and can be summarized by the following chemical equation:

$$C_6H_{12}O_6 + 6\ O_2 \xrightarrow[\text{CHLOROPHYLL AND ENZYMES}]{\text{ENZYMES}} 6\ CO_2 + 6\ H_2O + 36\ ATP$$

GLUCOSE + OXYGEN → CARBON DIOXIDE + WATER

Instead of being produced in the cells, the energy-rich glucose molecules are taken apart to release their stored energy. Oxygen is used, and $CO_2$ and water are released as wastes. (Because oxygen is used to produce ATP, this is referred to as an *aerobic* process.) Cells use

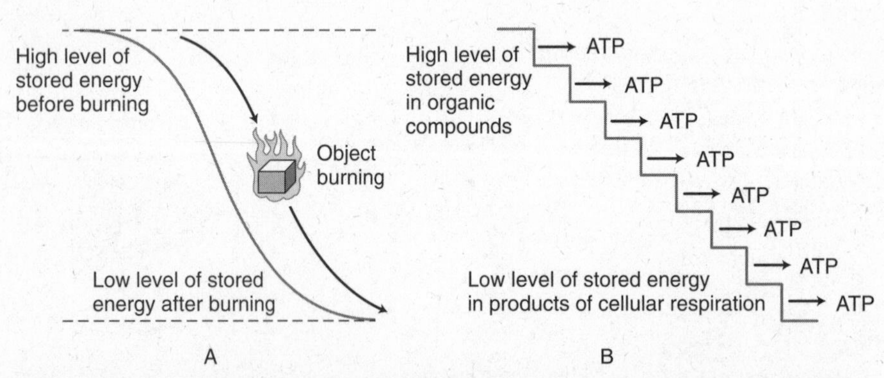


**Figure 5-5** The burning of an object (diagram A) is due to the sudden release in one step of the energy stored in that object. By contrast, in cellular respiration (diagram B), energy is released from organic compounds as ATP in a series of small, enzyme-controlled steps.

the energy from ATP to perform many functions, such as obtaining materials and eliminating wastes.

## A FINAL VISIT BACK TO PLANTS

A plant does not specifically go through the process of photosynthesis to make food for people and other animals. The apple tree, for example, is simply making food for itself to live long enough to reproduce successfully. The apples contain seeds, which may get carried away to new places by animals that eat the apples. This makes it possible for the apple tree to produce more apple trees in other places. However, most of the glucose made in the leaves of the tree does not go into storage in the form of apples. Rather, it gets taken to different parts of the tree and used by the tree to stay alive. In fact, the tree uses the same process as you do to get the energy it needs from the glucose it has made—cellular respiration.

To summarize: Plants are autotrophs; they are able to produce their own food by photosynthesis and use it for energy through cellular respiration. Animals are heterotrophs; they must obtain food energy from other organisms. Animals use cellular respiration, just as plants do, to obtain energy from the food they eat.

# Chapter 5 Review

### *Part A—Multiple Choice*

**1.** Which of the following is an autotroph?

1 lizard
2 cactus
3 shark
4 antelope

**2** In heterotrophs, energy for the life processes comes from the chemical energy stored in the bonds of

1 water molecules
2 organic compounds
3 oxygen molecules
4 inorganic compounds

**3.** During photosynthesis,

1 animals use sunlight to convert the starch in plants into food
2 animals use the oxygen released by plants to make carbon dioxide
3 plants use the energy of sunlight to convert carbon dioxide and water into glucose and oxygen
4 plants use the energy of sunlight to convert glucose and oxygen into carbon dioxide and water

**4.** The following equation represents an important biological process. This process is carried out within a cell's

$$\text{carbon dioxide} + \text{water} \longrightarrow \text{glucose} + \text{water} + \text{oxygen}$$

1 mitochondria
2 cell membranes
3 ribosomes
4 chloroplasts

**5.** The source of energy for photosynthesis is

1 oxygen
2 sunlight
3 carbon dioxide
4 glucose

**6.** To occur, photosynthesis requires the presence of the green substance

1 tree sap
2 glucose
3 chlorophyll
4 copper

**7.** The approximate mass of a field of corn plants at the end of its growth period was 3 tons per hectare. Most of this mass was produced from

1 water and organic compounds absorbed from the soil
2 minerals and organic materials absorbed from the soil
3 minerals from the soil and oxygen from the air
4 water from the soil and carbon dioxide from the air

**8.** The diagram below represents part of the life process that occurs inside a leaf chloroplast. If the process were to be interrupted by a chemical at point X, there would be an immediate effect on the release of which substance?

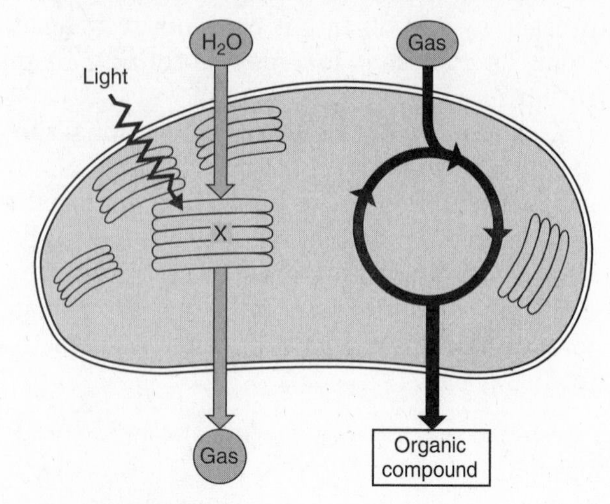


1 chlorophyll
2 carbon dioxide
3 nitrogen
4 oxygen

**9.** The food produced by plants during photosynthesis is used

1 by the plants themselves only
2 by animals that eat them only
3 by both the plants and the animals that eat them
4 up at the end of the reaction

**10.** If stored energy were to be released too quickly, a cell would

1 release too much heat
2 produce ATP molecules
3 become an autotroph
4 become a heterotroph

*Answer question 11 based on the following information and word diagram.*

The flow of energy through an ecosystem involves many energy transfers. The diagram below summarizes the transfer of energy that eventually powers muscle activity.

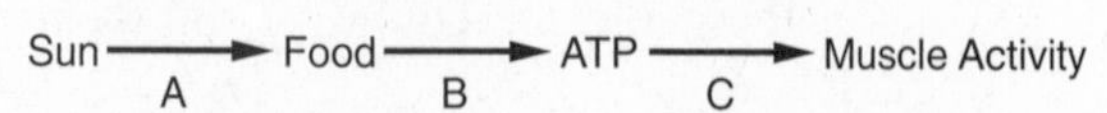


**11.** The process of cellular respiration is represented by

1 arrow A only
2 arrow B only
3 arrow C only
4 arrows A, B, and C

**12.** How do humans and plants interact in terms of the two gases involved in photosynthesis?

1 Humans take in the $CO_2$ released by plants and release $O_2$ to the plants.
2 Humans take in the $O_2$ released by plants and release $CO_2$ to the plants.
3 Plants and humans usually compete for the same $O_2$ available in the air.
4 Plants and humans usually compete for the same $CO_2$ available in the air.

**13.** Cellular respiration occurs in

1 autotrophs only
2 heterotrophs only
3 autotrophs and heterotrophs
4 humans only

**14.** Eating a sweet potato provides energy for human metabolic processes. The original source of this energy is the energy

1 in protein molecules stored within the potato
2 that is made available by photosynthesis
3 from starch molecules absorbed by the potato plant
4 in vitamins and minerals found in the soil

**15.** In nature, during a 24-hour period, green plants *continuously* use

1 carbon dioxide only
2 oxygen only
3 both carbon dioxide and oxygen
4 neither carbon dioxide nor oxygen

**16.** Plant leaves contain openings known as stomates, which are opened and closed by specialized cells, allowing for gas exchange between the leaf and the outside environment. Which phrase best describes the net flow of gases involved in photosynthesis into and out of the stomates on a sunny day?

1 carbon dioxide moves in, oxygen moves out
2 oxygen moves in, nitrogen moves out
3 carbon dioxide and oxygen move in, ozone moves out
4 water and ozone move in, carbon dioxide moves out

### Part B—Analysis and Open Ended

*Answer question 17 based on the following information and graph.*

As the depth of the ocean increases, the amount of light that penetrates to that depth decreases. At about 200 meters, there is almost no light present. The graph below illustrates the population size of four different species at different water depths.

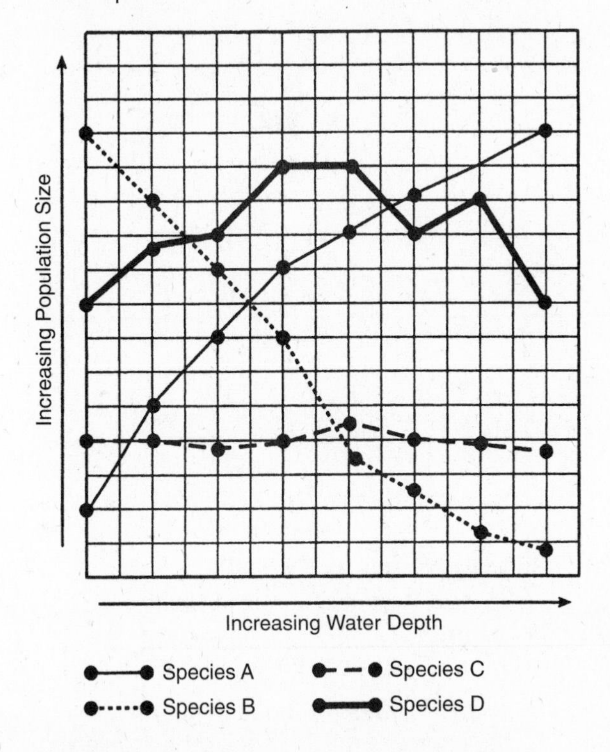


**17.** Which species most likely performs photosynthesis?

1 species A
2 species B
3 species C
4 species D

**18.** Explain why plants are defined as autotrophs and why animals are defined as heterotrophs.

**19.** Why might the process of photosynthesis be considered a "bridge" between the living and nonliving parts of the world?

**20.** Briefly describe *three* ways in which the structures of a leaf enable the process of photosynthesis to occur. Your answer should include the following factors:

- light
- water
- gases

*Base your answers to questions 21 and 22 on the summary equations of two processes shown below and on your knowledge of biology.*

**Photosynthesis**

Water + Carbon Dioxide —ENZYMES→ Glucose + Oxygen (+ Water)

**Respiration**

Glucose + Oxygen —ENZYMES→ Water + Carbon Dioxide

**21.** Choose *one* of the processes shown above and identify the following:

a. the source of the energy in the process you chose; and
b. where the energy ends up at the end of that process.

**22.** State *one* reason why *each* of the following processes is important for living things:

a. respiration; and
b. photosynthesis.

*Base your answers to questions 23 to 25 on the information and diagram below and on your knowledge of biology.*

The diagram represents a system in a space station that includes a tank containing algae. An astronaut from a spaceship boards the space station.

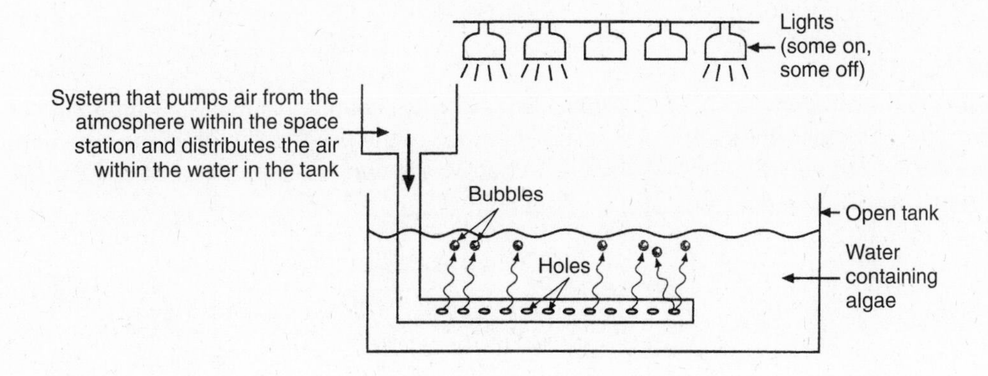

**23.** Identify *one* process that is being controlled in the setup shown in the diagram.

**24.** State *two* changes in the chemical composition of the space station atmosphere as a result of the astronaut coming on board the station.

**25.** State *two* changes in the chemical composition of the space station atmosphere that would result from turning on more lights.

*Base your answers to questions 26 and 27 on the information and diagram below.*

The diagram represents a single-celled organism known as *Euglena*. This organism is able to carry out both photosynthesis and cellular respiration.

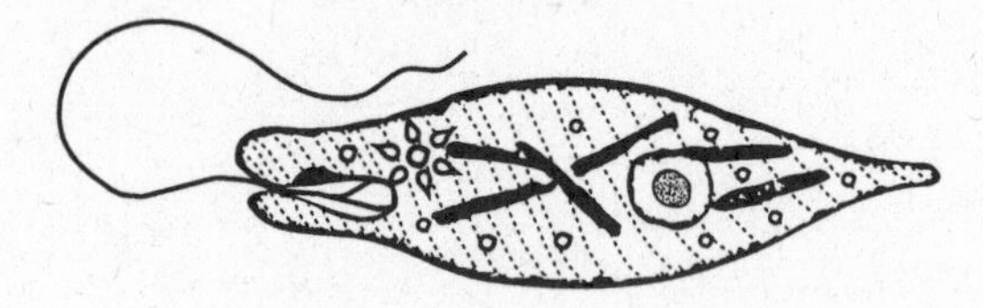

**26.** Choose *one* of the two processes that *Euglena* carries out. Write down the word for it; then use words or chemical symbols to summarize the reaction for the process you chose.

**27.** State *one* reason why the process you chose is essential for the survival of the *Euglena*.

**28.** Look at the chart below to answer this question. Which phrase would you choose to fill in the missing title?

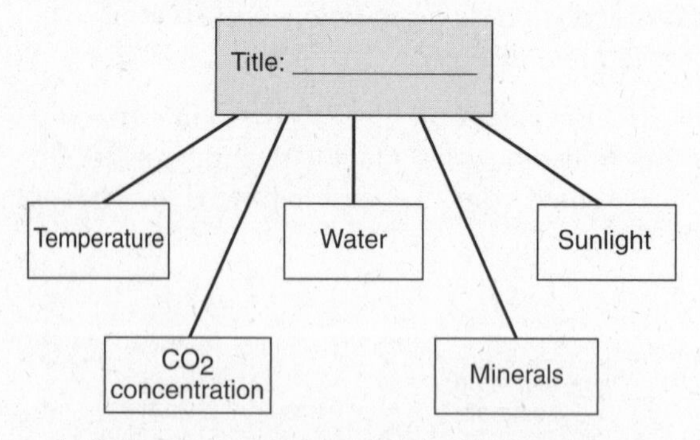


1 Some Living Factors in the Environment
2 The Chemical Process of Photosynthesis
3 Factors that Affect the Rate of Photosynthesis
4 The Nonliving Things that Make Up a Plant

**29.** Look at the following diagrams (A and B) to answer this question. The energy change in diagram B is different from the energy change in diagram A because, in diagram B,

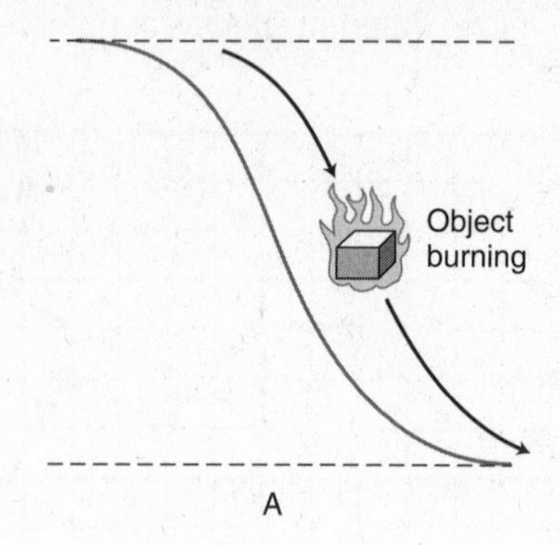


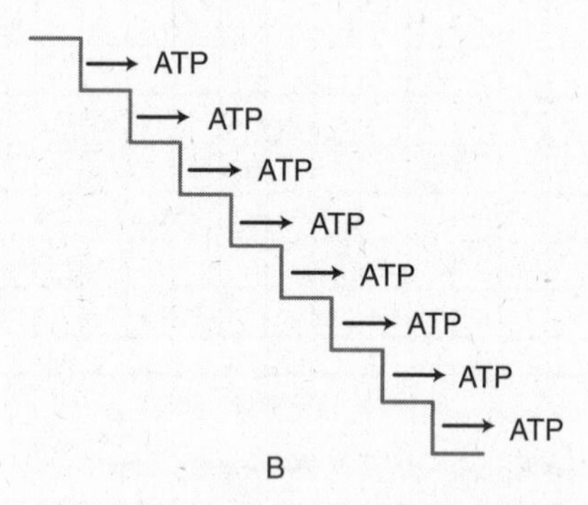


1 energy is released suddenly in one step
2 the energy is released in a series of steps
3 there is less stored energy at the beginning
4 there is less stored energy remaining at the end

*Refer to the chemical equation below to answer questions 30 and 31.*

$$CO_2 + H_2O \xrightarrow[\text{CHLOROPHYLL}]{\text{LIGHT ENERGY}} C_6H_{12}O_6 + O_2 \ (+ H_2O)$$

**30.** What important life process is described by this equation? What are the two vital products of this reaction?

**31.** Explain why "cellular respiration is basically the opposite" of the process shown in the equation. What are the two waste products of cellular respiration?

### *Part C—Reading Comprehension*

*Base your answers to questions 32 to 34 on the information below and on your knowledge of biology. Use one or more complete sentences to answer each question.*

We live on land. Even the very name *Earth* is used to mean land. But look at a world map and you will see a lot of blue space. In fact, more than 70 percent of Earth's surface is covered by water, mostly oceans. Unseen in these waters—drifting along with waves and currents—are countless numbers of tiny organisms. Photosynthetic bacteria, algae, and plants are included in these drifters. Some of these unicellular species are so small that if 12 million cells were lined up in a row, the line would be only about 1 centimeter long. In some places in the oceans, these microscopic organisms are so numerous that a cup of seawater may hold 24 million individuals of a single species, and that cup would contain other species as well!

These species are very small, but their importance to the overall life on the planet is huge. Tiny sea-dwelling organisms are the beginning food source for almost all living things in the oceans. It is easy for us land dwellers to understand that many animals eat plants to get food. We have seen cattle and sheep grazing on grasses in a pasture. The drifting cells in the ocean could be called the *fields* or *pastures* of the sea. Just like grass on land, the sea drifters capture energy from the sun and convert inorganic $CO_2$ and water into organic molecules, which become important foods for other organisms. On land, plants bloom with wild displays of colorful flowers in spring. The photosynthetic drifters in the sea are said to "bloom" in the spring, too, as the water warms and nutrients from ocean depths are brought to the surface by currents. A great deal has been learned recently about the seasonal explosive growth of these photosynthetic cells in the ocean from photographs taken by orbiting satellites.

**32.** Explain why the drifting cells in the ocean can be called the fields or pastures of the sea.

**33.** Describe *three* ways in which microscopic drifting cells in the ocean are similar to plants on land.

**34.** How has modern technology improved our ability to study life in the ocean?

# 6

# Getting Food to Cells: Nutrition

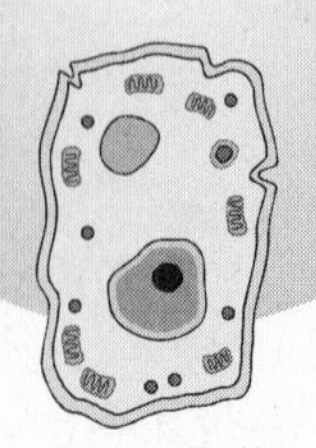

**Vocabulary**

| | | |
|---|---|---|
| absorption | digestion | nutrition |
| bacteria | nutrients | synthesis |
| bile | | |

## THE NEED FOR DIGESTION

The food you eat provides the matter your body needs to build cells, tissues, and organs. In addition to being used to build cells and tissues, food provides the energy an organism needs to remain alive. The process of taking in and utilizing food is called **nutrition**. (See Figure 6-1.)

For all organisms, food must be digested. Why? Every organism is either a single cell or a collection of cells. The cell theory states that cells are the basic units of structure and function of all living things. Therefore, what an organism needs is what its cells need. In order to nourish you, the food you eat must get into each of your cells.

**Figure 6-1** Food provides you with matter and energy to keep you alive.

That is why digestion is necessary. No matter what the food is—an earthworm for a bird or a hamburger for a person—the food is too large, and its molecules too complex, to get inside a single cell. To get inside the cells, food must be broken down into relatively simple molecules. **Digestion** is the process of breaking down food particles into molecules small enough to be absorbed by cells.

## THE HUMAN DIGESTIVE SYSTEM

The digestive system in humans has the same purpose as the digestive system in other organisms: to get **nutrients** (the vital molecules from food) into your cells. Food does not really enter your body when you put it into your mouth; not even when it goes into your stomach or intestine. These locations are actually connected to the outside world. Only

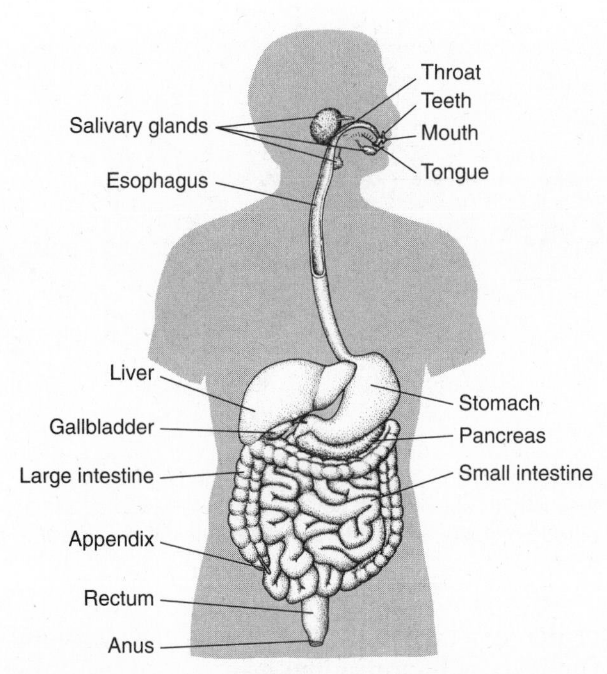


**Figure 6-2** The human digestive system—it functions to get the nutrients from your food into your cells.

when the food is absorbed across the membranes of cells that line your digestive system is it truly inside your body. (See Figure 6-2.)

Follow the path of a hamburger as it travels through the human digestive system. First, the hamburger must enter the mouth. This is the process of *ingestion*. Then teeth in the mouth provide *mechanical digestion*; they grind up the meat and bun into smaller and smaller pieces. The smaller pieces of food provide more surface area for the body's enzymes to work on. *Chemical digestion* of the bun begins when an enzyme in saliva, called *salivary amylase*, starts to break down the bun's starch into sugars.

The food is swallowed and pushed by the tongue to the back of the mouth where it enters the *esophagus*, a muscular tube that carries mouthfuls of hamburger down to the stomach. As this begins, the pathways up to the nose and down to the lungs are closed off. The *epiglottis*, a small flap of tissue in the throat, is pushed down by the food to close off the trachea. The *trachea* is the pathway for air to the lungs; if food enters it by mistake, coughing and difficult breathing occur. (See Figure 6-3.)

When the food arrives at the stomach, a muscular valve quickly opens and closes. The valve allows food to enter the stomach without letting out its acidic contents. Occasionally, acid backs up through the valve, causing a painful feeling in the chest, commonly called "heartburn." The *stomach* is a large muscular organ located in the abdomen. The muscular movement of the stomach churns the food and continues the process of mechanical digestion. As muscles in the stomach wall contract and relax, food gets mixed with *gastric juice*, which contains hydrochloric acid, pepsinogen, and water. The hydrochloric acid kills many bacteria that may be present in the food. The acid also turns the pepsinogen into the protein-digesting enzyme pepsin. Therefore, a second step in the chemical digestion of the hamburger begins in the stomach when pepsin breaks down proteins in the meat.

Muscular contractions of the stomach walls, called *peristalsis*, push the food contents into the *small intestine*, which is a narrow, muscular tube about seven meters in length. Most of the chemical digestion of food occurs in the small intestine, not in the stomach. Large quantities of different enzymes accomplish this task. These enzymes come from two main sources: the lining of the small intestine itself and the pancreas. The *liver* (the

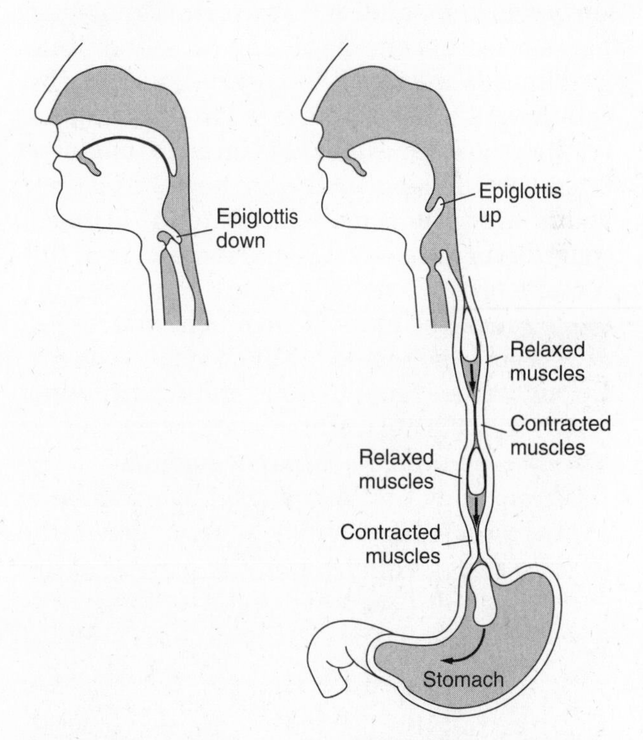


**Figure 6-3** The process of swallowing food—muscular contractions move food down the esophagus to the stomach.

largest organ in the abdomen) helps the body digest fat by producing **bile**, which is stored in the gallbladder until it is needed for digestion. Then it is emptied into the small intestine through a duct. When it comes into contact with fat, the bile acts like a detergent, breaking down the lipids into smaller droplets of fat. This process, called *emulsification*, allows the fat droplets to be more effectively digested by enzymes.

Now the hamburger is very much changed. The starch that was in the bun has been changed into simple sugar molecules, the meat protein exists as amino acids, and the fat is fatty acid molecules and glycerol. Yet the food is still outside the body. Only now are its molecules tiny enough to pass through the wall of the small intestine and into the blood vessels. This process is known as **absorption**. Tiny projections called *villi* on the lining of the small intestine increase its surface area and aid absorption of nutrients. Once they are in the blood, the food molecules are carried to all the cells and then are, finally, really inside the body. Once inside the cells, the nutrients are used in the building, or **synthesis**, of compounds needed for life. (See Figure 6-4.)

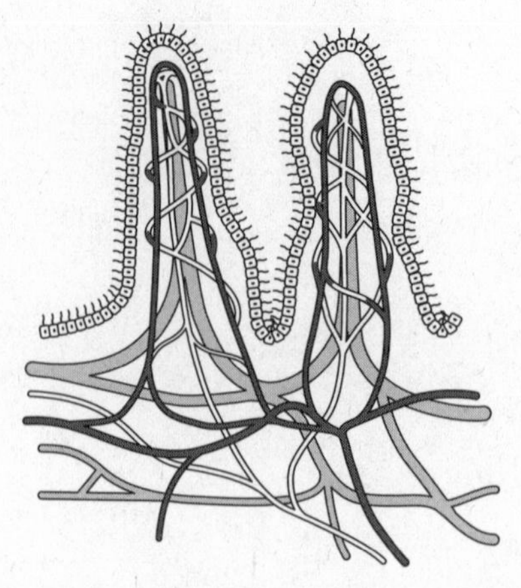

**Figure 6-4** Tiny projections on the lining of the small intestine increase its surface area. This makes absorption of food molecules into the blood vessels more efficient.

Almost all of the useful nutrients get absorbed in the small intestine. Anything that passes on from there into the large intestine is primarily indigestible material, such as the cellulose in lettuce. Compared to the small intestine, the *large intestine* is large in diameter but short in length. In the large intestine, water from the remaining material is reabsorbed into the body. Then, the muscles of the rectum force feces (the solids that remain) out through the anus. If too much water is reabsorbed, the feces cannot move easily through the large intestine, and constipation occurs. If too little water is reabsorbed, the feces are too liquid, and diarrhea occurs.

Throughout the intestines, huge numbers of microscopic **bacteria** help break down the food. In addition to helping the process of digestion, these intestinal bacteria make several important vitamins. They also help rid the body of harmful bacteria.

## WHEN THINGS GO WRONG: DISEASES OF THE DIGESTIVE SYSTEM

To prevent the stomach from digesting itself, cells in the wall of the stomach produce a thick layer of mucus. Sometimes, the layer of mucus protection fails. Gastric juice reaches the wall of the stomach and begins to break it down. The eating away of tissues produces an *ulcer*, a painful and serious condition. When this happens in the stomach, it is called a peptic ulcer. Infection by acid-resistant bacteria is the main cause of ulcers. Treatment with antibiotics is often most effective for eliminating ulcers.

One of the most common types of cancer in North America is *colon cancer*. The large intestine is made up of the colon and the very end of the intestine, the rectum. In North America, the typical diet contains low levels of fiber. As a result, the feces move too slowly through the colon. This digestive problem may be related to the high incidence of colon cancer. Physicians also suggest that there may be a strong hereditary predisposition to colon cancer.

# Chapter 6 Review

## Part A—Multiple Choice

**1.** The main function of the human digestive system is to

1 rid the body of cellular waste materials
2 break down glucose in order to release energy
3 process organic molecules so they can enter cells
4 change amino acids into proteins and carbohydrates

**2.** At what point does food really enter your body?

1 when you bite into the food
2 when you chew the food
3 when it enters your stomach
4 when it is absorbed into your cells

**3.** The teeth play an important role in digestion because they

1 begin chemical digestion by releasing an enzyme in the mouth
2 begin mechanical digestion by grinding food into smaller pieces
3 break the food down so that it can be absorbed by the cells
4 force the epiglottis to close off the trachea

**4.** The epiglottis is important because it

1 pushes food to the back of the mouth
2 prevents food from entering the esophagus
3 lets food into the stomach without letting acid out
4 prevents food from entering the trachea

**5.** What causes heartburn?

1 stomach acid backing up to the esophagus
2 food blocking the trachea
3 undigested food reaching the heart
4 stomach acid reaching the heart

**6.** Which statement is true of the stomach's role in digestion?

1 The process of digestion is completed in the stomach.
2 Aside from the mouth, all chemical digestion takes place in the stomach.
3 Both mechanical and chemical digestion occur in the stomach.
4 Nutrients from food are absorbed by the cells of the stomach.

**7.** What happens when the muscles in the stomach wall relax and contract?

1 Food is pulled through the esophagus.
2 Food is mixed with gastric juice.
3 Nutrients in food are moved into body cells.
4 Wastes are removed from the body.

**8.** The stomach kills bacteria that may be present in food by

1 churning the food repeatedly
2 releasing hydrochloric acid
3 producing pepsinogen
4 pushing the food into the small intestine

*Base your answer to question 9 on the information below and on your knowledge of biology.*

A student completed a series of experiments and found that a protein-digesting enzyme (intestinal protease) functions best when the pH is 8.0 and the temperature is 37°C. During an experiment, the student used some of the procedures listed below:

a adding more protease
b adding more protein
c decreasing the pH to 6.0
d increasing the temperature to 45°C
e decreasing the amount of light

**9.** Which procedure would have the *least* effect on the rate of protein digestion?

1 procedure a
2 procedure b
3 procedure d
4 procedure e

**10.** Most of the chemical digestion of food takes place in the

1 stomach
2 small intestine
3 pancreas
4 large intestine

**11.** A sample of food containing one type of a large molecule was treated with a specific digestive enzyme. Nutrient tests performed on the resulting products showed the presence of simple sugars only. Based on these test results, you could determine that the original large molecules contained in the sample were molecules of

1 protein
2 starch
3 glucose
4 DNA

**12.** The pancreas is an organ connected to the digestive tract of humans by a duct (tube) through which digestive enzymes flow. These enzymes are important to the digestive system because they

1 form proteins needed in the stomach
2 change the food into molecules that can pass into the bloodstream
3 form the acids that break down the food
4 change food materials into wastes that can be passed out of the body

**13.** During the process of emulsification,

1 gastric juices break down proteins
2 bile breaks down fat into tiny droplets
3 enzymes break down fat into droplets
4 bacteria present in food are killed

**14.** The liver and the gallbladder are involved in digestion because the

1 food from the small intestine is passed to the liver and then to the gallbladder
2 food from the small intestine is passed to the gallbladder and then to the liver
3 gallbladder produces bile that is stored in the liver until needed
4 liver produces bile that is stored in the gallbladder until needed

**15.** During digestion, fat is changed into

1 fatty acid molecules and glycerol
2 amino acids
3 monosaccharides
4 bile

*Base your answer to question 16 on the diagram and chart below. The diagram represents one metabolic activity of a human.*

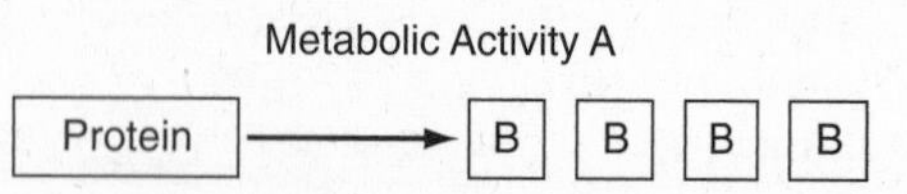


| Row | Metabolic Activity A | B |
|---|---|---|
| (1) | Respiration | Oxygen molecules |
| (2) | Reproduction | Hormone molecules |
| (3) | Excretion | Simple sugar molecules |
| (4) | Digestion | Amino acid molecules |

**16.** Letters A and B in the diagram are best represented by which row in the chart?

1 row 1
2 row 2
3 row 3
4 row 4

**17.** Food that reaches the large intestine is

1 basically indigestible
2 small enough to be absorbed by cells
3 carried by the blood to the cells
4 broken down by bile

**18.** Solids that remain in the large intestine are

1 returned to the small intestine for further digestion
2 forced out of the body through the anus
3 diluted with water until they can be digested
4 stored until they break down further

**19.** Bacteria that live in the intestines

1 help break down food and get rid of harmful bacteria
2 are all harmful and are killed by gastric juices
3 cause illness unless they are released with feces
4 exist in small enough numbers to be ignored

**20.** A peptic ulcer may result when

1 helpful bacteria are removed from the digestive system
2 too little fiber passes through the digestive system
3 too much water is removed from the large intestine
4 gastric juice eats away at the tissue of the stomach lining

### *Part B—Analysis and Open Ended*

**21.** Distinguish between the processes of ingestion and digestion.

**22.** According to the cell theory, the cell is the basic unit of structure and function in the body. In two or more sentences, explain how this fact relates to the need for digestion.

**23.** Identify and describe the two types of digestion that take place in the mouth. Your answer should include:

- the role of the teeth
- the role of saliva

*Use the figure below and your knowledge of digestion to answer the following question.*

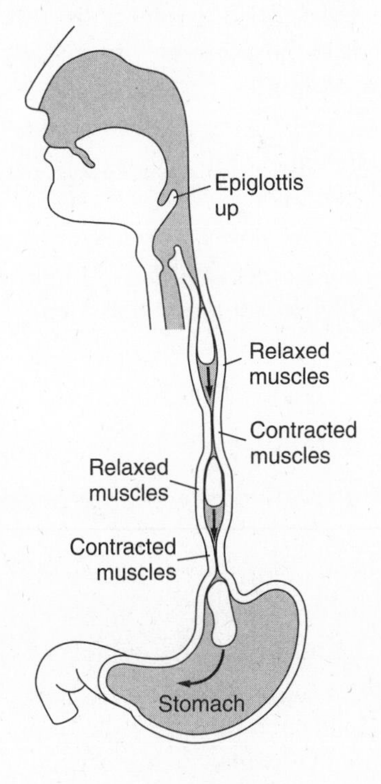


**24.** How is food moved from the esophagus to the stomach?

1 The tongue pushes the food down.
2 The epiglottis pushes the food down.
3 The esophagus muscles move the food down.
4 The stomach muscles pull the food down.

**25.** Describe the ways that muscles are important to the functioning of our digestive system. Your answer should include at least:

- *two* ways muscles function in the stomach
- *two* ways muscles function in the intestines

**26.** Complete the following chart by indicating where digestion begins for each of the following nutrients: proteins; fats; and starches.

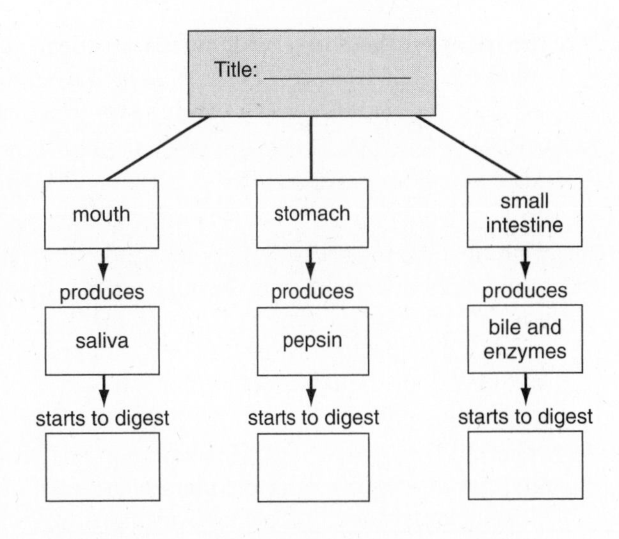


**27.** The best title for this concept map probably would be

1 The Basic Food Groups
2 The Types of Enzymes
3 Chemical Digestion
4 Mechanical Digestion

*Base your answers to questions 28 to 30 on the information and diagram below and on your knowledge of biology.*

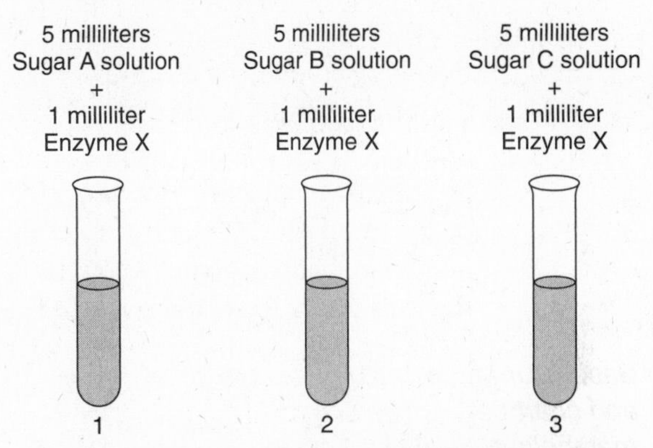


An investigation was performed to determine the effects of enzyme *X* on three different disaccharides at 37°C. Three test tubes were set up as shown in the diagram. At the end of five minutes, the solution in each test tube was tested for the presence of disaccharides (double sugars) and monosaccharides (simple sugars).

| | Test Tube 1 | Test Tube 2 | Test Tube 3 |
|---|---|---|---|
| **Monosaccharide** | not present | not present | present |
| **Disaccharide** | present | present | not present |

The results of these tests are shown in the data table.

**28.** What can be concluded about the activity of enzyme X from the data table?

**29.** With only the materials listed below and common laboratory equipment, design an investigation that could show how a change in pH would affect the activity of enzyme X. Your design need only include a detailed procedure and a data table.

Materials: enzyme X; sugar C solution; indicators; substances of various pH values (vinegar [acidic], water [neutral], baking soda [basic]); data table.

**30.** State *one* safety precaution that should be used during this investigation.

**31.** How is the human digestive system similar to a factory assembly line? In what important way is it *not* similar (aside from its being a living system)?

**32.** Explain why the liver, pancreas, and gallbladder are all considered to be part of the digestive system, even though food does not pass through them.

**33.** Briefly describe the roles of the following three liquids in the digestive system: saliva; gastric juice; and bile.

**34.** Explain why absorption is considered to be the final step in digestion.

**35.** Why are some bacteria considered to be "friendly" to our digestive system?

### Part C—Reading Comprehension

*Base your answers to questions 36 to 38 on the information below and on your knowledge of biology. Use one or more complete sentences to answer each question.*

People try to lose weight by dieting. Researchers now know that dieting may reduce the amount of fat in the body, but it also changes how the body functions. When people diet they usually limit the amount of food calories they take in. A dieter's body reacts to protect itself. Because of our evolutionary history, the body thinks that there is an actual shortage of food—that starvation is imminent. The body does not know that the dieting person is intentionally limiting the amount of food taken in, and it reacts by slowing down to survive the food shortage. The result? Fewer calories are burned during normal activities, and not much weight loss occurs. The fat cells that normally store fat are being emptied, but they still remain in the body. The person feels hungry and may end the diet. The "starved" fat cells quickly refill their reserves and a type of on again, off again dieting may result. The fluctuating weight loss and gain that occurs can be dangerous.

Most researchers now realize that the best way to avoid becoming overweight is to reduce the amount you eat somewhat and to increase physical activity. Exercise increases the amount of energy used by the body. It also increases the amount of muscle tissue, which even when resting burns more calories than other types of body tissues.

There are other serious health risks involved in severe weight loss that is caused by a refusal to eat. The disorder called *anorexia nervosa* is most common in young women. Abnormal fears of being overweight, as well as other fears, may lead to anorexia nervosa. An anorexic person appears unhealthy. This disorder can be fatal.

*Bulimia* is another eating disorder. Unlike most anorexics, a person with bulimia might appear healthy. However, this person swings between overeating and getting rid of the food, often by taking laxatives or inducing vomiting. Some studies show that as many as 20 percent of college-age women suffer from some form of bulimia. This disorder can be dangerous. It can damage the heart, kidneys, or digestive system. Counseling to help a person understand the reasons behind these eating disorders is important. It is also important to learn how to make wise choices about what one eats. In some severe cases, hospital treatment may be necessary.

**36.** In a few sentences, describe how evolution explains the unintended effects of dieting.

**37.** Explain why exercising is a healthier way to lose weight than dieting is.

**38.** Compare and contrast the eating disorders *anorexia nervosa* and *bulimia*.

# 7
# Gas Exchange and Transport

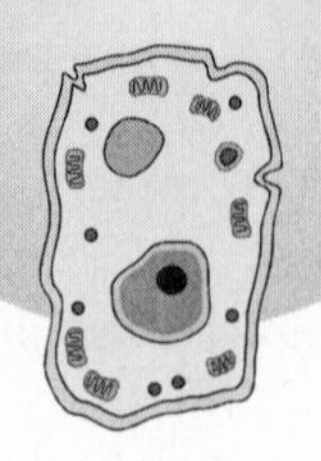

## Vocabulary

| | | |
|---|---|---|
| alveoli | circulation | platelets |
| arteries | diaphragm | respiration |
| capillaries | plasma | veins |

## THE RESPIRATORY SYSTEM, BREATHING, AND CELLULAR RESPIRATION

Animals move oxygen and carbon dioxide into and out of their bodies with a respiratory system. Although cellular respiration refers to the energy-releasing chemical reactions that occur in cells, the word **respiration** also means the process of *exchanging gases*. In many animals, air is physically pumped into and out of the body by the process of breathing. In multicellular animals, breathing or respiration that involves a respiratory system is necessary to allow the life-sustaining activities of cellular respiration to occur.

## GAS EXCHANGE SURFACES

All aerobic organisms, both plants and animals, exchange oxygen and carbon dioxide with their environment. (The oxygen is used by living things to produce ATP during cellular respiration.) Respiration in all organisms involves the diffusion of gases across cell membranes. This occurs for plants in the spongy layer of cells within the plants' leaves. Although gas exchange in animals may involve a complex respiratory system, the actual process of taking in and getting rid of gases is identical to that of the ameba. Gases must cross a barrier to be moved in or out of the animal. This barrier, a part of the animal's body, is known as the *respiratory surface*. Respiratory surfaces in different animal species vary in shape and size. However, they all share certain requirements:

- The respiratory surface has to remain moist at all times so that gases can diffuse across the cell membranes.
- The respiratory surface must be very thin so that gases are able to pass through it.
- There must be a source of oxygen, either in the air or dissolved in the water.
- The respiratory surface must be closely connected to the transport system that delivers gases to and from cells.

## THE HUMAN RESPIRATORY SYSTEM

The human respiratory system is similar to the respiratory system of other mammals. (See Figure 7-1.) Air moves through the nostrils into the nasal cavity, where dirt and other particles in the air are trapped by tiny hairs and mucus. The air is also warmed, humidified, and tested for odors. The nasal cavity leads to the *pharynx*, where it meets air and food arriving from the mouth. Air continues flowing down, passing by the vocal cords in the larynx. Farther down the tube is the *trachea*, which is surrounded by rings of stiff cartilage that help maintain its tubelike shape, keeping it open for airflow.

The trachea branches into two *bronchi*. Each bronchus leads to a lung. In the two lungs, the bronchi continue branching into smaller and smaller tubes called *bronchioles*. Most of these tubes are covered on the inside by mucus and by tiny hairlike extensions called *cilia*. The cilia help keep the delicate tissues that line the lungs clean. When cilia are paralyzed by, for example, the harmful effects of cigarette smoking, the lungs lose much of their ability to keep themselves clean.

The tiniest bronchioles end in bunches of microscopic air sacs, the **alveoli**. It is the lining of the alveoli that acts as the respiratory surface. Blood vessels surround the alveoli. It is only when oxygen molecules diffuse across

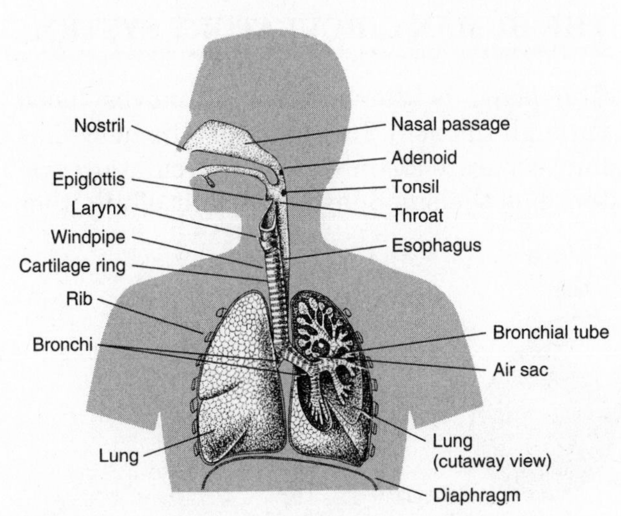


**Figure 7-1** The human respiratory system (showing cutaway view of a lung).

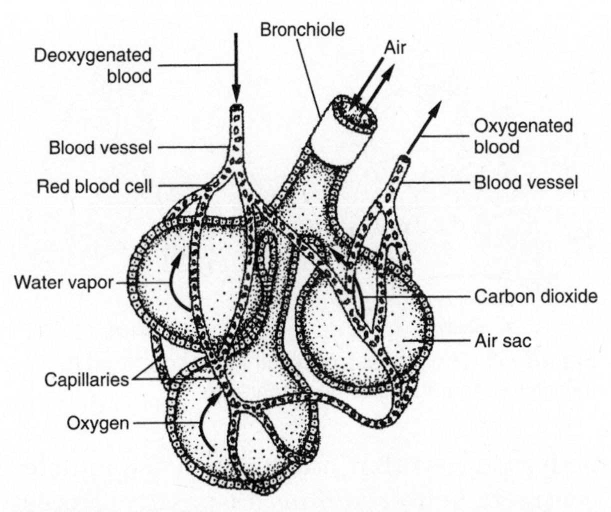


**Figure 7-2** The lining of the alveoli acts as our respiratory surface—it is where gas exchange occurs in our bodies.

this lining—from the alveoli into the blood vessels—that they really enter the body. (See Figure 7-2.)

## BREATHING: MAKING THE AIR MOVE

On their own, the lungs cannot move air into or out of the body. Lungs have no muscles. All mammals, including humans, move air into their lungs by lowering the air pressure in their lungs. Two sets of structures are involved: the ribs and the muscles between them. The muscles move the ribs. When the muscles move the ribs upward and outward, the rib cage expands. At the same time, the **diaphragm**, a large flat muscle that lies across the bottom of the chest cavity, contracts and moves down. This movement also increases the size of the chest cavity. The air pressure decreases in the lungs because the same amount of air suddenly has more space to fill. Inhalation occurs when air rushes into the lungs through the respiratory tubes, and the lungs fill with air. (See Figure 7-3.)

To move air out, the diaphragm and the muscles between the ribs relax. When these return to their original positions, the volume of the chest cavity decreases. With less space to fill, the air pressure in the lungs increases and air moves out. Exhalation occurs. Breathing is the physical process of *inhalation* (an

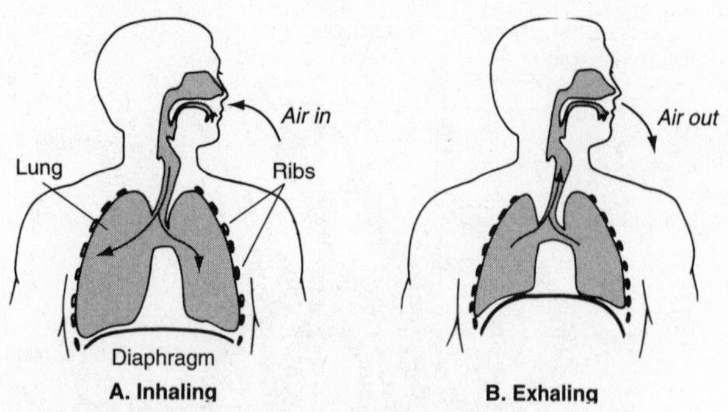


**Figure 7-3** The process of breathing is controlled by muscles of the ribs and the diaphragm.

active process that occurs when the muscles contract) and *exhalation* (a passive process that occurs when the muscles relax). Humans inhale and exhale at a rate of about 12 times a minute, automatically, 24 hours a day.

## WHEN THINGS GO WRONG: DISEASES OF THE RESPIRATORY SYSTEM

Cigarette smoking causes many deaths from respiratory diseases. Evidence has shown that the chemicals in tobacco smoke can cause cancer in the lungs, esophagus, larynx, and mouth. Other diseases of the respiratory system, which smokers are more likely to get, include:

- *Bronchitis*. This is an inflammation and swelling of the inside of the bronchial tubes.
- *Asthma*. The walls of the bronchi contract, restricting the flow of air. As a result, the lungs do not fill or empty normally.
- *Emphysema*. This is a very serious chronic disease in which the alveoli break down, greatly reducing the total area of the respiratory surface.
- *Pneumonia*. This disease results from a bacterial or viral infection that causes the alveoli to fill with fluid.

## WANTED: A TRANSPORT SYSTEM

For an aerobic organism to live and grow, matter in the form of food molecules must be delivered to every cell. To release the energy stored in those molecules, oxygen must make that same trip to all cells. Finally, to prevent harmful buildups, carbon dioxide gas and other wastes must be taken away from all cells. An organism needs a *transport system* to accomplish these tasks. Transport involves two processes: absorption and circulation. *Absorption* occurs when materials cross cell membranes from the outside environment into the body (into the cells). Then, during the process of **circulation**, materials are circulated throughout the body to where they are needed.

All organisms transport materials inside of them within a liquid, and the transport fluid must reach every cell in the body. Thus, there have to be vessels that are able to deliver materials around the body—similar to the water pipes in a house. Finally, there must be something that forces the fluid through the vessels—the transport system must have a pump.

## THE HUMAN CIRCULATORY SYSTEM

The *heart* is the pump that moves blood through the body. In mammals, the heart has four separate chambers. Our circulatory system has the same layout as that of all other

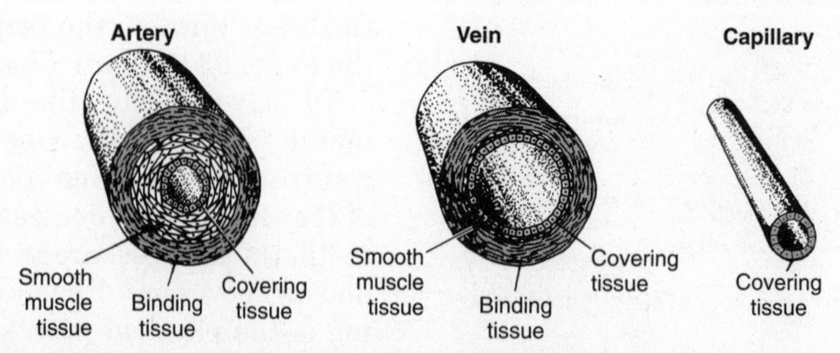


**Figure 7-4** Cross sections of the three kinds of blood vessels in humans.

mammals and is highly efficient at the job it does.

The circulatory system accomplishes the vital task of transport through the many thousands of kilometers of *blood vessels* in the body. (See Figure 7-4.) Blood from the lungs enters the left atrium of the heart and then passes to the left ventricle. This oxygen-rich blood then begins its journey throughout the body by passing through the aorta (the body's largest artery) to the other arteries.

The **arteries** are vessels that have thick, muscular walls. They are very elastic, expanding and contracting as blood from the heart is pumped through them in pulses—one pulse for every heartbeat. The elasticity of the arteries' walls exerts pressure against the blood inside. This pressure is measured as your *blood pressure*.

The arteries become smaller and smaller as they get farther away from the heart. Eventually the blood enters **capillaries**, the smallest vessels. Capillaries are close to every body cell. It is at the capillaries that the exchange of nutrients and gases between the blood and the cells takes place. After moving through capillaries, blood returns through the thin-walled **veins**, which get larger and larger closer to the heart. Unlike the walls of arteries, the walls of veins have little elasticity; blood is under low pressure in them. It would be very easy for blood to flow backward in the veins of the legs. To prevent the blood from moving backward, one-way valves work to trap it. The blood remains stationary in the veins until the beating of the heart and other muscular activities, such as leg muscle contractions, force the blood back up toward the heart.

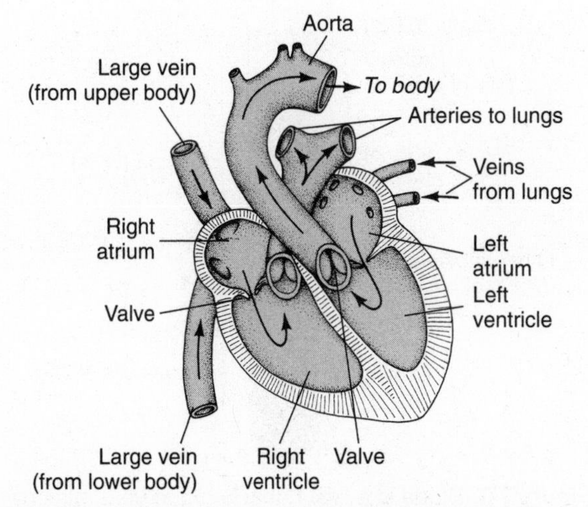


**Figure 7-5** All mammals, including humans, have an efficient four-chambered heart to pump blood through the body.

The oxygen-poor blood from the body enters the right atrium of the heart, passes through the right ventricle, and then enters the arteries that take it to the lungs. After the exchange of gases that occurs in the lungs, oxygen-rich blood once again returns by veins to the left atrium of the heart. (See Figure 7-5.)

## BLOOD TISSUE AND BLOOD FLOW

The entire purpose of the transport system is to move materials to and from cells. A capillary is so small that red blood cells must travel through it in single file. Molecules, including water, diffuse through the capillary walls and enter the spaces around body cells. These intercellular spaces are filled with a fluid that surrounds all cells. Molecules diffuse between this *intercellular fluid* (ICF) and the body cells. (See Figure 7-6.)

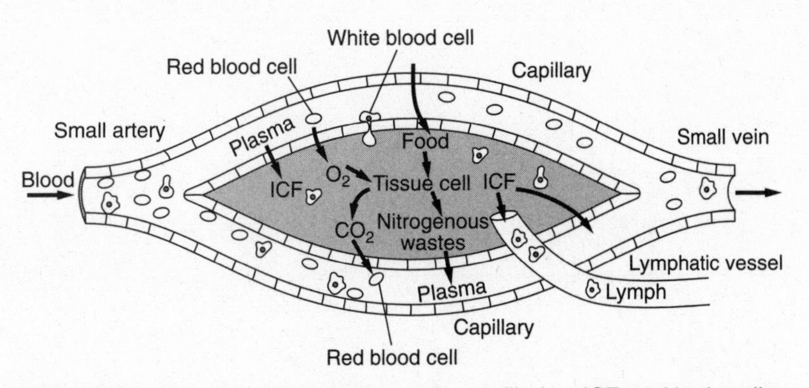


**Figure 7-6** Molecules diffuse between the capillaries, ICF, and body cells.

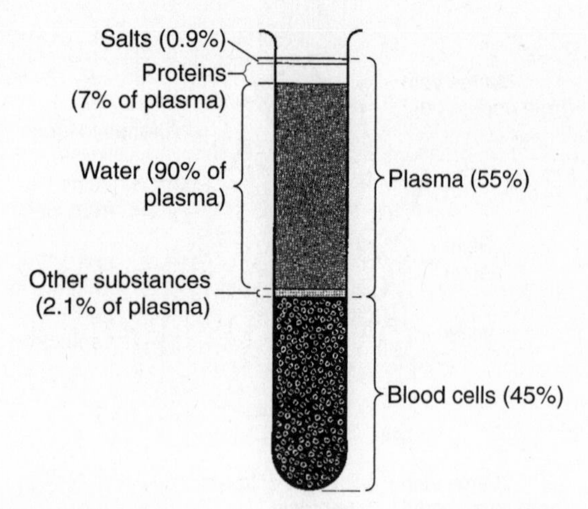


**Figure 7-7** Blood is a tissue that is made up of plasma and blood cells.

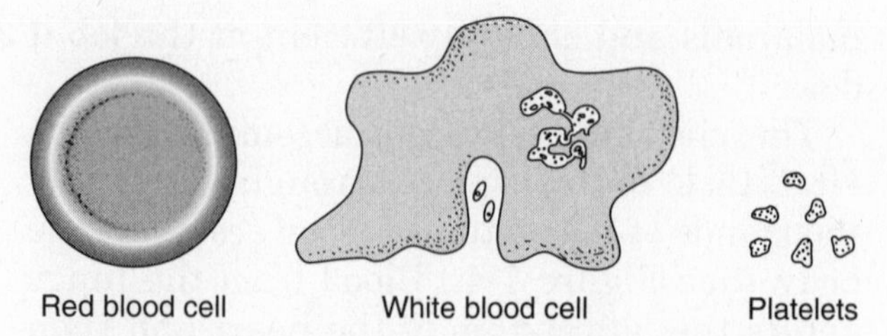


**Figure 7-8** Red blood cells carry oxygen, white blood cells protect us from disease, and platelets begin the clotting process.

Blood itself is a tissue. But unlike any other tissue in the body, blood is a liquid. Blood is made up of cells, cell parts, and a clear, light-yellow-colored liquid called *plasma*. The **plasma** is 90 percent water, plus many important proteins, salts, vitamins, hormones, gases, sugars, and other nutrients. One of the proteins, called *fibrinogen*, helps in the clotting process that stops bleeding caused by an injury. (See Figure 7-7.)

The cells in the blood include *red blood cells*, which contain the oxygen-carrying protein hemoglobin, and *white blood cells*. There are five types of white blood cells, all of which are involved in protecting the body from disease-causing foreign substances. Blood also contains **platelets**—fragments of cells that plug "leaks" when an injury occurs. Platelets also begin the complex chemical process that results in formation of a blood clot. A clot stops the flow of blood out of a damaged blood vessel. (See Figure 7-8.)

## WHEN THINGS GO WRONG: CARDIOVASCULAR DISEASE

The human transport system is called the *cardio* (heart) *vascular* (vessels) *system*. Cardiovascular disease includes several important, and potentially fatal, conditions. A *heart attack* occurs when the vessels that bring blood to the heart get blocked and the heart tissue beyond that point is not supplied with blood. The muscle tissue in that area of the heart then dies.

Unlike a heart attack, which often occurs suddenly, some forms of cardiovascular disease develop slowly over a long period of time. Clogged arteries result from the gradual buildup of layers of fatty deposits inside the arteries. Cardiovascular disease also includes strokes, which may block an artery in the brain. Depending on what area of the brain is affected, strokes can damage a person's ability to feel things or to speak and move. Scientists think that both diet and heredity play a part in the development of cardiovascular disease.

# Chapter 7 Review

## *Part A—Multiple Choice*

**1.** What happens during respiration in all animals?

1 Air must be physically pumped into and out of the body.
2 Gases ($O_2$ and $CO_2$) must move into and out of the body.
3 Blood must move gases throughout the body.
4 The diaphragm must relax and contract.

**2.** Which is *not* a requirement of a respiratory surface?

1 It must be thin.
2 It must be moist.
3 It must be thick and solid.
4 It must be closely connected to the system that delivers gases.

**3.** How are dirt and other small particles removed from the air that humans breathe?

1 They are trapped by hairs and mucus in the nasal cavity.
2 They are filtered by the pharynx at the back of the mouth.
3 They become stuck on the rings of the trachea.
4 They are caught by the lining of the alveoli.

**4.** Moving from the nasal cavity to the lungs, which sequence is correct?

1 larynx, bronchi, bronchioles, trachea, alveoli
2 trachea, larynx, alveoli, bronchioles, bronchi
3 larynx, trachea, bronchi, bronchioles, alveoli
4 pharynx, trachea, alveoli, bronchi, bronchioles

**5.** What is the purpose of the rings extending along the trachea?

1 They allow the trachea to rotate sideways.
2 They maintain its tubelike shape to allow for airflow.
3 They relax and contract to force air in and out of the lungs.
4 They allow both food and air to pass through the same tube.

**6.** Cilia are important to the lungs because the tiny hairs

1 help clean the tissues that line the lungs
2 allow air to pass from the trachea into the lungs
3 act as our respiratory surface
4 hold the end of the trachea open

**7.** Oxygen in the air we breathe enters our blood through the lining of the

1 bronchi
2 bronchioles
3 alveoli
4 cilia

**8.** You physically inhale when the

1 muscles of the lungs relax, increasing the size of the lungs
2 diaphragm relaxes, increasing the size of the chest cavity
3 diaphragm contracts, increasing the size of the chest cavity
4 diaphragm contracts, decreasing the size of the chest cavity

**9.** During inhalation, the air pressure in the chest cavity is

1 greater than the pressure outside the body
2 lower than the pressure outside the body
3 the same as the pressure outside the body
4 exactly balanced by the pressure inside the lungs

**10.** The process of circulation serves to

1 break down gases absorbed in the lungs
2 move air into and out of the body
3 absorb gases inhaled through the mouth or nose
4 transport absorbed materials throughout the body

**11.** The pump that pushes blood through the body is the

1 heart
2 artery
3 brain
4 plasma

**12.** Nutrients and gases are exchanged between the cells and blood across the

1 capillaries
2 ventricles
3 aorta
4 veins

**13.** When compared with the blood in the arteries, the blood in the veins

1 does not move at all
2 is under lower pressure
3 contains more gases
4 contains more water

**14.** Which sequence states the path of blood after it leaves the lungs?

1 aorta → left atrium → left ventricle
2 left atrium → right atrium → aorta
3 aorta → right atrium → right ventricle
4 left atrium → left ventricle → aorta

**15.** Molecules diffuse from capillaries to body cells by moving through the

1 veins
2 arteries
3 intercellular fluid
4 plasma

**16.** The main component of plasma is

1 oxygen
2 water
3 protein
4 gases

**17.** Red blood cells are important because they

1 repair damaged blood vessels
2 contain hemoglobin
3 protect the body from disease
4 produce blood clots

**18.** In the blood, both fibrinogen and platelets work to

1 pump blood through the veins
2 carry blood to the heart
3 help in the clotting process
4 transport nutrients into the cells

**19.** The immediate cause of a heart attack is

1 blood pressure drops in the veins
2 layers of fatty deposits build up slowly
3 an artery in the brain becomes blocked by a clot
4 vessels that bring blood to the heart get blocked

**20.** In an investigation to determine the change in heart rate with increased activity, a biology teacher asked students to take their pulses immediately before and immediately after exercising for 2 minutes. The data showed an average heart rate of 72 beats per minute before exercising and 90 beats per minute after exercising. If a valid conclusion is to be made from the results of this investigation, which assumption must be made?

1 In most students, the average heart rate is not affected by exercise.
2 Each student exercised with the same intensity.
3 Exercise causes the heart rate to slow down.
4 The heart rate of each student goes up 18 beats after jogging for 2 minutes.

### *Part B—Analysis and Open Ended*

**21.** Briefly explain the difference between "breathing" and "respiration."

**22.** Describe the *four* main characteristics of an animal's respiratory surface.

**23.** Trace the pathway of air through the human respiratory system, from the nostrils to the lungs. At what point within the lungs do the oxygen molecules really enter the body?

**24.** Refer to the diagram below to explain the process of breathing. Your answer should include the following:

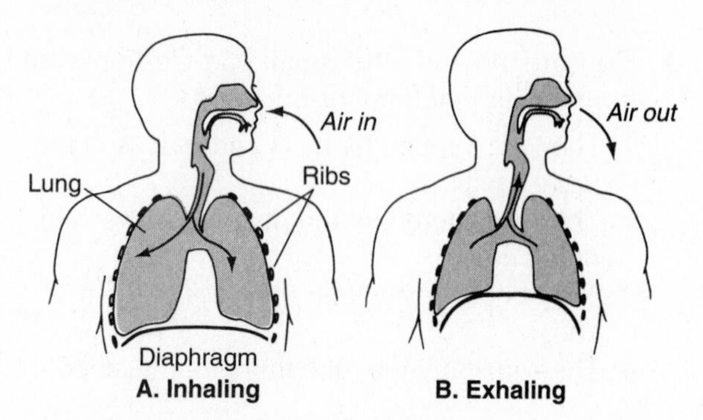


- what happens to the rib cage during inhalation and exhalation
- what happens to the diaphragm during inhalation and exhalation
- how these movements affect the volume of the chest cavity
- how these movements affect the air pressure in the lungs

**25.** Why is inhalation an active process whereas exhalation is a passive process?

**26.** Transport involves absorption and circulation. Define these two processes.

**27.** In general, what are *three* main components of a transport system?

**28.** Identify and describe the *three* types of blood vessels in the human body. Your answer should explain the following: (*Hint:* You may want to refer to Figure 7-4 on page 56.)

- the differences in their size (width) and structure
- the differences in their main functions and direction of flow
- if they usually carry oxygen-rich and/or oxygen-poor blood

**29.** In what way do the lungs depend upon the circulatory system?

**30.** In Texas, researchers gave a cholesterol-reducing drug to 2335 people and an inactive substitute (placebo) to 2081. Most of the volunteers were men who had normal cholesterol levels and no history of heart disease. After 5 years, 97 people getting the placebo had suffered heart attacks compared to only 57 people who had received the actual drug. The researchers are recommending that to help prevent heart attacks, all people (even those without high cholesterol) take these cholesterol-reducing drugs. In addition to the information above, which piece of information should the researchers have *before* support for the recommendation can be justified?

1 Were the eating habits of the two groups similar?
2 Did the heart attacks result in deaths?
3 How does a heart attack affect cholesterol levels?
4 What chemical is in the placebo?

**31.** According to the diagram, how many chambers does the human heart have?

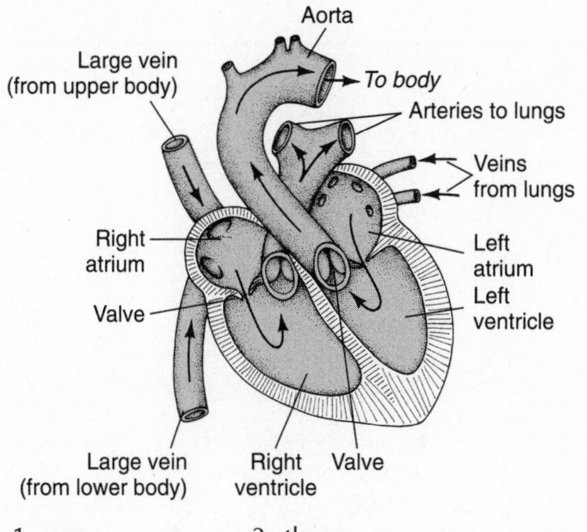


1 one
2 two
3 three
4 four

**32.** Why is intercellular fluid so important to the functioning of the transport system? (Explain its role in relation to capillaries and body cells.)

*Base your answers to questions 33 and 34 on the diagram of blood cells shown below.*

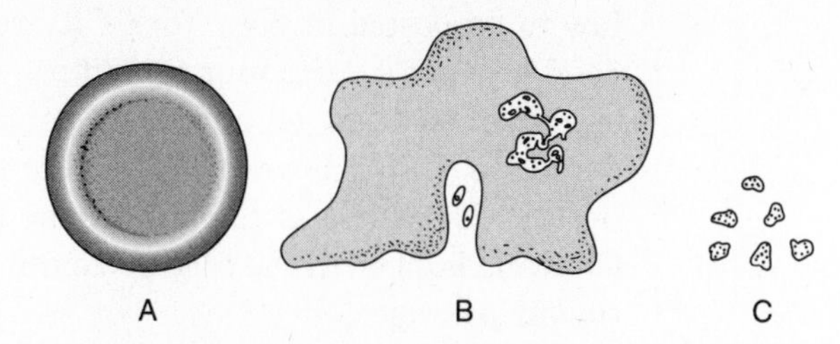


**33.** Identify the *three* types of blood cells shown in the diagram. What is the main function of each type of blood cell?

**34.** How does each type of blood cell help to maintain homeostasis?

**35.** Imagine that you are a red blood cell. Describe your trip through the human body, going from one foot, through the circulatory system, to the lungs, through the circulatory system, then back to the foot. Identify which parts of the heart you would pass through and how often you would pass through the heart on one round-trip.

### *Part C—Reading Comprehension*

*Base your answers to questions 36 to 38 on the information below and on your knowledge of biology. Use one or more complete sentences to answer each question.*

While it may seem out of place, certain health-related posters are required by law to be posted in New York City restaurants. People eating in restaurants need to know a particular procedure—called the *Heimlich maneuver*—because a piece of food can get stuck in a person's throat and block airflow, causing the person to choke. However, food is not the only cause of choking. It is just as important for people at home to know how to perform the Heimlich maneuver. Chewing gum or, in the case of a child, small toys or a piece of balloon can block the air passage.

The Heimlich maneuver uses the air that is already present in the lungs to push an object up and out of the throat. The Heimlich maneuver is actually quite simple. First, you must determine if a person has an object blocking the throat or is, in fact, experiencing another condition, such as a heart attack. Ask the person, "Are you choking?" If so, he or she will not be able to speak to answer you. For a person who is standing, you administer the Heimlich maneuver by standing behind the choking person and grasping him or her with both arms around the waist. Place one hand, now closed as a fist, just below the bottom of the rib cage and above the navel. Grasp the closed hand with your other hand and make an upward, not inward, thrust. This thrusting motion puts pressure on the victim's diaphragm. The increased pressure forces air up and out of the lungs in order to force the piece of food or foreign object out of the throat.

**36.** Describe how the Heimlich maneuver works to push an object out of the throat.

**37.** Write a list that shows each of the steps in the Heimlich maneuver.

**38.** Why is it necessary in this procedure to make an upward, rather than an inward, thrust?

# THEME III

## Maintaining a Dynamic Equilibrium

# 8
# The Need for Homeostasis

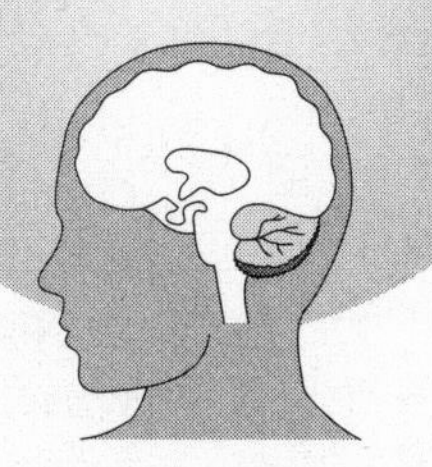

### Vocabulary

dynamic equilibrium
feedback mechanism
homeostasis
stomata

## HOMEOSTASIS

Organisms live in a world of changing conditions. But, to remain alive, every organism needs to keep the conditions inside itself fairly constant. An organism must have ways to keep its internal conditions from changing when its external environment changes. This ability of all living things to detect external changes and to maintain a constant internal environment is known as **homeostasis**.

An obvious change that has occurred in the course of evolution is the development of larger *multicellular* organisms from microscopic, single-celled ones. Is there an advantage to being multicellular? Being microscopic and single-celled makes it difficult for an organism to maintain homeostasis. Having a multicellular body makes possible many types of protection against changes in the environment. In other words, an organism with many cells is able to have structures and systems that protect its individual cells from external changes, thus helping it to stay alive. (See Figure 8-1.)

To maintain homeostasis, organisms actually must make constant changes. That is why homeostasis is often referred to as maintaining a **dynamic equilibrium**. *Dynamic* means "active," and *equilibrium* means "balanced." Homeostasis requires active balancing.

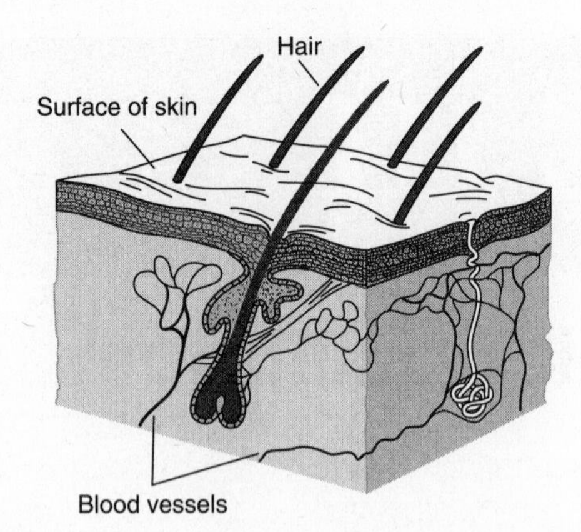


**Figure 8-1** Multicellular organisms have systems and structures that help them maintain homeostasis. For example, our skin has features that detect and respond to changes in external temperature.

## THE CELL AND ITS ENVIRONMENT

The smallest blood vessels in our bodies, the capillaries, are close to every cell. There is a small amount of space between the capillaries and the body cells, and this space is filled with fluid. In fact, each of our many cells is surrounded by liquid. This fluid that surrounds cells is made up mostly of water, with many substances dissolved in it. The intercellular fluid is important in helping to maintain stable conditions inside each of our cells. Many materials are exchanged between the cells and the fluid. In turn, materials may be exchanged between the fluid and the blood in the capillaries. All of this is done to make sure that each and every body cell is able to maintain homeostasis and remain healthy. (See Figure 8-2.)

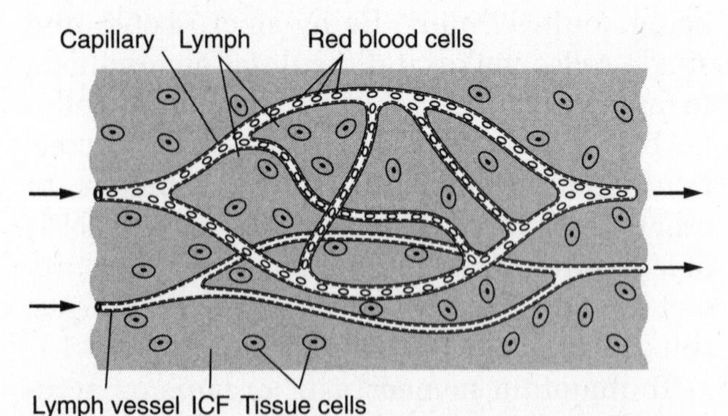


**Figure 8-2** All the cells in our body are surrounded by intercellular fluid (ICF). Materials are exchanged between the cells and the fluid, which helps to maintain stable conditions inside each of the cells.

## MAINTAINING HOMEOSTASIS WHEN WE EXERCISE

Exercise involves increased muscle activity. This activity creates changes within the body. To maintain homeostasis, the body needs to be able to respond to these changes.

An example of a change that occurs when we exercise is the increase in the body of carbon dioxide ($CO_2$), produced by muscle cells as a result of cellular respiration. The level of $CO_2$ increases in both the intercellular fluid and the blood. To maintain homeostasis, the body first must be able to detect this change and then respond to the change.

A structure in the brain detects the increased $CO_2$ level in the blood passing through the brain and in the fluid around the brain cells. As a result, this part of the brain sends signals to the chest to increase the rate of breathing and the amount of air taken in on each breath. These changes in breathing increase the exchange of gases in the lungs, lowering the $CO_2$ levels in the body. These lower levels are then detected in the brain, which in turn sends a signal to reduce the breathing rate. This process is an example of a **feedback mechanism**. Feedback mecha-

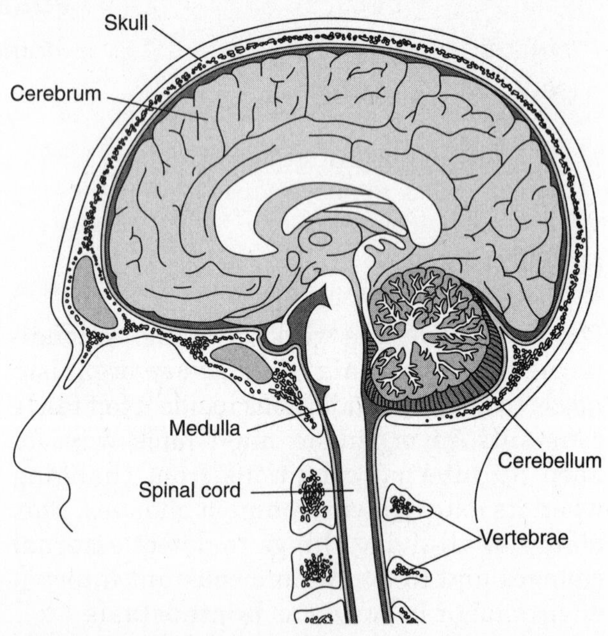


**Figure 8-3** A structure in the brain (the medulla) monitors the amount of $CO_2$ in the body, adjusting the breathing rate to maintain proper levels.

nisms are important in maintaining homeostasis. (See Figure 8-3.)

## FEEDBACK MECHANISMS

Carbon dioxide levels in your body are regulated somewhat as a thermostat regulates the temperature of your house. A thermostat measures the temperature of the air in a room. When the air temperature in the house falls below a preset figure, the thermostat turns a furnace on. The furnace produces heat, and the temperature of the air in the house increases. When the temperature of the air rises above the preset temperature, the thermostat tells the furnace to shut down. The temperature in the house stops rising, the air begins to cool, and the thermostat continues the cycle of telling the furnace to produce heat or to shut down. (See Figure 8-4.)

In this type of feedback mechanism, a change occurs that produces another change, which in turn reverses the first change. This is an important process in maintaining homeostasis.

The following are parts of a feedback mechanism used in maintaining homeostasis:

- *Sensor.* Something must be able to detect a change. A thermometer attached to a thermostat is a sensor. In the body, structures in the brain detect changes in $CO_2$ levels.
- *Control unit.* Something must know what the correct level should be. A thermostat in a house is set to a particular comfort level. Information in the brain is preset at the correct $CO_2$ level.
- *Effector.* Something must take instructions from the control unit and make the necessary changes. In a house, the effector would be a furnace or an air conditioner. In the body, the effector for $CO_2$ levels would be the muscles in the chest that are used for breathing.

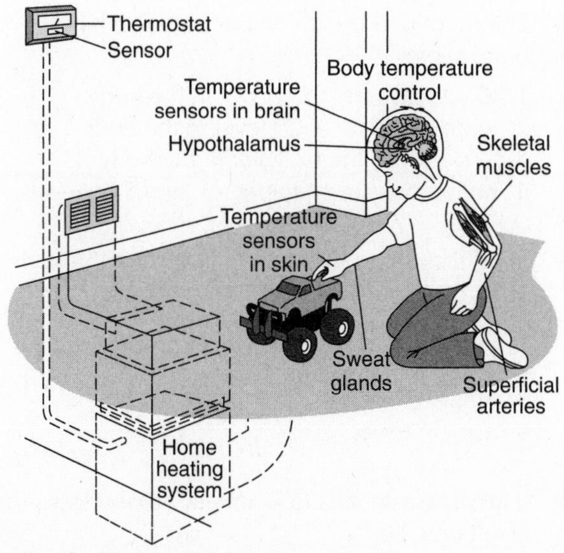


**Figure 8-4** Both $CO_2$ levels and body temperature are regulated by feedback mechanisms, much as a thermostat controls the temperature in a room.

## MAINTAINING HOMEOSTASIS: WATER BALANCE IN PLANTS

Maintaining water balance is a major concern for all living things. Plants as well as animals must maintain water balance. Openings, called **stomata** (singular, *stoma*), in the surface of a leaf are adapted to control the loss of water. Each stoma is surrounded by two *guard cells*. These guard cells, like any cells, allow water to diffuse through their cell membrane. When water is abundant, it moves into the guard cells. The increased quantity of water increases the pressure within the cells. Guard cells are somewhat curved in shape; when they are filled with water, they become even more curved. The space between them expands, the stoma widens, and excess water is allowed to evaporate out of the air spaces inside the leaf to the air that surrounds the plant. (See Figure 8-5.)

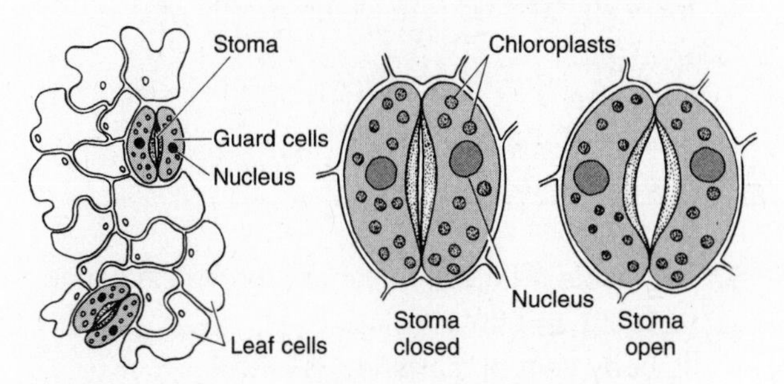


**Figure 8-5** Special openings in the surface of a leaf function to maintain water balance in plants.

When water becomes scarce, the leaf's guard cells become less curved in shape and the stomata close. Water loss is reduced and the plant is able to maintain its water balance.

## SYSTEMS FOR MAINTAINING HOMEOSTASIS

Multicellular animals have evolved highly organized, complex organ systems especially suited to maintaining a relatively constant internal environment. These organ systems include the excretory system, which regulates the chemistry of the body's fluids while removing harmful wastes; the nervous system, which uses electrochemical impulses to regulate body functions; the endocrine system, which produces hormones—chemical messengers essential in regulating the functions and behavior of the body; and finally, the immune system, which uses a set of defenses to protect the body from dangerous substances and microorganisms that could upset the internal balance on which life itself depends.

# Chapter 8 Review

### *Part A—Multiple Choice*

**1.** Organisms undergo constant chemical changes as they maintain an internal balance known as

1 interdependence
2 synthesis
3 homeostasis
4 recombination

**2.** Which description fits an organism that is better able to maintain homeostasis?

1 a taller body with larger cells
2 a shorter body with fewer cells
3 a multicellular body with many cells
4 a multicellular body with fewer cells

**3.** A system in dynamic equilibrium

1 makes constant changes
2 changes in intervals or steps
3 changes very infrequently
4 never changes at all

**4.** Intercellular fluid is made up mostly of

1 water
2 blood
3 mineral salts
4 cytoplasm

**5.** Intercellular fluid is important for the exchange of materials between

1 body cells and arteries
2 body cells and veins
3 veins and capillaries
4 body cells and capillaries

**6.** As a result of exercise, $CO_2$ levels increase in the

1 blood only
2 intercellular fluid only
3 blood and intercellular fluid
4 muscles only

**7.** The brain sends a signal to increase the breathing rate when the $CO_2$ level has

1 not changed for a while
2 decreased
3 increased
4 increased, then decreased

**8.** The increased breathing rate signaled by the brain serves

1 to increase the $CO_2$ level in the body
2 to decrease the $CO_2$ level in the body
3 to decrease the $O_2$ level in the body
4 no function in changing $O_2$ and $CO_2$ levels

**9.** In adjusting the $CO_2$ level, the part of the body that acts like a thermostat in the home is the

1 brain
2 chest
3 lungs
4 muscle tissue

**10.** If an organism fails to maintain homeostasis, the result may be

1 disease only
2 death only
3 disease or death
4 none of the above

**11.** A change in the body results in another change. This second change reverses the first change to maintain homeostasis. This describes a type of

1 control mechanism
2 feedback controller
3 feedback mechanism
4 effector mechanism

**12.** The effector for adjusting the $CO_2$ level in the body would be the

1 blood tissue
2 brain
3 lungs
4 chest muscles

**13.** Why might a blood clot be important to maintaining homeostasis?

1 It slows the flow of blood through the body.
2 It prevents the loss of blood from the body.
3 It increases the amount of water in the blood.
4 It adds more cells to the blood tissue.

**14.** The changing shape of a plant's guard cells helps to

1 allow the plant to grow stronger
2 prevent the plant from losing food
3 regulate the temperature of the plant
4 maintain the plant's water balance

*Base your answer to question 15 on the table below, which shows the rate of water loss in three different plants.*

| Plant | Liters of Water Lost Per Day |
|---|---|
| Cactus | 0.02 |
| Potato plant | 1.00 |
| Apple tree | 19.00 |

**15.** One reason each plant loses a different amount of water from the other plants is that each has

1 different regulation of water loss by its guard cells to maintain homeostasis
2 the same number of chloroplasts but different rates of photosynthesis
3 different types of insulin-secreting cells that regulate water levels
4 the same rate of photosynthesis but different numbers of chloroplasts

**16.** The nervous system helps to maintain homeostasis by

1 using electrochemical impulses to regulate functions
2 regulating the chemistry of the body's fluids
3 releasing hormones directly into the bloodstream
4 protecting the body from harmful bacteria

**17.** Which homeostatic adjustment does the human body make in response to an increase in environmental temperatures?

1 a decrease in glucose levels
2 an increase in perspiration
3 a decrease in fat storage
4 an increase in urine production

**18.** Which situation is *not* an example of the maintenance of a dynamic equilibrium in an organism?

1 Guard cells contribute to the regulation of water content in a geranium plant.
2 The release of insulin lowers the blood sugar level in a human after eating a big meal.
3 Water passes into an animal cell, causing it to swell.
4 A runner perspires while running a race on a hot summer day.

### *Part B—Analysis and Open Ended*

*Base your answer to question 19 on the photograph below, which shows a microscopic view of the underside (lower surface) of a leaf.*

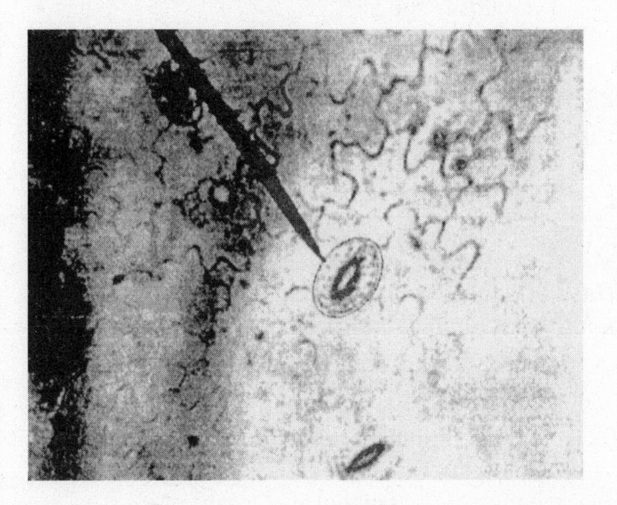

**19.** What is the main function of the cells indicated by the black pointer?

1 to regulate the rate of gas exchange
2 to store food for winter dormancy
3 to undergo mitotic cell division
4 to give support to the leaf's veins

**20.** How does being multicellular increase an organism's ability to maintain homeostasis and survive?

**21.** Write a brief paragraph comparing the life of a cell in your body with that of an ameba in the soil. Why is it more likely that the body cell will survive for a long time, but the ameba will not?

*Refer to the diagram below to answer questions 22 and 23.*

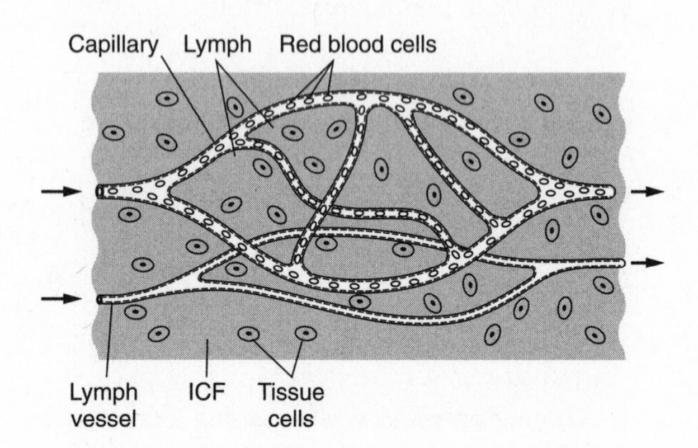


**22.** Which analogy most accurately describes the location of the body's tissue cells?

1 cities within states
2 islands within oceans
3 chains of mountains
4 clouds in the air

**23.** Use your knowledge of biology and the diagram above to explain the purpose of intercellular fluid (ICF). Why is it so important for homeostasis?

*Base your answer to question 24 on the information and diagrams below.*

To survive, an organism must maintain the health of its cells. The normal internal environment of a human's cells would include a temperature of 37°C, a pH of 7, and a water/salt balance of 0.1 percent.

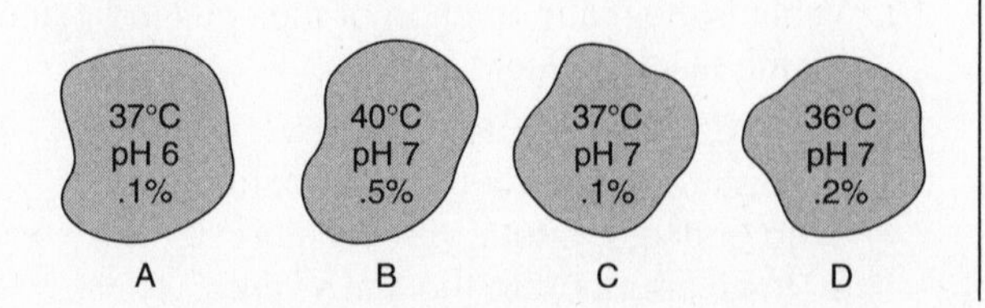


**24.** Which of the cells shown in the diagrams would belong to someone who is *not* maintaining homeostasis?

**25.** List, and describe the roles of, the three components of a homeostatic process.

**26.** Use the diagram below to explain how feedback mechanisms maintain homeostasis.

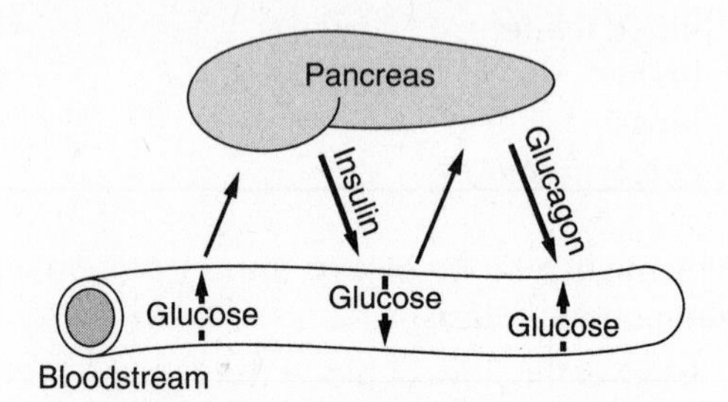


**27.** Use the following terms to replace the definitions given within the boxes in the following chart: *Higher $CO_2$ levels; Lower $CO_2$ levels; Muscles in the chest; Structures in the brain (with preset information).*

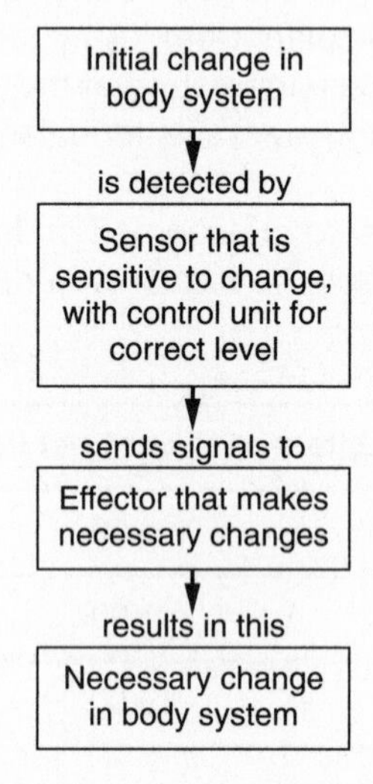


**28.** The best title for this concept map probably would be:

1 The Respiratory System and $CO_2$ Levels
2 The Circulatory System and $CO_2$ Levels
3 Feedback Mechanisms and $CO_2$ Levels
4 The Bloodstream and Its $CO_2$ Levels

**29.** Briefly explain the way our bodies adjust our breathing rates to maintain homeostasis.

*Base your answer to question 30 on the data in the following graph.*

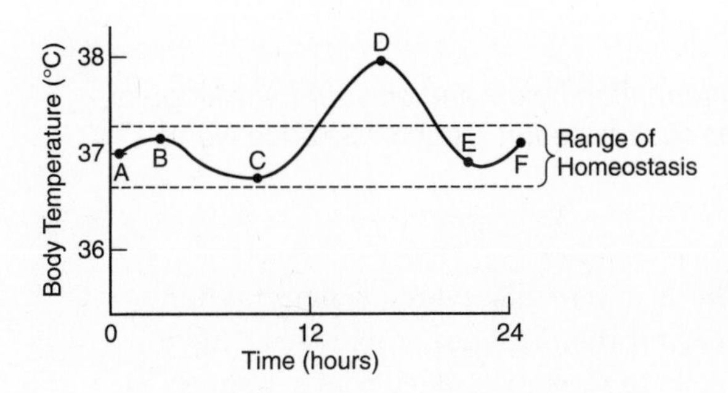


**30.** The graph shows evidence of disease in the human body. A disruption in the dynamic equilibrium is indicated by the temperature change that occurs between points

1 A and B
2 B and C
3 C and D
4 E and F

*Study the following graph to answer questions 31 and 32.*

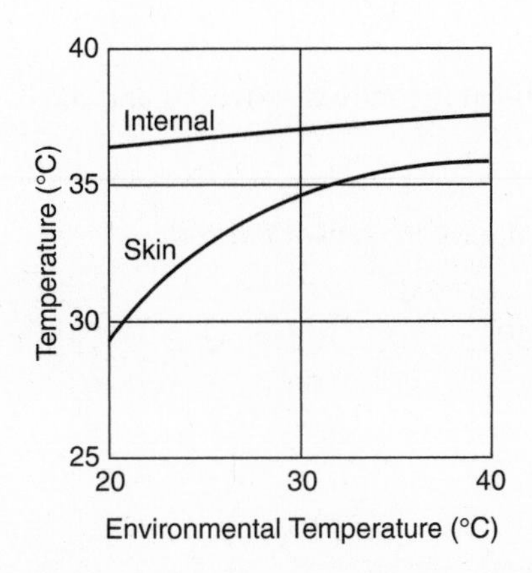


**31.** The graph shows the effect of external (environmental) temperatures on a student's skin and internal temperatures. Which statement best describes what happens as the environmental temperature increases?

1 The skin temperature increases, then decreases to 20°C.
2 The internal temperature increases abruptly to about 30°C.
3 The skin temperature decreases, due to sweating, to 30°C.
4 The skin temperature increases, then levels off at about 36°C.

**32.** What is the difference between the effects of rising external temperatures on the student's internal temperature and skin temperature? Explain how homeostatic processes are responsible for the effects seen in the graph.

**33.** In desert environments, organisms that cannot maintain a constant internal body temperature, such as snakes and lizards, rarely go out during the hottest daylight hours. Instead, they stay in the shade, under rocks, or in burrows. Explain how this behavior helps these organisms to maintain homeostasis.

**34.** Describe how plants maintain their water balance. Your answer should include the following:

- *one* reason why water balance is important to plants
- the structure that plants have to perform this function
- how this structure works to maintain water balance

**35.** In what way are the functions of the contractile vacuoles of an ameba and the guard cells of a plant similar?

**36.** Identify the *four* main organ systems that are involved in maintaining homeostasis. Briefly describe each of their roles in this process.

### *Part C—Reading Comprehension*

*Base your answers to questions 37 through 39 on the information below and on your knowledge of biology. Use one or more complete sentences to answer each question.* Source: *Science News*, Vol. 163, p. 301.

> They're on a hot streak. Researchers who last year discovered a mammalian cell-surface protein that senses coolness—and the presence of menthol—have now found a protein that enables nerve cells to recognize much colder temperatures (*SN:* 2/16/02, p.101).
>
> Whereas the cool-menthol receptor kicks in around 25°C, the newly identified receptor doesn't trigger nerve cells until the thermometer falls below 15°C, Ardem Patapoutian of the Scripps Research Institute in La Jolla, Calif., and his colleagues report in the March 21 *Cell*. They had suspected that an additional temperature sensor exists because other scientists had recently documented nerve cells that respond to cold temperatures but not to menthol.
>
> The two cold-activated receptors are related in structure, but their amino acid makeup is very different, the researchers report. Curiously, the new cold receptor is found on nerve cells that also sport a receptor for hot temperatures and capsaicin, the chemical that gives chilies and other foods their fiery kick.

**37.** State the difference between the cold-temperature sensor that has been recently discovered and the one that scientists already knew about.

**38.** Explain the evidence scientists knew of that suggested a second temperature sensor existed.

**39.** Explain what is remarkable about the newly discovered cold receptor.

# 9

# Nervous and Hormonal Regulation

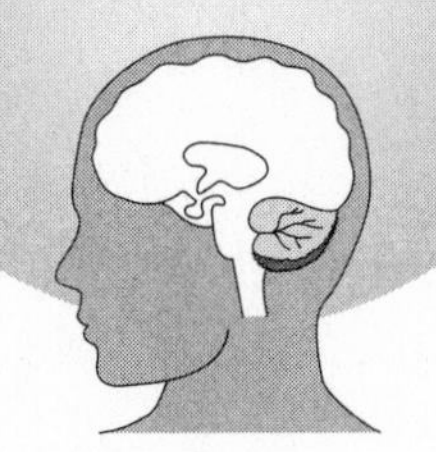

## Vocabulary

coordination
endocrine system
hormones
insulin
nerve cells
nervous system
receptor molecules
response
stimulus
synapse

## COMMUNICATION BETWEEN CELLS

All living things interact with their environment in many ways. Conditions outside and inside an organism are constantly being checked and, when needed, adjustments are made to maintain homeostasis. Whatever the interaction is—finding food, maintaining body temperature, or protecting oneself from disease—communication is required. Information must be received from the environment, processed, and responded to. Organisms, particularly complex multicellular ones, must organize the information they receive and respond to it. This makes it necessary for all parts of an organism to work in a *coordinated* fashion. The means by which body systems work together to maintain homeostasis is called **coordination**. To do so, an organism must have a means for *integration*—making all of its body parts work together—and a means for *control*—acting in an organized fashion.

Every function of an organism must involve cells. This includes the communication of an organism with its environment and the communication within an organism between all of its parts. Most important, the only way that cells communicate is chemically. Communication for a cell means having chemicals moving into and out of it. The work of the two organ systems responsible for integration and control—the nervous system and the endocrine system—is based on the chemical communication between cells.

## THE NERVE CELL: A CELL FOR RAPID COMMUNICATION

How does a message travel through the nervous system? The cell theory tells us that the messages must travel along pathways composed of cells. The specialized cells that make up these pathways are the **nerve cells**. The message itself is a *nerve impulse*. Every nerve cell does three things: it receives, conducts, and sends impulses. The most important part of a nerve cell involved in conducting an impulse is the cell membrane. Through the rapid movement of positive ions across the cell membrane, an electrical voltage is created. Electrical voltage is a form of energy. In a nerve cell, the voltage changes that occur at

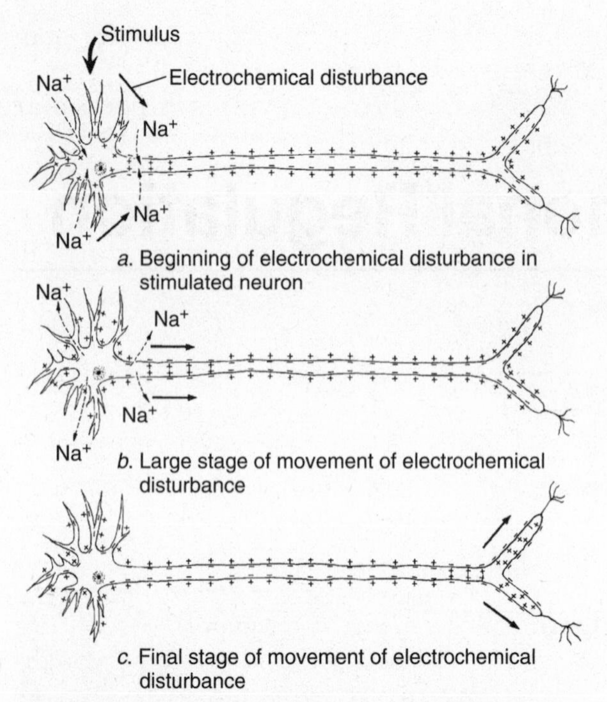


**Figure 9-1** A nerve impulse is the movement of cell membrane voltage changes along the length of a nerve cell (neuron).

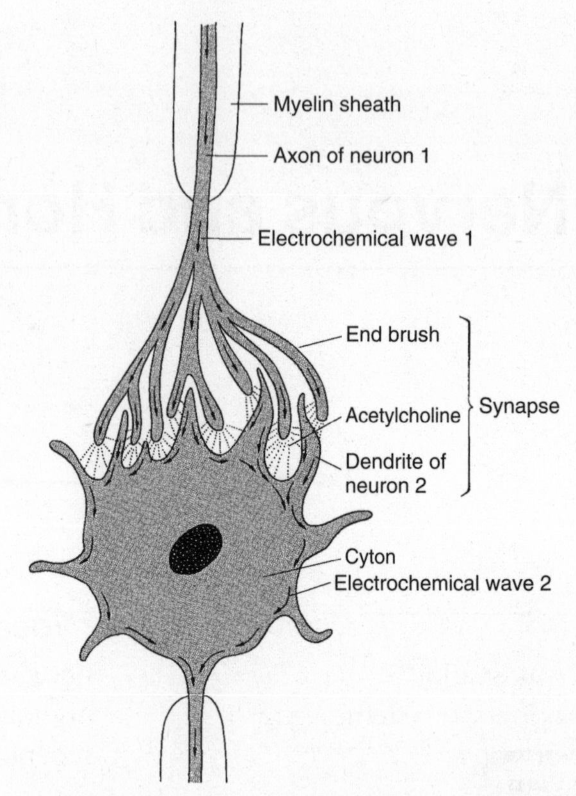


**Figure 9-2** Nerve impulses travel from one nerve cell (neuron) to the next by way of chemicals that diffuse across the gap (at the synapse).

one place on the membrane trigger the same kind of changes at the next spot on the membrane. This movement of voltage changes along the length of the nerve cell, or *neuron,* is the nerve impulse. (See Figure 9-1.)

## CROSSING THE GAP BETWEEN NERVE CELLS

If you accidentally touch a hot pan on a stove, you immediately pull your hand away. The nerve pathway from your finger to your spinal cord and brain, and then back to your finger, consists of many nerve cells. Close examination shows that nerve cells do not touch each other. They are separated by a gap, at a region known as the **synapse**. How does the impulse get from one neuron to another? Extremely important chemicals called *neurotransmitters* (such as acetylcholine, dopamine, melatonin, and epinephrine) are released by the ends of one nerve cell. The chemicals are released as the impulse arrives at the nerve cell endings. These chemicals diffuse across the gap to the next nerve cell. Once received by the next neuron, the chemicals make a new nerve impulse possible. In this way, the message continues along the entire nerve pathway, moving from one nerve cell to another. (See Figure 9-2.)

## THE HUMAN NERVOUS SYSTEM

Any event, change, or condition in the environment that may cause an organism to react is called a **stimulus** (plural, *stimuli*). A hot pan on the stove is a stimulus; pulling your hand away is a **response**. The **nervous system** of an organism makes possible both detection of a stimulus and response to a stimulus. In vertebrates, such as humans, the nervous system is a complex organization of cells and organs and includes both a central and a peripheral nervous system.

The *central nervous system* consists of the brain and spinal cord. The spinal cord runs from the base of the brain to the lower portion of the back. In most vertebrates, the spinal cord is surrounded by hollow, bony

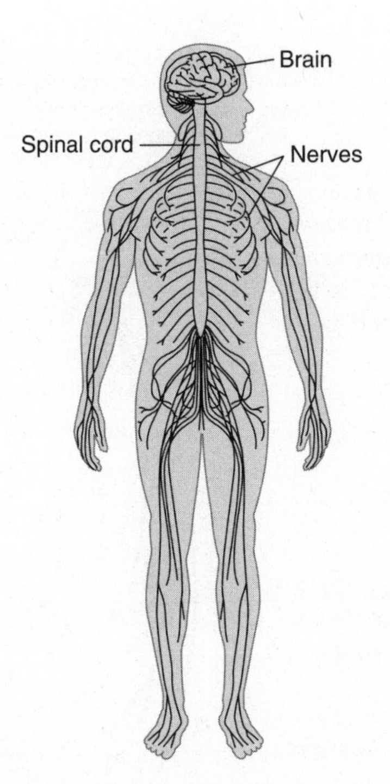


**Figure 9-3** The human nervous system: the central nervous system consists of the brain and spinal cord; the peripheral nervous system consists of the nerve cells that connect the spinal cord to all parts of the body.

vertebrae that make up the backbone. The *peripheral nervous system* consists of nerve cells that travel out of the central nervous system to all parts of the body. Signals are carried by these nerve cells to muscles or glands. Other nerve cells carry signals to the central nervous system from *sensory receptors*, such as those in the ears and in the eyes. (See Figure 9-3.)

## DISEASES THAT AFFECT THE NERVOUS SYSTEM

Cerebral palsy results from brain damage and is the collective name for a group of disorders that affect a person's ability to control body movements.

Multiple sclerosis occurs when cells in the brain and spinal cord do not function normally. A wide variety of symptoms, including shaking of the hands, blurred vision, and slurred speech, occur in people with multiple sclerosis.

Alzheimer's disease is a progressive degenerative disease. Eventually, memory loss and the inability to think, speak, or care for oneself occur.

Parkinson's disease involves a particular group of nerve cells in the brain. Loss of function in these nerve cells produces the typical shaking motion, poor balance, lack of coordination, and stiffening of the muscles that occur with this disease.

## THE ENDOCRINE SYSTEM: ANOTHER COMMUNICATION NETWORK

A system of glands and hormones makes up the **endocrine system**. The key function of this important system is maintaining homeostasis. The *endocrine glands* produce **hormones**, chemical messengers that are released into the blood and carried throughout the body by the circulatory system. At some place in the body—often far from the gland that made it—a hormone arrives at its special *target cells* and puts into effect whatever changes it has been designed to produce.

How do hormones do their work? Some hormones bind to specific **receptor molecules** (proteins) found in the cell membrane. The binding of a hormone with a receptor protein then causes a change inside the cell, usually involving the cell's enzymes. Other hormones pass right through the cell membrane and bind to receptor proteins in the cytoplasm. The hormone-receptor complex may then move to the nucleus and interact with the cell's DNA, affecting gene activity.

Both the nervous system and the endocrine system are communication networks. However, there are important differences in the two systems. Impulses sent by the nervous system usually produce rapid responses, frequently produced by the actions of muscles. Hormones generally produce slower, more long-lasting changes, which often involve metabolic activity within the target cells.

An important characteristic of hormones is that only small amounts are usually needed to produce the required effect. The group of target cells for a hormone is usually very sensitive to its particular hormone. Feedback mechanisms work to control the amounts of many hormones that are released. In this way,

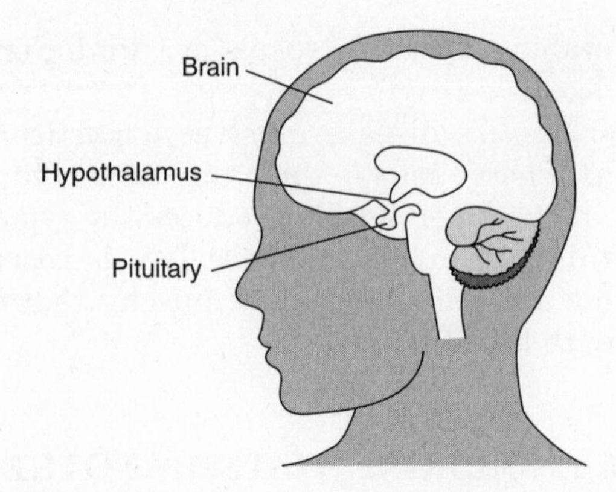


**Figure 9-4** The actions of the hypothalamus connect the nervous and endocrine systems—it receives information from the body and uses it to control hormones that are released from the pituitary gland.

the release of the hormone has the effect of stopping any further release of it until the hormone is needed once again.

## THE HUMAN ENDOCRINE SYSTEM

There is a close link between the human nervous and endocrine systems. The *hypothalamus* is a part of the brain. The hypothalamus receives information about conditions in the body as blood passes through it. It also receives information from nerve impulses that are carried to it by nerve cells. In turn, the hypothalamus uses the information it receives to control hormones that are released from its neighbor in the brain, the pituitary gland. (See Figure 9-4.)

The *pituitary gland*, only about the size of a pea, is sometimes called the "master gland" because it controls the activities of so many other glands of the endocrine system. When the hypothalamus detects a need for one of the pituitary hormones in the body, it sends a tiny amount of a releasing factor to the pituitary to secrete the correct hormone into the blood.

Attached to the top of each kidney is an *adrenal gland*. The most important hormone released from the adrenal gland is cortisol. This hormone is released only after the hormone ACTH, from the pituitary gland, triggers it to do so. This process of the pituitary hormone causing another gland to release its hormone occurs throughout the body.

Other major glands of the endocrine system are the thyroid gland (thyroxin) in the neck, the four small parathyroids (parathormone) connected to the thyroid, the pancreas (insulin), the ovaries (estrogen and progesterone), and the testes (testosterone). (See Figure 9-5.)

Key: TSH – Thyroid Stimulating Hormone
ACTH – Adrenocorticotropic Hormone
LH – Luteinizing Hormone
FSH – Follicle Stimulating Hormone

Hormones that control the release of anterior pituitary hormones
Hypothalamus
Anterior Pituitary
Posterior Pituitary
TSH
Thyroid
Parathyroids
LH
FSH
ACTH
Adrenal Cortex
Adrenal Medulla
Pancreas
LH
Ovaries ♀
Testes ♂
FSH

**Figure 9-5** The human endocrine system—the pituitary gland controls the release of hormones by many other glands in the body.

## WHEN THINGS GO WRONG: DIABETES—A DISEASE OF THE ENDOCRINE SYSTEM

Diabetes is a disease that occurs for a variety of reasons. However, all cases of diabetes are caused by a malfunction in the metabolism of carbohydrates. Carbohydrate metabolism involves the hormone **insulin**, which is released from the *pancreas*. Normally, when an increase in blood sugar level is detected, insulin is released from the pancreas to help metabolize the sugar. The result is a drop in the blood sugar level and then the pancreas stops releasing more insulin. This is, therefore, a feedback mechanism. Interruption of this feedback process may result in insulin shock and, possibly, even a coma.

# Chapter 9 Review

## Part A—Multiple Choice

**1.** Nerve cells are essential to an animal because they directly provide

1 communication between cells
2 regulation of reproductive rates within other cells
3 transport of nutrients to various organs
4 an exchange of gases within the body

**2.** A nerve impulse results from

1 the removal of fluid from between cells
2 electrical voltage changes on the cells
3 the motion of groups of cells
4 a collision between two cells

**3.** A nerve impulse is transmitted along the length of a cell's

1 membrane
2 nucleus
3 endoplasmic reticulum
4 Golgi complex

**4.** How do nerve impulses cross the gap between nerve cells?

1 The impulse pulls the two nerve cells together to close the gap.
2 The impulse forms a bridge across the gap.
3 One nerve cell releases chemicals that diffuse across the gap.
4 One nerve cell releases chemicals that build cells to fill the gap.

**5.** When you pull your finger away after touching a sharp pin, the stimulus is the

1 sharp pin itself
2 message your finger sends to your brain
3 motion of your finger away from the pin
4 motion of your finger toward the pin

**6.** The central nervous system consists of the

1 stomach and chest cavity
2 brain and spinal cord
3 eyes, ears, and mouth
4 legs and arms

**7.** The endocrine system maintains homeostasis by

1 maintaining physical coordination
2 controlling the size of blood vessels
3 sending nerve impulses to the brain
4 releasing hormones into the blood

**8.** Hormones begin their work by

1 replacing the nucleus of a cell
2 breaking down the membrane of a cell
3 binding to receptor proteins in a cell's membrane or cytoplasm
4 changing the shape of a cell and then becoming part of the cell

**9.** The endocrine system differs from the nervous system in that the endocrine system

1 produces faster, short-term changes
2 produces slower, long-term changes
3 operates in isolated regions of the body
4 produces changes in the brain only

**10.** Hormones and secretions of the nervous system are chemical messengers that

1 store genetic information
2 extract energy from nutrients
3 carry out the circulation of materials
4 coordinate system interactions

**11.** Which of the following statements is true?

1 Hormones are carried by the respiratory system.
2 Hormones are produced by the nervous system.
3 Hormones are generally needed in small amounts.
4 Hormones are generally needed in large amounts.

**12.** The body does not produce too much of most hormones because

1 the excretory system controls the amount released
2 feedback mechanisms control the amount released
3 cells can store any excess hormones they receive
4 the body can produce only fixed amounts of each hormone

**13.** The hypothalamus controls the release of hormones directly from the

1 pituitary gland
2 adrenal gland
3 thyroid gland
4 ovaries

**14.** Which statement describes a feedback mechanism involving the human pancreas?

1 The production of estrogen stimulates the formation of gametes for sexual reproduction.
2 The level of sugar in the blood is affected by the amount of insulin in the blood.
3 The level of oxygen in the blood is related to heart rate.
4 The production of urine allows for excretion of cell waste.

**15.** The pancreas produces one hormone that lowers blood sugar level and another that increases blood sugar level. The interaction of these two hormones most directly helps humans to

1 maintain a balanced internal environment
2 dispose of wastes formed in other body organs
3 digest needed substances for other body organs
4 increase the rate of cellular communication

**16.** What process is represented by the boxed sequence below?

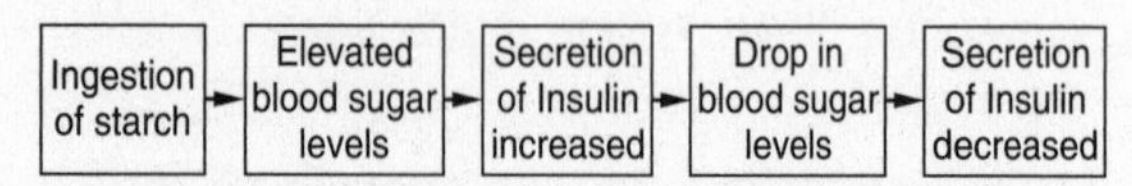


1 a feedback mechanism in multicellular organisms
2 the differentiation of organic molecules
3 an immune response by cells of the pancreas
4 the disruption of cellular communication

## *Part B—Analysis and Open Ended*

**17.** Explain the importance of both "integration" and "control" for maintaining homeostasis in an organism. What *two* systems accomplish this in humans?

**18.** Briefly describe the means by which cells communicate with each other.

**19.** What are the *three* main functions of every nerve cell?

**20.** Describe how a nerve impulse travels along the length of a nerve cell. Your answer should explain the following:

- what the nerve impulse actually is
- what the role of the cell membrane is
- what happens to the impulse at the end of the cell

**21.** How does a nerve impulse travel from one nerve cell to the next?

***Base your answers to questions 22 through 24 on the diagram below, which illustrates one type of cellular communication, and on your knowledge of biology.***

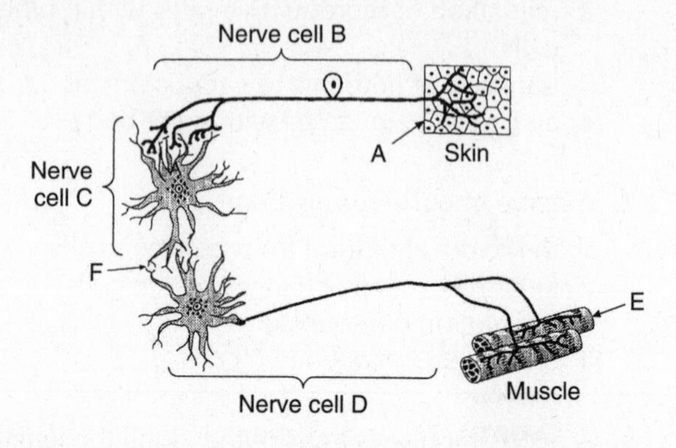


**22.** In region F, there is a space between nerve cells C and D. Nerve cell D is usually stimulated to respond by

1 a chemical produced by cell C moving to cell D
2 the movement of a virus from cell C to cell D
3 the flow of blood out of cell C to cell D
4 the movement of material through a blood vessel that forms between cell C and cell D

**23.** If a stimulus is received by the skin cells in area A, the muscle cells in area E will most likely use energy obtained from a reaction between

1 fats and enzymes
2 glucose and oxygen
3 ATP and pathogens
4 water and carbon dioxide

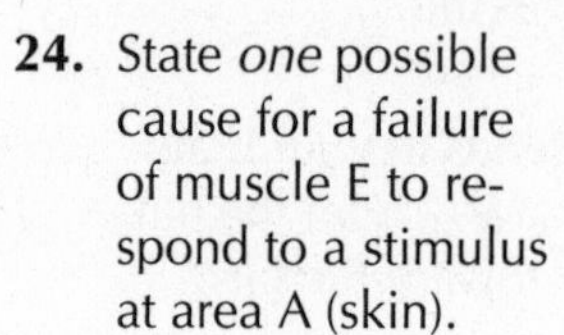

**24.** State *one* possible cause for a failure of muscle E to respond to a stimulus at area A (skin).

**25.** The vertebrate nervous system consists of the central nervous system and peripheral nervous system. The main components of the human nervous system are shown in the diagram at right.

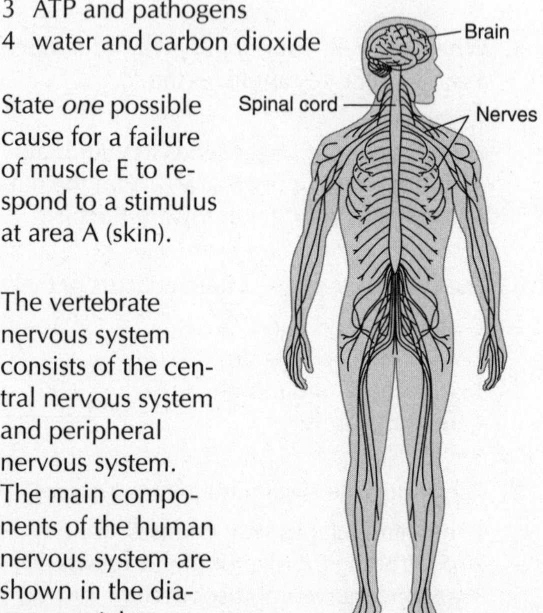

To which part of the system do the labeled parts belong?

**26.** Compare the central nervous system and the peripheral nervous system. Where do they carry their signals to, and from, in the body?

**27.** An important method of communication between cells in an organism is shown in the flowchart below. What type of chemical is referred to in the diagram?

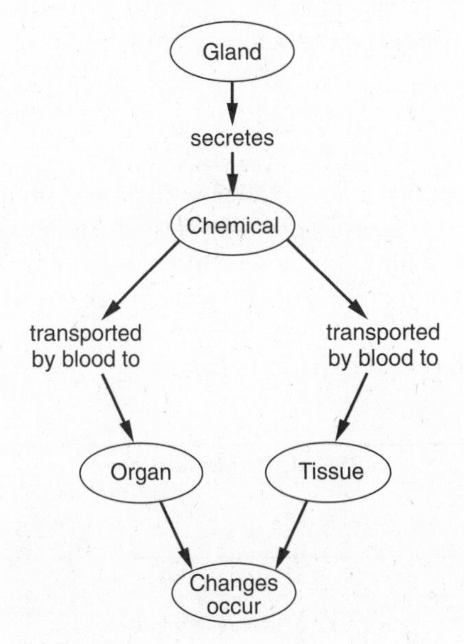


1 a hormone important in maintaining homeostasis
2 DNA necessary for regulating cell functions
3 an enzyme detected by a cell membrane receptor
4 a food molecule taken in by an organism

**28.** Explain what hormones are and tell the following facts about them:

- what body system releases them into the blood
- what body system carries them throughout the body
- what body system controls the release of hormones

**29.** Use the following terms to fill in the boxes of the following chart, which compares the endocrine system and nervous system: *electrical impulse; chemical messenger; nerve cells; bloodstream; to muscles and glands; from glands to target cells; rapid (muscle) responses; slow (metabolic) changes.*

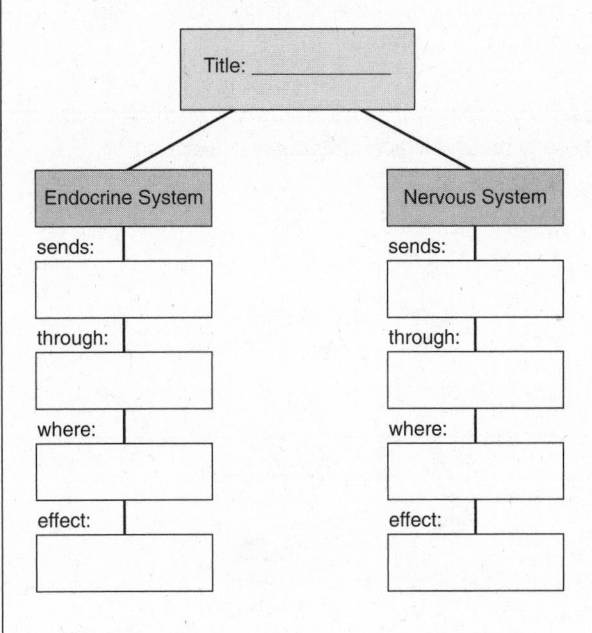


**30.** The best title for this concept map probably would be

1 Why the Endocrine System Is Better than the Nervous System
2 Why the Nervous System Is Better than the Endocrine System
3 Contrasting Features of the Body's Communication Networks
4 The Role of Chemicals in the Body's Feedback Mechanisms

**31.** Which graph of blood sugar level over a 12-hour period best illustrates the concept of dynamic equilibrium in the body?

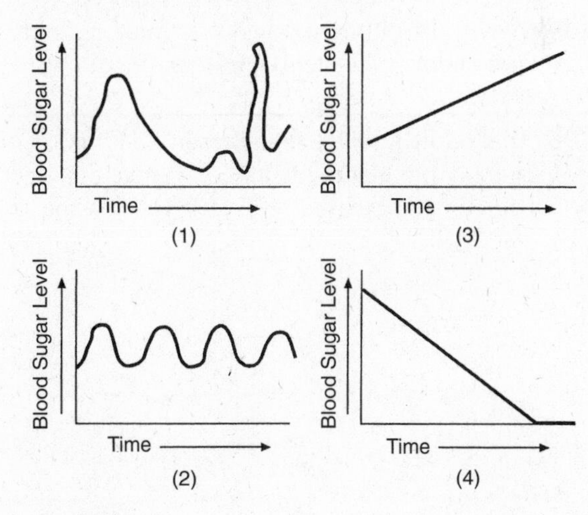


1 graph 1
2 graph 2
3 graph 3
4 graph 4

**32.** Why are only small amounts of a hormone usually needed to produce an effect? By what means does the body stop the release of too much of a hormone?

*Base your answers to questions 33 and 34 on the diagram below, which illustrates a function of hormones.*

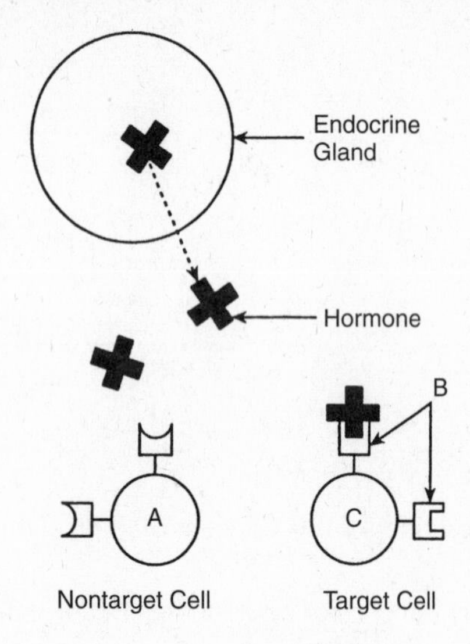


**33.** The structures identified by letter B, which are attached to cell C, represent
1 ribosomes
2 receptor molecules
3 specialized tissues
4 inorganic substances

**34.** How can you tell that cell A is *not* a target cell for the hormone shown in the diagram?

**35.** Why is the pituitary called the "master gland" of the endocrine system?

**36.** The following diagram represents a function of the thyroid gland. Identify *one* effect that occurs due to an increased level of TSH-releasing factor.

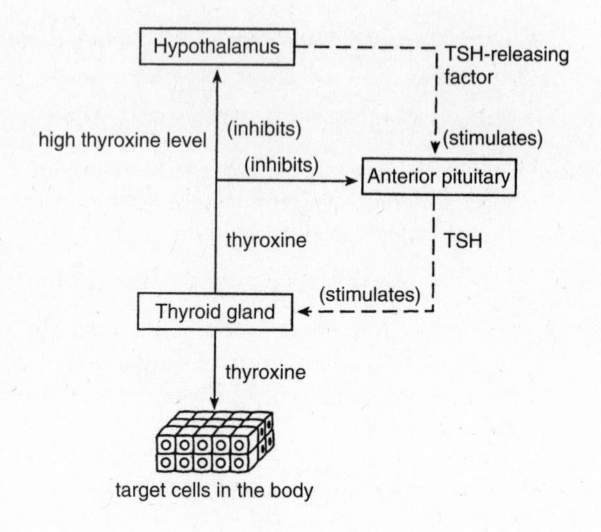


**37.** Parts of the brain and the endocrine system interact to release hormones. Use the following terms to fill in the flowchart below so that it shows which two body parts control the release of cortisol by another body part: *adrenal gland; pituitary gland; hypothalamus.*

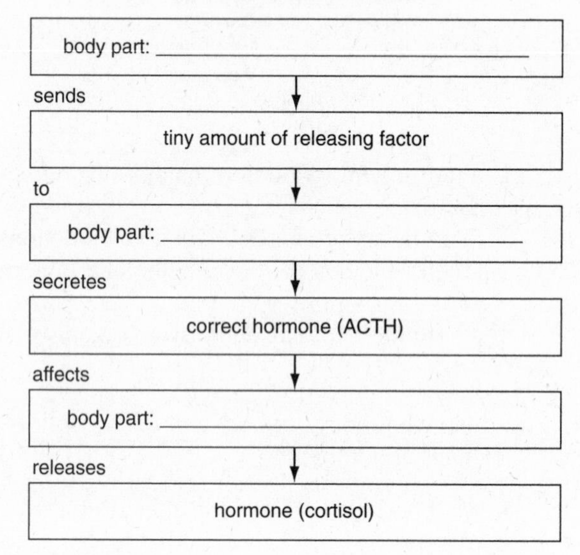

### *Part C—Reading Comprehension*

*Base your answers to questions 38 to 40 on the information below and on your knowledge of biology. Use one or more complete sentences to answer each question.* Source: *Science News*, Vol. 163, p. 206.

> A protein related to oxygen-carrying hemoglobin in blood cells may protect the brain during strokes. Scientists discovered the hemoglobin cousin several years ago and dubbed it *neuroglobin* because only nerve cells in the brain of vertebrates make it.
>
> Seeking to uncover neuroglobin's role, David A. Greenberg of the Buck Institute for Age Research in Novato, Calif., and his colleagues recently induced strokes in rats whose brains had been injected with viruses genetically engineered to churn out the protein. The amount of brain tissue damaged by the strokes was significantly less in those animals than in rats not given the virus, or in rats whose brains had less-than-normal amounts of neuroglobin, the investigators report in the March 18 *Proceedings of the National Academy of Sciences*.
>
> Greenberg and his colleagues conclude that neuroglobin naturally protects brain cells faced with too little oxygen. They speculate that drugs that increase the production of neuroglobin could become a new stroke therapy.

**38.** State two facts about the brain-protecting protein that led to its being given the name *neuroglobin*.

**39.** Describe the basic design of the experiment that was conducted to study the role of neuroglobin.

**40.** Explain how this understanding of the role of neuroglobin could be used to help stroke victims.

# 10
# Excretion and Water Balance

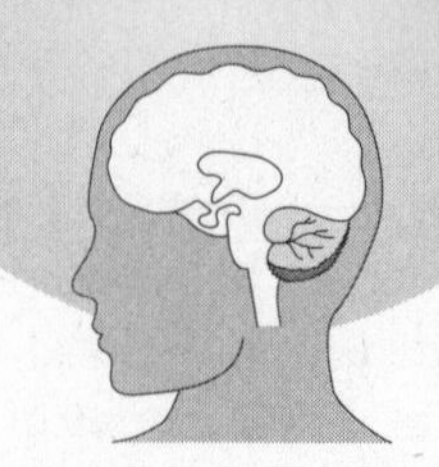

## Vocabulary

excretion
kidneys
liver
nitrogenous wastes
sweat glands
water balance

## REGULATION OF BODY CHEMISTRY

The human body is about 70 percent water. This water contains many types of solutes, including table salt. Yet we cannot drink salt water, only fresh water. Drinking salt water disrupts the careful balance that our bodies need internally, and the results could be very dangerous. That is because the amounts of these solutes in our bodies must remain constant day to day and hour to hour. In other words, one of the most important functions of homeostasis is to maintain a constant internal chemical environment.

The real meaning of the cell theory is that we are only as healthy as our cells are. Therefore, the chemical contents of our cells are regulated carefully. All body cells are surrounded by intercellular fluid, which is mostly water. (Refer to Figure 8-2 on page 64.) The level of chemicals in cells must remain within very narrow limits, so it is essential that the composition of intercellular fluid remains within narrow limits, too. How is its chemical composition regulated? Capillaries deliver blood to every cell in the body and exchange materials with the intercellular fluid that surrounds each cell. The blood controls the levels of substances in the fluid that surround each cell. So, one of the most important jobs of homeostasis is to regulate the chemical composition of blood. (See Figure 10-1.)

In vertebrates that live on land, the organ primarily responsible for regulating the chemical composition of blood is the *kidney*. When the **kidneys** do not work properly, the result is blood that does not have the proper balance of chemicals. In time, this can cause death, unless the kidneys' functions are taken over by mechanical means (dialysis).

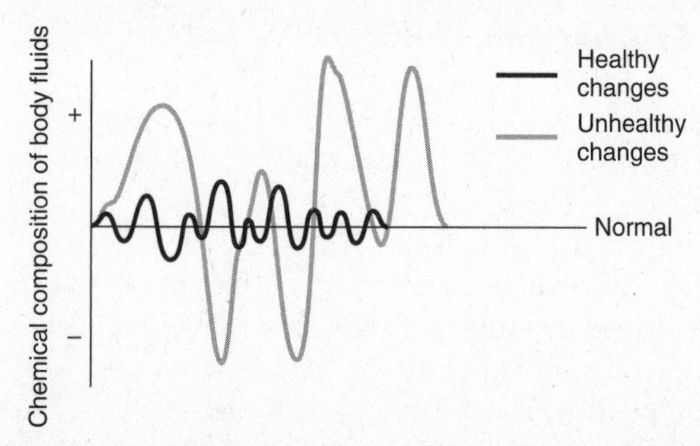


**Figure 10-1** If our cells are to remain healthy, the chemical composition of our body fluids (blood and intercellular) must remain within very narrow limits.

## THE KIDNEYS

The job of regulating both salt levels and water levels is often called maintaining a **water balance** in the body. This is one of the functions of the kidneys.

The kidneys also function as the cleansing system of the body. All metabolic activities in the body produce wastes. These wastes are *toxic* and must not be allowed to accumulate in the body. So another important function of the kidneys is to carefully select the chemical wastes for removal while keeping useful nutrients. This process of cleansing the body of metabolic wastes is called **excretion**.

By far the most important and potentially dangerous metabolic wastes produced in the body contain the element *nitrogen*. These **nitrogenous wastes** result when amino acids, the building blocks of proteins, are broken down. In the case of one-celled aquatic organisms, such as the ameba, the nitrogen wastes diffuse directly into the water through the cell membrane. But most animals, which are multicellular, have some type of internal organ that contains tubelike structures to carry out chemical regulation. As body fluids pass around and inside these tubes, materials are exchanged between the tubes and the blood in capillaries that surround the tubes. Useful materials stay in the blood; wastes are collected in the tubes and passed to the external environment.

## THE HUMAN EXCRETORY SYSTEM

The *kidneys* are the most important part of the excretory system in humans. There are two kidneys, located in the lower rear portion of the abdominal cavity.

A large artery brings blood to each of the kidneys. A large vein leaves each kidney with the newly cleansed blood. As blood passes through the kidneys, metabolic wastes are removed, and the correct balance of salt and water in the blood is maintained. As a result, the kidneys produce *urine*. Urine leaves each kidney through a tube, called the *ureter*, which drips the urine into the *bladder*. Here the urine is stored until it is passed from the body through another tube, called the *urethra*. (See Figure 10-2.)

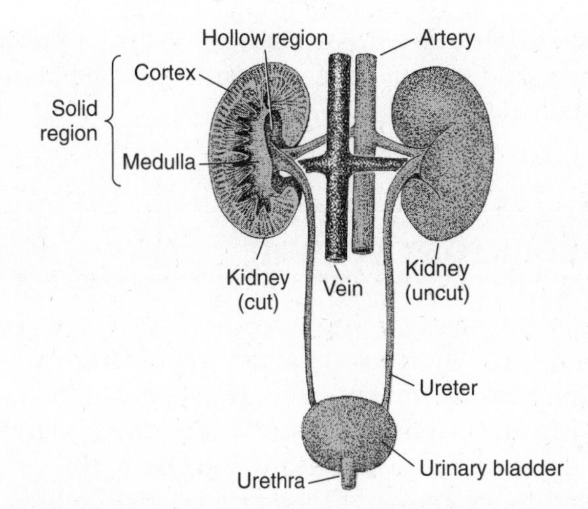


**Figure 10-2** The human excretory system—the kidneys, which are the most important part of this system, cleanse the blood of wastes and maintain water balance.

There are about one million tiny, tubular structures in each kidney. Around each structure, a complex network of capillaries is wrapped. Blood is cleansed, water balance is maintained, and urine is produced through the exchange of materials between the capillaries and the tubular structures. (See Figure 10-3.)

The result of all this activity in the kidneys is that the blood returns through veins to the body, cleansed of wastes and properly balanced with salts and water. In addition, urine,

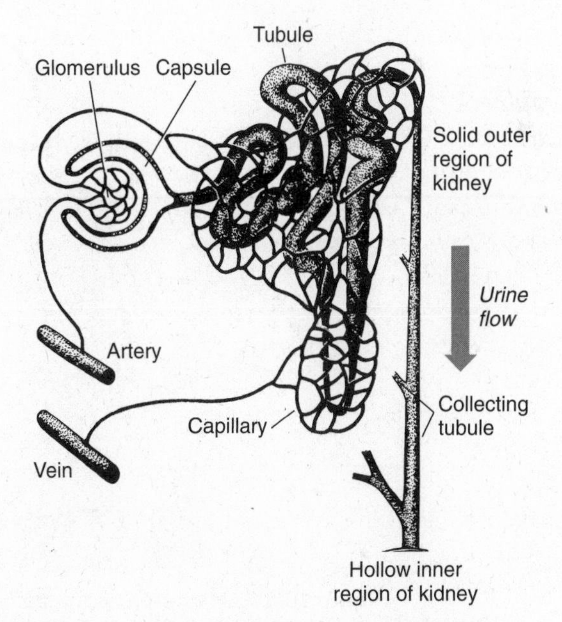


**Figure 10-3** Within each kidney, materials are exchanged between about one million tiny tubular structures and the network of capillaries that surrounds them.

which contains wastes as well as any excess salts and water, is collected and removed from the body.

## REGULATION OF THE EXCRETORY SYSTEM

The purpose of the excretory system is to maintain homeostasis by regulating the chemical composition of the blood and, in turn, all the body's cells. At times, to carry out this extremely important job, both the endocrine system and the nervous system have to help.

Regulation of the level of water in the body is very important. For example, when you are exercising a great deal and sweating a lot, your body loses water. The result is that the volume of your blood decreases, since it consists mostly of water. It is the blood, and not your cells, that loses the water. Our bodies "know" that it is absolutely necessary to keep the right amount of water inside our cells.

The lower volume of the blood acts as a stimulus to the endocrine and nervous systems. Glands in the brain detect the lower blood volume and send a hormone to the kidneys. The hormone signals the kidneys to return more water from the urine back into the blood. The blood volume returns to normal. Meanwhile, wastes become more concentrated in the urine. The urine becomes darker, which shows that the body is automatically readjusting itself by not releasing as much water. This is another example of a feedback mechanism.

## HOMEOSTASIS AND THE SKIN

The *skin* is the largest organ of the body. It is made up of a variety of different types of cells and tissues. This organ, which surrounds and protects the body, also has an important role in keeping the conditions that exist within the body fairly constant.

There are two separate layers in the skin: an outer layer and an inner layer. The outer layer is a very thin layer where new skin cells push their way toward the surface, replacing the dead skin cells there. This process occurs all the time. These skin cells are very tough and form a protective layer that prevents bacteria and harmful chemicals from entering, and water from leaving, the body. Beneath this layer are cells that produce the pigments that give skin its color.

The inner layer is a thick, complex layer that contains a variety of structures that help maintain homeostasis: nerve endings that can detect temperature and touch; the hair roots; capillaries that help regulate body tempera-

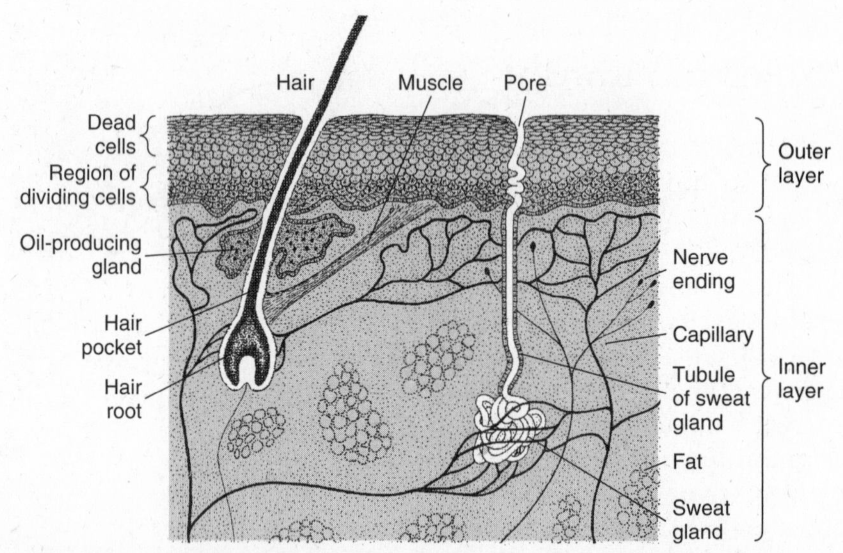


**Figure 10-4** Skin, the largest organ of the body, helps us maintain homeostasis through a variety of structures. The sweat glands help carry out excretion, since perspiration contains some nitrogen wastes.

ture; glands that keep skin and hair soft by releasing oils; and **sweat glands**. Perspiration from sweat glands contains some nitrogenous wastes, so these glands can be considered excretory structures. However, the main role of perspiration is to assist in regulating temperature by cooling the body through evaporation. (See Figure 10-4.)

## HOMEOSTASIS AND THE LIVER

No theme on homeostasis and excretion would be complete without a discussion of the **liver**. This organ is involved in homeostasis by assisting most of the important systems of the body. The liver helps in excretion by removing ammonia, a nitrogenous waste that results from the digestion of proteins; it breaks down old red blood cells; and chemical poisons, such as alcohol (which is considered a poison by the body) are also made harmless by the liver. However, an excessive intake of alcohol makes the liver work very hard and can, in time, cause the disease *cirrhosis*, which is a scarring of the liver tissue.

An example of homeostasis is the need to keep blood sugar levels constant. The liver helps the body do this by storing excess glucose in the form of a starch, called *glycogen*, and then releasing the sugar into the bloodstream when it is needed. In addition, the liver helps the digestive process by making bile, the substance used to digest fats; packages fats for transport in the blood; controls the level of cholesterol in the blood; stores vitamins A, D, and E; and helps maintain the body's water balance. (See Figure 10-5.)

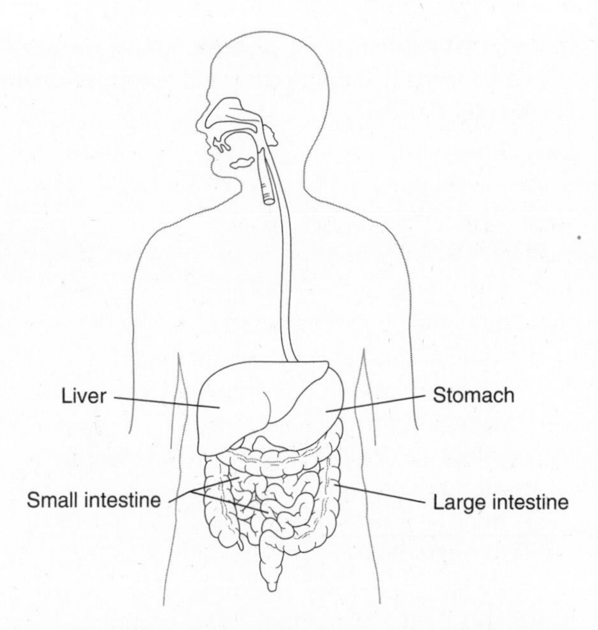


**Figure 10-5** The liver performs many functions that maintain homeostasis: it removes nitrogen from waste amino acids; breaks down old red blood cells; detoxifies chemical poisons; stores excess glucose; and helps maintain the body's water balance.

## WHEN THINGS GO WRONG: DISEASES OF THE EXCRETORY SYSTEM

One of the most common diseases of the kidney, a condition called *nephritis*, can be detected by a microscopic examination of the urine. Cells and chemicals that normally should not be in the urine are indicators of the disease. It is caused by a bacterial infection and can usually be treated with antibiotics. Kidney stones occur when a chemical compound that contains calcium builds up in the kidney. Gout, a condition caused by high levels of nitrogenous wastes in the blood, produces severe pain in the joints.

# Chapter 10 Review

### *Part A—Multiple Choice*

**1.** Humans cannot drink salt water because

1 it tastes so bad
2 the human body does not contain any salt
3 even small amounts of salt cause cells to die
4 it disrupts the balance of salt in the body

**2.** The levels of substances in the intercellular fluid are controlled by the

1 blood
2 lungs
3 heart
4 skin

**3.** In land vertebrates, the organ primarily responsible for regulating the chemical composition of the blood is the

1 liver
2 kidney
3 pancreas
4 heart

**4.** During excretion, the kidneys function to

1 remove solid wastes from the body
2 cause the skin to perspire, thereby removing excess water
3 select wastes for removal while keeping useful nutrients
4 remove water from the blood that passes through them

**5.** Wastes that contain nitrogen are produced when

1 polluted air is breathed in
2 proteins are formed
3 fats are digested
4 amino acids are broken down

**6.** In the ameba, nitrogen wastes

1 are released by the kidneys
2 are pumped out through specialized tubes
3 diffuse out through the cell membrane
4 are stored in a special saclike structure

**7.** In humans, wastes and useful materials are *directly* exchanged between the kidneys'

1 tiny tubes and the skin surface
2 tiny tubes and the bladder
3 veins and arteries and the bladder
4 tiny tubes and the blood in surrounding capillaries

**8.** Blood that returns to the body through veins from the kidneys has

1 a higher salt content than when it entered the kidneys
2 been cleansed of wastes and has a proper salt balance
3 lower levels of oxygen and carbon dioxide in it
4 more waste products, including a high level of urine

**9.** In humans, urine that is darker than normal indicates

1 excess water and a high volume of blood in the body
2 loss of water and a low volume of blood in the body
3 a high amount of nitrogen wastes in the body
4 a low amount of nitrogen wastes in the body

**10.** Perspiration from sweat glands in the skin helps maintain homeostasis by

1 excreting nitrogen wastes and salts only
2 cooling the body and regulating its temperature only
3 adjusting the salt and water balance of the body
4 excreting nitrogen wastes and regulating body temperature

**11.** When a person does strenuous exercise, small blood vessels (capillaries) near the surface of the skin increase in diameter. This change allows the body to be cooled. These statements best illustrate

1 synthesis
2 excretion
3 homeostasis
4 locomotion

**12.** The liver helps the body maintain homeostasis by doing all of the following *except*

1 removing nitrogen from waste amino acids
2 producing red blood cells and vitamins for the body
3 making bile, packaging fats, and controlling cholesterol levels
4 storing excess sugar and then releasing it when it is needed

**13.** The diagram below represents one metabolic activity of a human. Which row in the table under it correctly represents the letters A and B in the diagram?

Metabolic Activity A

Protein → B B B B

| Row | Metabolic Activity A | Product B |
|---|---|---|
| 1 | Respiration | Oxygen molecules |
| 2 | Reproduction | Hormone molecules |
| 3 | Excretion | Simple sugar molecules |
| 4 | Digestion | Amino acid molecules |

1 row 1
2 row 2
3 row 3
4 row 4

**14.** The diagram below represents events involved as energy is ultimately released from food.

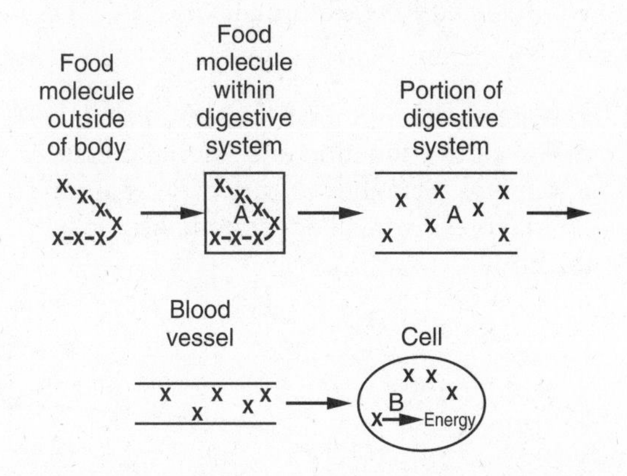


Which row in the table below correctly represents the X's and letters A and B in the diagram?

| Row | X-X-X-X-X-X-X-X | Substances A and B |
|---|---|---|
| 1 | Nutrient | Antibodies |
| 2 | Nutrient | Enzymes |
| 3 | Hemoglobin | Wastes |
| 4 | Hemoglobin | Hormones |

1 row 1
2 row 2
3 row 3
4 row 4

### *Part B—Analysis and Open Ended*

**15.** Explain why it is harmful for a person to drink salt water.

**16.** Why is it necessary for the body to regulate the chemical composition of the blood?

**17.** Describe the two main functions of the kidneys.

**18.** What is the source of nitrogenous wastes in the body?

**19.** Briefly compare the removal of nitrogenous wastes by one-celled and multicellular organisms. Your answer should include the following:

- the basic structures involved for each type of organism
- the process by which each removes the nitrogen wastes
- the term that describes the removal of metabolic wastes

**20.** List the *four* main structures of the human excretory system. In what sequence are the parts connected to carry out waste (urine) removal?

*Refer to the following figure to answer questions 21 and 22.*

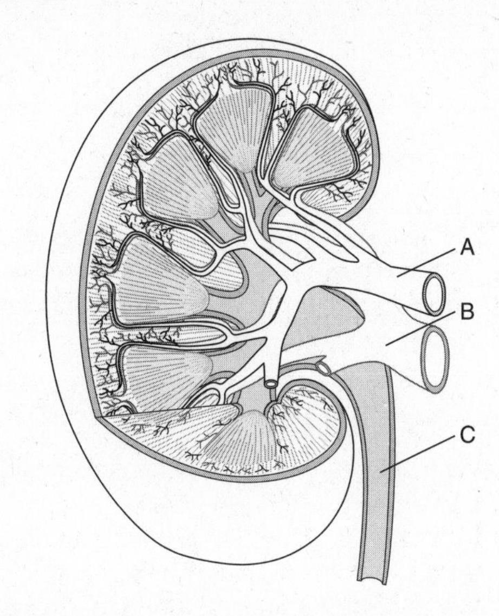


**21.** The diagram represents a cross section of what excretory organ? Identify the three main tubes—labeled A, B, and C—that are connected to it.

**22.** Briefly describe the function of each of these tubes. Your answer should explain the following:

- what kind of body fluid is transported in each of these tubes
- which way the fluid is flowing in relation to the organ
- what process occurs as each fluid passes through the organ

**23.** What kind of blood vessel surrounds the millions of tiny tubules in the kidneys? Describe what occurs between these blood vessels and the tubules.

**24.** Use the following terms to replace the phrases in the chart on page 86 so that they show how the endocrine, nervous, and excretory systems work together to maintain a healthy water balance in the body: *hormone to the kidneys; normal blood volume; lower blood volume; concentrated wastes in urine; glands in the brain; return water from urine to blood.*

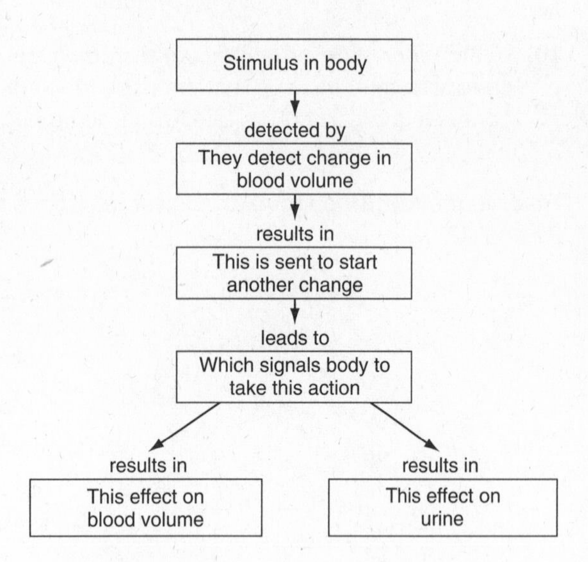


**25.** Refer to Figure 10-4 on page 82 to answer the following questions.

a. What structure in the skin can be considered part of the excretory system?
b. In what part of the skin is this structure found—the inner layer or outer layer?
c. What job does this structure perform for the excretory system? Is this the structure's *main* function? (If not, what is?)

**26.** The liver carries out many functions that help to maintain homeostasis. Briefly describe the two ways that it helps in (a) waste excretion and (b) water balance.

**27.** Why is excessive alcohol consumption harmful to a person's health?

*Base your answers to questions 28 to 30 on the diagram below, which shows some of the specialized organelles in a single-celled organism, and on your knowledge of biology.*

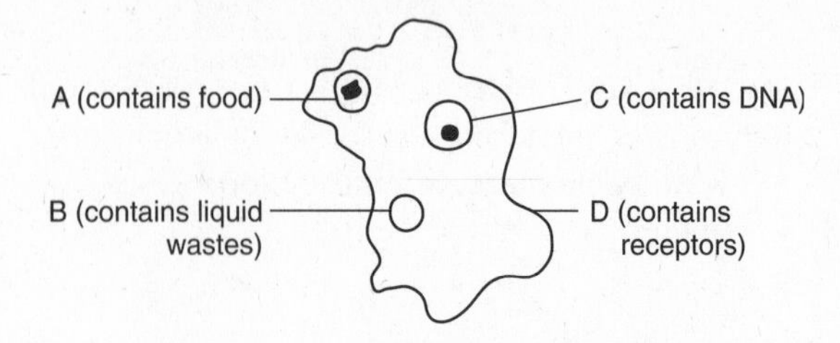


**28.** Write the letter of *one* of the labeled organelles and state the name of that organelle.

**29.** Explain how the function of the organelle you selected in question 28 assists in the maintenance of homeostasis.

**30.** Identify a system in the human body that performs a function similar to that of the organelle you selected in question 28.

**31.** The graph below shows the relationship between kidney function and arterial pressure in humans. State how a steady decrease in arterial pressure will affect homeostasis in the body.

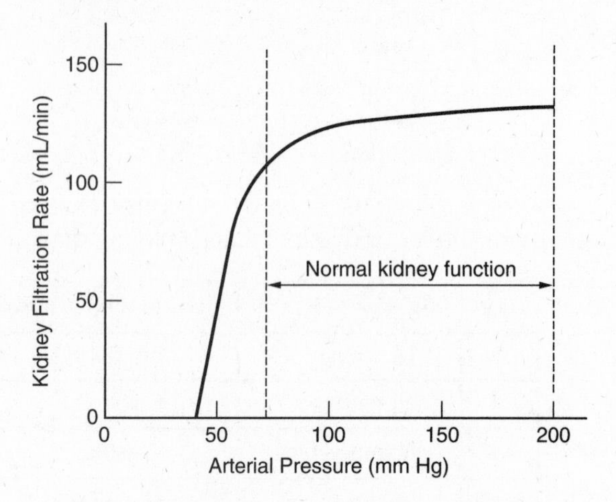


**32.** The skeletal system of an animal is shown in the photograph below. List *three* systems (other than the skeletal system) that the animal had when alive, which helped it to survive. Describe how each of these three systems contributed to maintaining homeostasis.

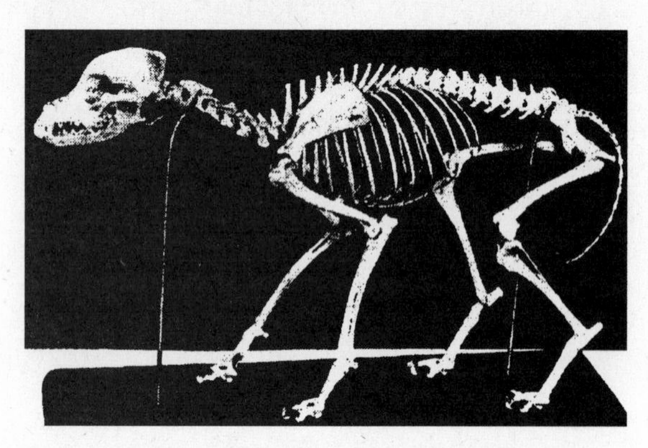

**33.** Several systems interact in the human body to maintain homeostasis. Four of these systems are the circulatory, the digestive, the respiratory, and the excretory. Based on these, answer the following:

a. Select *two* of the systems listed. Identify each system selected and state its function in helping to maintain homeostasis in the body.
b. Explain how a malfunction of *one* of the four systems listed disrupts homeostasis and how

that malfunction could be prevented or treated. In your answer be sure to:

- name the system and state *one* possible malfunction of that system
- explain how the malfunction disrupts homeostasis
- describe *one* way the malfunction could be prevented or treated

### *Part C—Reading Comprehension*

*Base your answers to questions 34 to 36 on the information below and on your knowledge of biology. Use one or more complete sentences to answer each question.*

Kidney failure, now called *ESRD* (*e*nd-*s*tage *r*enal *d*isease), leads to a buildup of wastes in the blood, higher concentrations of salt and water, and unbalanced pH levels. Homeostasis is seriously disrupted. Without kidney function, life cannot continue. Fortunately, however, treatments have been developed to artificially cleanse the blood. The treatments involve the exchange of substances across membranes in a process called *dialysis*. In the 1960s, when the first artificial kidney was developed, a patient had to be connected to a dialysis machine for 8 to 10 hours at a time, several times a week. There were also serious side effects from the treatment, such as severe infections and inflammation around the heart.

Today, the many options that ESRD patients can choose from are remarkable. The two basic choices are hemodialysis or peritoneal dialysis. With *hemodialysis*, the patient's blood passes through tubes into a machine placed next to him or her, where materials are exchanged between the blood and fluids in the machine. The patient has the option to undergo hemodialysis at home or in the hospital, and at different times during the week. Some people have dialysis treatment every day, others just two or three times a week.

The other kind of treatment, *peritoneal dialysis*, amazingly uses the lining of the patient's own abdomen to filter the blood. A cleansing solution is transported into the abdomen, where it stays for awhile. As the person's blood passes in capillaries through the solution, the blood gets cleansed. Then the solution is drained out, and a fresh solution is put into the patient's abdomen to start the process again. Peritoneal dialysis can be done in several ways. One method uses no machine; it just goes on continuously, even as the patient walks around. A second method uses a machine to move the cleansing solution through a tube, in and out of the person's abdomen. The machine does this work at night, while the person sleeps. A third method also uses a machine, but only at certain times, and usually in the hospital.

So there are now several treatment choices for people with kidney failure choices that allow individuals who no longer have normal use of their kidneys to go on living normal lives nevertheless.

**34.** In what ways does kidney failure disrupt the body's homeostasis?

**35.** Describe the options available to a person undergoing hemodialysis.

**36.** How does peritoneal dialysis use the patient's body to cleanse the blood?

# 11
# Disease and Immunity

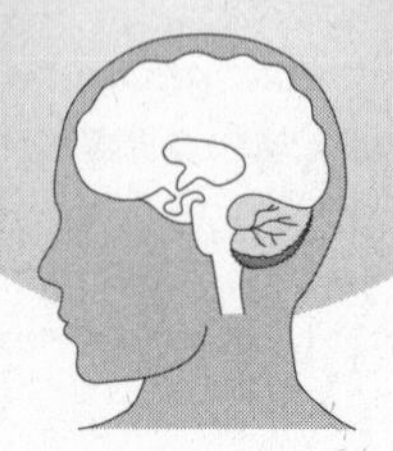

## Vocabulary

AIDS
allergic reactions
antibodies
antigens
autoimmune diseases
bacteria
fungi
immune system
immunodeficiency diseases
immunity
malfunction
microbes
pathogens
vaccinations
viruses
white blood cells

## DISEASE: A LACK OF HOMEOSTASIS

The theme of homeostasis emphasizes the need for organisms to maintain a carefully controlled internal set of conditions, that is, a dynamic equilibrium. Maintaining these conditions—including pH, temperature, water and salt balance, and levels of carbon dioxide and oxygen—allows an organism's cells to function normally. Organisms can tolerate changes that occur within very definite limits. But changes outside normal limits disrupt homeostasis, producing illness, disease, and even death.

There are many reasons why the body may be pushed beyond its normal limits. These reasons, or factors, are often the causes of disease. An inherited defect in a genetic trait might be the cause of a disease. The disruption of homeostasis in such a disease would be caused, in a sense, by a factor inside the body. Many other diseases result from some influence outside the body, that is, from the environment.

## FACTORS THAT CAUSE DISEASES

Diseases may be caused by one of the following factors, or by a combination of several of these.

- *Inheritance.* Defective genetic traits can be passed from parents to offspring. Often, neither parent has the disease, but both may carry a single form of a gene for the disease. It is the combination of these two defective, or mutated, genes in the child that gives him or her the disease. A well-known example of an inherited disease is sickle-cell anemia, in which the protein that carries oxygen in red blood cells is flawed. (See Figure 11-1.)
- *Microorganisms.* Microscopic organisms that cause diseases are called **pathogens**; they include certain **fungi, bacteria, viruses**, and protozoa. Some diseases caused by pathogenic microorganisms may be passed in a variety of ways from one person to another. These are called *in-*

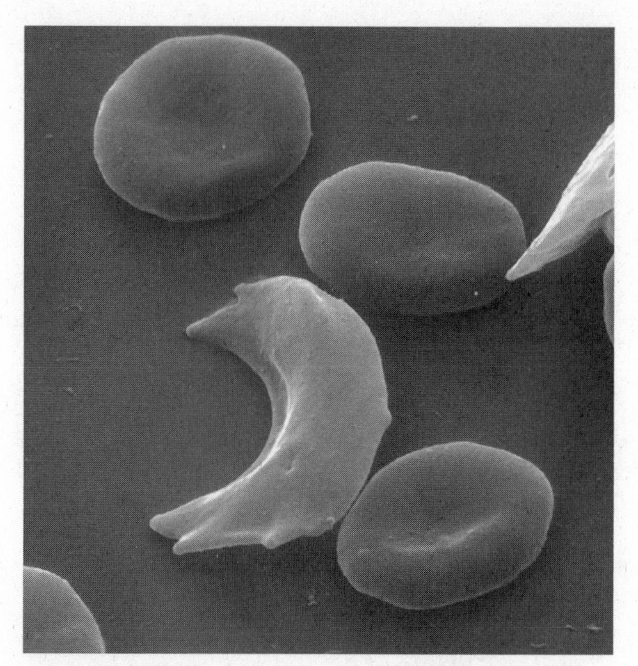

**Figure 11-1** Sickle-cell anemia is an inherited disease that disrupts homeostasis because the protein that carries oxygen in red blood cells is flawed.

*fectious diseases*. Microorganisms most often enter the body through respiratory pathways, the digestive system, or pathways of the excretory system. Infections may also occur through breaks in the skin. Tuberculosis is an infectious disease caused by certain bacteria. (See Figure 11-2.)

- *Pollutants* and *poisons*. Chemical agents present in the environment may upset the body's normal functioning and produce disease. These pollutants and poisons include coal dust, asbestos, lead, mercury, and many others. For example, when asbestos fibers enter the respiratory system, they cause asbestosis, a disease of the lungs; years later, this may result in cancer in the lungs and chest.

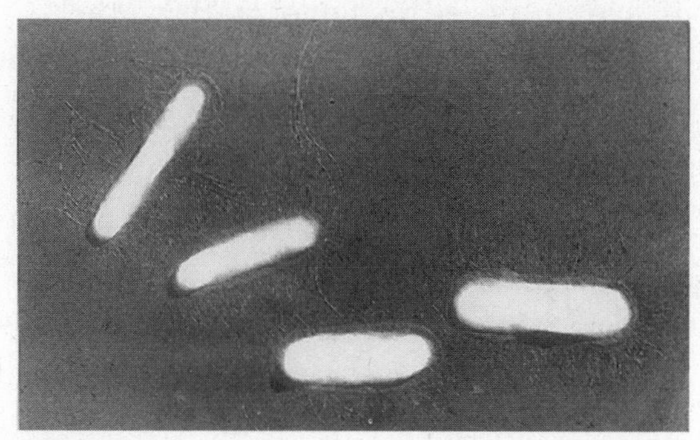

**Figure 11-2** These bacteria are an example of microorganisms that can cause infectious diseases, such as tuberculosis.

- *Organ malfunction*. A disease may develop when one or more of the body's organs **malfunction**. When an organ such as the liver, lung, heart, stomach, or kidney is damaged by disease or injury and cannot function properly, serious effects on the body result.
- *Harmful lifestyles*. The way one lives can also be an important factor in causing disease. Specifically, the use of tobacco, alcohol, and drugs can disrupt homeostasis in the body, producing illness. In addition, overeating, poor nutrition, lack of exercise, unsafe sexual practices, and living with stress can lead to certain diseases. Hypertension, or high blood pressure, is one such disease.

## THE BODY'S DEFENSES AGAINST DISEASE

Our bodies are surrounded by countless microorganisms. Some of them succeed in entering our bodies through the nose, through cuts in our skin, or along with the food we eat. Many of these microorganisms cause serious problems if they survive and reproduce inside us without challenge. Controlling these microscopic invaders is as important to homeostasis as is regulating body temperature and chemistry.

The first line of defense against infection consists of *physical barriers* that block the entry of microorganisms. The skin is the main physical barrier in our body. A second line of defense, called *inflammation*, is present when microorganisms get through our physical barriers. For example, when we get a cut or scrape on the skin, the injured area may become warm, reddened, and perhaps swollen with pus. (See Figure 11-3.) Chemicals released by the damaged tissues are acting like an alarm, causing an increase in blood flow to the site of the injury. Special white blood cells that arrive engulf microorganisms, destroying them by ingesting them. All of this activity

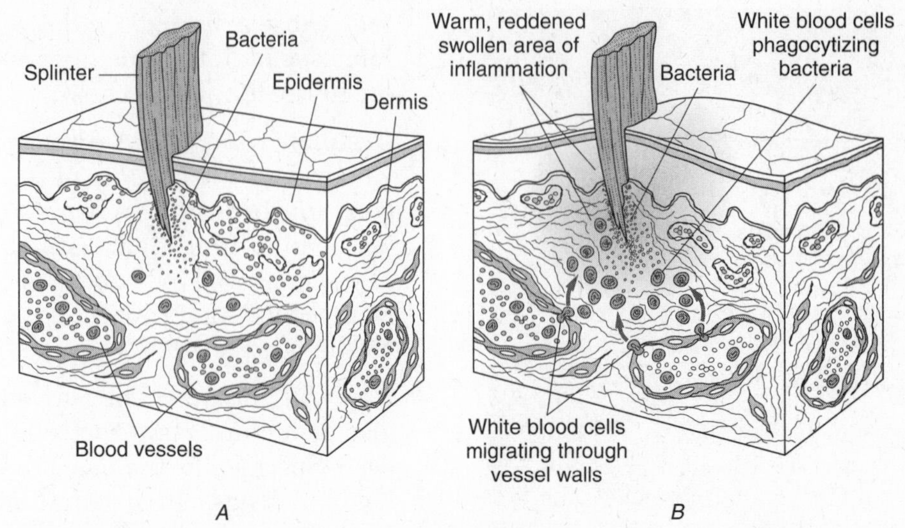


**Figure 11-3** Inflammation is the body's second line of defense against infectious bacteria that get through the skin's first defense.

helps prevent a more serious infection from developing. (See Figure 11-4.)

Vertebrates have evolved a very important system that fights specific invaders. This is the **immune system**. The immune system recognizes who the invaders are and attacks these microorganisms to try to keep them from disrupting normal body functions.

## THE HUMAN IMMUNE SYSTEM

The immune system defends our bodies against very specific microscopic invaders. Each invader—usually a bacterium or virus—has specific protein molecules attached to its surface. These protein molecules are called **antigens**. It is these molecules that are detected by the body's immune system.

When the immune system detects a foreign antigen, it produces **antibodies**, molecules that bind to that antigen. Once the antibodies bind to the antigen, the invader can be destroyed by the body. (See Figure 11-5.) **Vaccinations** use weakened microorganisms, or parts of them, to stimulate the immune system to react by recognizing the specific antigens. This reaction provides the body with **immunity**, the ability to resist an infection. It does so by preparing the body to fight subsequent invasions by the same microorganisms, or **microbes**, because it can recognize the invaders and produce specific antibodies against them. People are now given harmless antigens in vaccines to offer protection against a number of diseases.

## B CELLS AND T CELLS

The immune system also includes B cells and T cells, special kinds of **white blood cells** that are produced in bone marrow, the thymus gland, the spleen, the lymph nodes, and

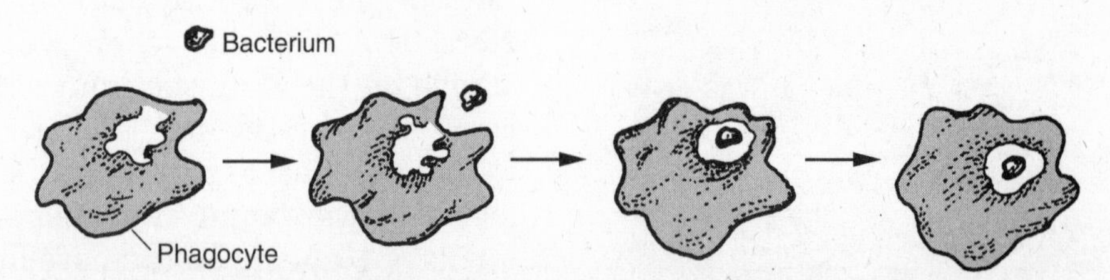


**Figure 11-4** During inflammation, the white blood cells engulf and destroy the invading bacteria, which prevents the development of a more serious infection.

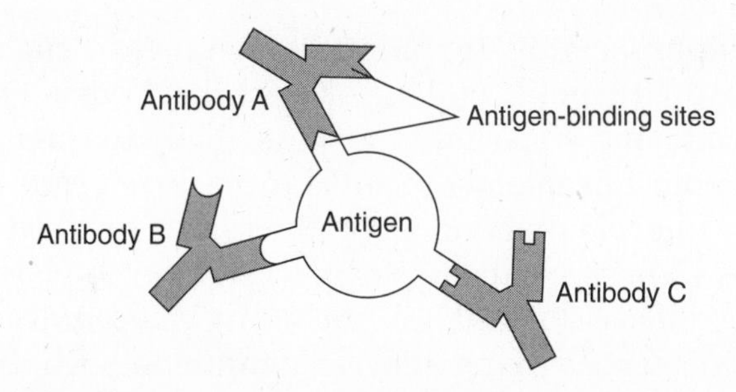


**Figure 11-5** Antibodies, produced by the immune system, bind to antigens, which are specific protein molecules on an invading microbe's surface.

the tonsils. (See Figure 11-6.) *B cells* are the ones that respond to specific antigens by beginning to produce antibody proteins that will bind only with that antigen.

As time goes on, the body comes to have many different types of B cells, each producing antibodies for one specific antigen. After having been invaded once by an antigen, some special B cells that recognize that antigen remain in the body for the rest of one's life. These are called memory B cells. Because they are already present in the body, you instantly start making antibodies the moment you encounter the same invading microorganisms again. That is why individuals usually do not get measles or chicken pox a second time. The immune system remembers the first exposure to the disease and is ready to defend the body. (See Figure 11-7.)

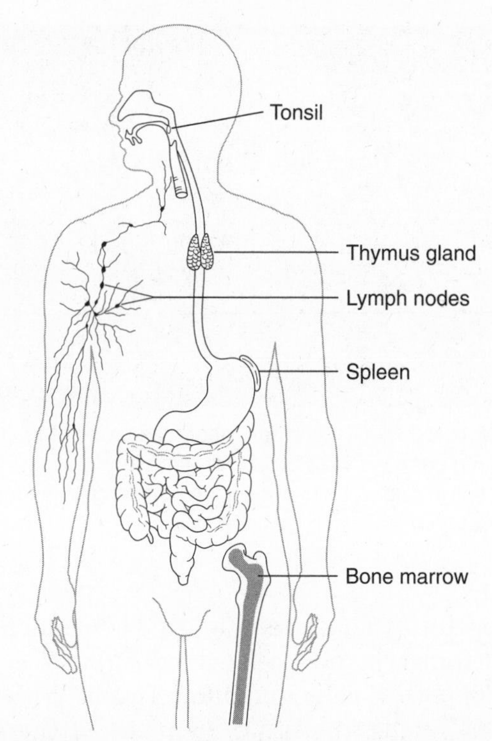


**Figure 11-6** Special white blood cells, called B cells and T cells, are produced in the tonsils, thymus gland, lymph nodes, spleen, and bone marrow.

One type of T cell is called *killer T cells.* Through protein receptors on their surface, they can recognize cells in the body that have been infected with invading microorganisms. The killer T cells punch holes in the membranes of the infected cells, sometimes injecting

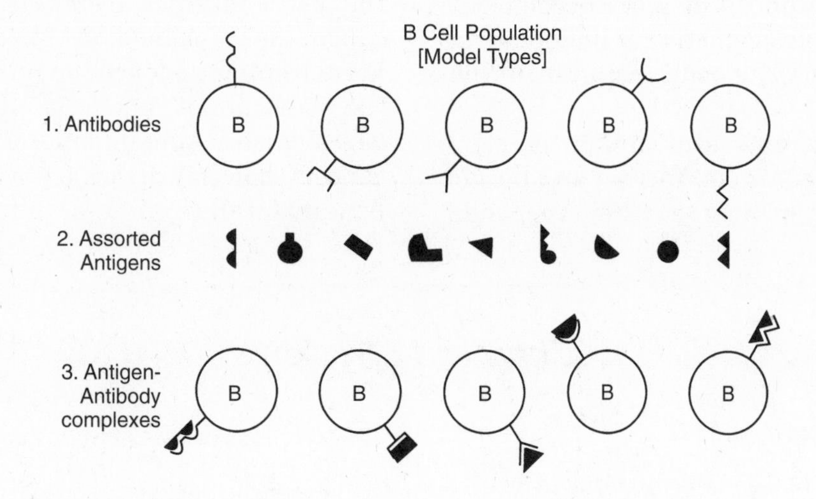


**Figure 11-7** Over time, a person's body comes to have many different B cells. The memory B cells, which remain in the body after their first exposure to an antigen, can instantly make antibodies when they encounter the same antigen again.

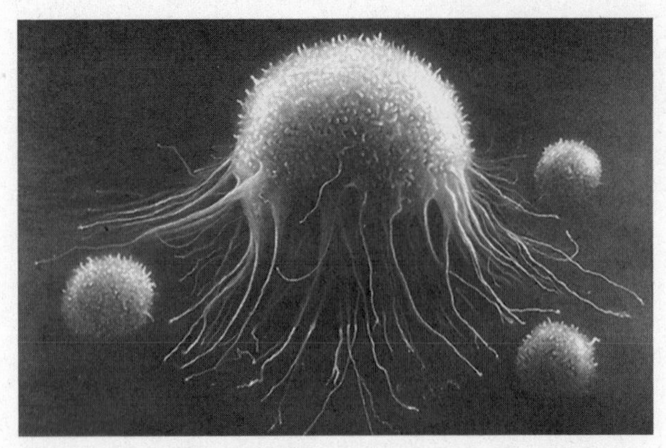

**Figure 11-8** Killer T cells can recognize cells in the body that have been infected by invading microorganisms. Here, some killer T cells are shown attacking an infected body cell.

poison into them. (See Figure 11-8.) Another important type of T cell, called *helper T cells*, assists both B cells and killer T cells. Without helper T cells, the other members of the immune system cannot do their job. The importance of helper T cells is shown by the fact that they are the cells that are destroyed by the *h*uman *i*mmunodeficiency *v*irus (HIV), which results in the disease called **AIDS**.

## WHEN THINGS GO WRONG: DISEASES OF THE IMMUNE SYSTEM

The immune system helps maintain the internal dynamic equilibrium necessary for life. However, the immune system can become out of balance. It can be overactive or underactive, and in either case the body's equilibrium is upset.

**Allergic reactions** result from overactivity of the immune system. During an allergic reaction, the body's immune system responds inappropriately to common substances—such as dust, mold, pollen, or certain foods—by producing a special type of antibody to them. (In most people, these substances do not cause an allergic reaction or production of antibodies.) These antibodies cause cells in the body to release substances, including *histamines*, which cause many allergic symptoms, such as extra fluid in the nasal pathways, difficulty breathing, or hives. The allergies are often treated with *antihistamines*, drugs that stop the release of histamines.

Sometimes an overactive immune system begins to attack its own normal body tissues. These are called **autoimmune diseases** and they are very serious. They include rheumatoid arthritis and lupus erythematosus. In all autoimmune diseases, the body is literally rejecting its own tissues. The immune system may also attack transplanted organs; medications are then taken to try to prevent organ rejection.

Also inflammation, which protects us when we are young, may actually be contributing to crippling diseases when we get older. For example, researchers now suspect that many heart attacks occur when a rupture develops in the wall of an artery, brought on by overactive immune system cells causing inflammation.

The illness known as *AIDS* (*a*cquired *i*mmuno*d*eficiency *s*yndrome) is a type of **immunodeficiency disease**, which means that the body's immune system is underactive because it is weakened, in this case by HIV, the human immunodeficiency virus. As a result, the body cannot protect itself from other diseases (such as pneumonia, tuberculosis, and cancer) that may attack it—a condition that is usually fatal.

## *Chapter 11 Review*

### *Part A—Multiple Choice*

**1.** A disruption of homeostasis can result in all of the following except

1 illness
2 death
3 disease
4 stability

**2.** Infectious diseases result from

1 genetic defects
2 microorganisms
3 pollutants
4 organ malfunctions

**3.** The inhalation of particles such as asbestos fibers and coal dust can result in respiratory diseases. In such a case, the main cause of the disease would be

1 microorganisms
2 pollutants
3 genetic defects
4 organ malfunction

**4.** The body's main physical barrier against infection is

1 the skin
2 white blood cells
3 red blood cells
4 inflammation

**5.** Scientific studies have indicated that there is a higher percentage of allergies in babies fed formula containing cow's milk than in breastfed babies. Which statement represents a valid inference made from these studies?

1 Milk from cows causes allergic reactions in all human infants.
2 There is no relationship between drinking cow's milk and having allergies.
3 Breastfeeding by humans prevents all allergies from occurring.
4 Breast milk most likely contains fewer substances that trigger allergies.

**6.** Allergic reactions are most closely associated with

1 the action of circulating hormones
2 immune responses to usually harmless substances
3 a low blood sugar level
4 the shape of red blood cells

**7.** White blood cells can prevent a serious infection by

1 filling the damaged tissues with pus
2 repairing the skin after it has been cut
3 ingesting the harmful microorganisms
4 constructing physical barriers against microorganisms

**8.** Certain microbes, foreign tissues, and some cancerous cells can cause immune responses in the human body because all three contain

1 antigens
2 lipids
3 enzymes
4 cytoplasm

**9.** Which activity would stimulate the human immune system to provide protection against an invasion by a microbe?

1 receiving antibiotic injections after surgery
2 being vaccinated against chicken pox
3 choosing a well-balanced diet and following it throughout life
4 receiving hormones contained in mother's milk while nursing

**10.** When the immune system detects an antigen, it

1 pushes it out of the body immediately
2 produces antibodies that bind to the antigen
3 produces antigens that cancel the effects of it
4 destroys the antigen by cutting it in half

**11.** Many vaccinations stimulate the immune system by exposing it to

1 antibodies
2 mutated genes
3 enzymes
4 weakened microbes

**12.** A part of the hepatitis B virus can be synthesized in the laboratory. This viral particle can be identified by the immune system as a foreign substance, but the viral particle is not capable of causing the disease. Immediately after this viral particle is injected into a human, it

1 stimulates the production of enzymes that are able to digest the hepatitis B virus
2 synthesizes specific hormones that provide immunity against the hepatitis B virus
3 triggers the formation of antibodies that protect against the hepatitis B virus
4 breaks down key receptor molecules so that the hepatitis B virus can enter body cells

**13.** The following diagram represents one possible immune response that can occur in the human body. The structures that are part of the immune system are represented by

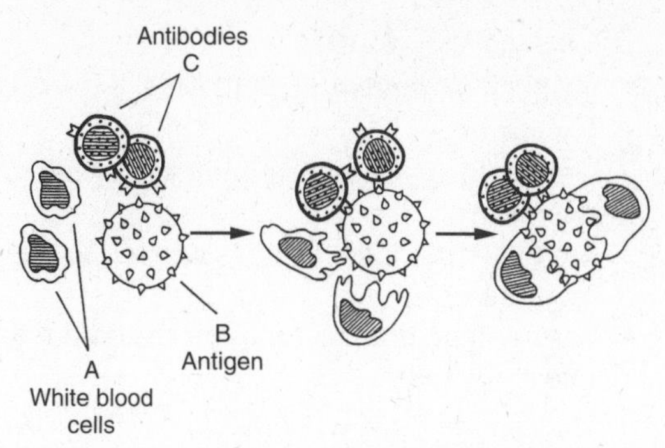


1 A, only
2 A and C only
3 B and C only
4 A, B, and C

**14.** Which of the following is *not* a characteristic of white blood cells?

1 They destroy some microbes by engulfing them.
2 They carry oxygen atoms throughout the body.
3 They make antibodies that bind with antigens.
4 They punch holes in membranes of infected cells.

**15.** Which statement does *not* identify a characteristic of antibodies?

1 They are produced in response to the presence of foreign substances.
2 They are nonspecific, acting against any foreign substance in the body.
3 They may be produced in response to an antigen.
4 They may be produced by the white blood cells.

**16.** Which is the best procedure for determining if a vaccine for a disease in a particular bird species is effective?

1 Vaccinate 100 birds; then expose all 100 birds to the disease.
2 Vaccinate 50 birds, do not vaccinate 50 other birds; expose all 100 birds to the disease.
3 Vaccinate 100 birds and expose only 50 of them to the disease.
4 Vaccinate 50 birds, do not vaccinate 50 other birds; expose only the vaccinated birds to the disease.

**17.** Produced in several parts of the body, B cells and T cells are special kinds of

1 blood platelets
2 red blood cells
3 white blood cells
4 microorganisms

**18.** The killer T cells function to

1 produce antibodies that kill invading microorganisms
2 destroy the cells that are infected by microorganisms
3 bind with the infected cells and repair their membranes
4 destroy invading microorganisms before they infect any cells

**19.** Which condition would most likely result in a human body being unable to defend itself against pathogens and cancerous cells?

1 a genetic tendency toward a disorder such as diabetes
2 the production of antibodies in response to an infection in the body
3 a parasitic infestation of ringworm on the body
4 the presence in the body of the virus that causes AIDS

**20.** The human immunodeficiency virus (HIV) is particularly devastating to the immune system because it destroys

1 all the white blood cells in the body
2 all the red blood cells in the body
3 the special B cells and killer T cells only
4 helper T cells, which assist B cells and killer T cells

**21.** Overactivity of the immune system due to a common substance can lead to an

1 equilibrium
2 allergic reaction
3 antihistamine
4 immunodeficiency

**22.** When an overactive immune system starts to attack its own body tissues, it causes serious conditions known as

1 antihistamine diseases
2 allergic reaction diseases
3 autoimmune diseases
4 immunodeficiency diseases

**23.** A characteristic shared by all enzymes, hormones, and antibodies is that their function is determined by the

1 shape of their protein molecules
2 inorganic molecules they contain
3 age of the organism that makes them
4 organelles present in their structure

**24.** A researcher needs information about antigen–antibody reactions. He could best find information on this topic by searching for which phrase on the Internet?

1 Protein Synthesis
2 Energy Sources in Nature
3 White Blood Cell Activity
4 DNA Replication

## Part B—Analysis and Open Ended

**25.** How can a child inherit a disease if neither parent appears to have it?

**26.** Explain what a pathogen is. Your answer should include the following information:

- what the *four* types of pathogens are that can cause diseases
- the term for such diseases when they are passed from one person to another
- the *three* common ways that pathogens can enter the body

**27.** What is the difference between an inherited disease and an infectious disease?

**28.** Briefly describe the relationship between organ malfunction and disease.

**29.** Explain why a harmful lifestyle can lead to disease. Give *one* example. In what way are the factors that cause such a disease preventable or controllable?

**30.** Describe the first line of defense against infection in the body. Include the following:

- what *kind* of defense it consists of
- what *organ* carries out this function
- what it is defending *against* and how

**31.** Refer to the diagram below to answer the following question: When bacteria enter a cut, what process occurs as part of the body's second line of defense?

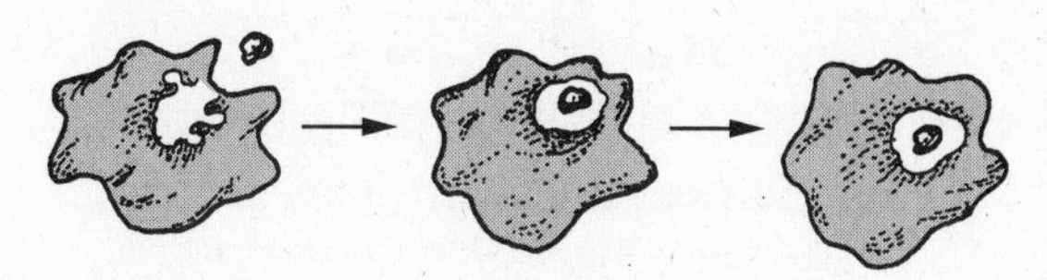

1 infection
2 inflammation
3 invasion
4 immunity

**32.** Use the terms listed below to fill in the missing words in the following flowchart, which describes the immune system's reaction to microscopic invaders: *body develops immunity; microbes destroyed by body; antibodies; antigens.*

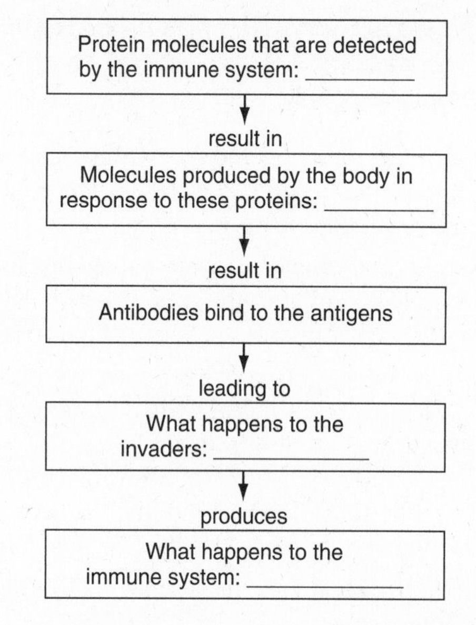


**33.** How does the immune system help to maintain homeostasis?

**34.** Briefly define the term *immunity.* How do vaccinations provide people with immunity?

*Base your answers to questions 35 and 36 on the table below and on your knowledge of biology.*

| Volunteer | Injected with Dead Chicken Pox Virus | Injected with Dead Mumps Virus | Injected with Distilled Water |
|---|---|---|---|
| A | ✔ | | |
| B | | ✔ | |
| C | | | ✔ |
| D | ✔ | ✔ | |

**35.** None of these volunteers ever had chicken pox. After the injection, there most likely would be antibodies to chicken pox in the bloodstream of

1 volunteers A and D only
2 volunteers A, B, and D
3 volunteer C only
4 volunteer D only

**36.** Volunteers A, B, and D underwent a medical procedure known as

1 cloning
2 vaccination
3 electrophoresis
4 chromatography

*Refer to the following list, which describes three ways of controlling viral diseases in humans, to answer question 37:*

- administer a vaccine containing dead or weakened viruses, which stimulates the body to form antibodies against the virus
- use chemotherapy (chemical agents) to kill viruses, similar to the way in which sulfa drugs or antibiotics act against bacteria
- rely on the action of interferon, which is produced by cells in the body and protects it against pathogenic viruses

**37.** Based on this information, which activity would provide the greatest protection against viruses?

1 producing a vaccine that is effective against interferon
2 using interferon to treat a number of diseases caused by bacteria
3 developing a method to stimulate the production of interferon in cells
4 synthesizing a drug that prevents the destruction of bacteria by viruses

**38.** Compare the functions of B cells with those of the killer T cells.

**39.** Explain how HIV affects the immune system. Your answer should include the following:

- what "HIV" stands for and what the term *immunodeficiency* means
- which cells in the immune system are affected and in what way
- how this affects the rest of the immune system's functioning

*Base your answers to questions 40 and 41 on the following image, which shows a slide of normal human blood cells (magnified several times), and on your knowledge of biology.*

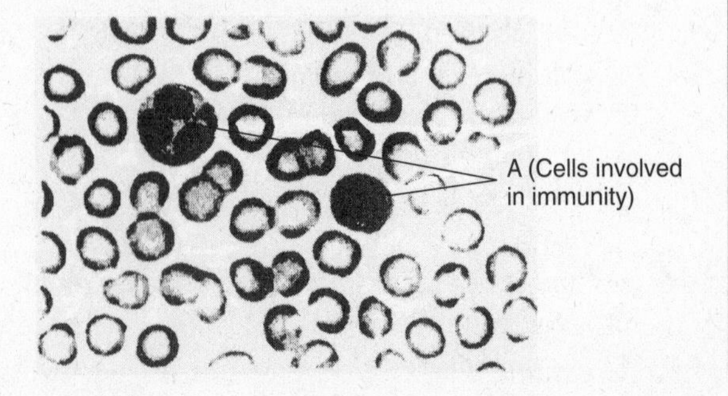


**40.** An increase in the production of the cells labeled *A* occurs in response to an internal environmental change. State *one* possible change that could cause this response.

**41.** Describe *one* possible immune response, other than an increase in number, that one of the cells labeled *A* could carry out.

**42.** In what way are allergic reactions and autoimmune diseases similar to one another? In what important way are they different?

*Base your answers to questions 43 and 44 on the information below and on your knowledge of biology.*

Immunization protects the human body from disease. The United States is now committed to the goal of immunizing all children against common childhood diseases. In fact, children must be vaccinated against certain diseases before they can enter school. However, some parents feel that vaccinations may be dangerous and do not want their children to receive them. Many parents are choosing not to immunize their children against childhood diseases such as diphtheria, whooping cough, and polio. For example, the mother of a newborn baby is concerned about having her child receive the DPT (diphtheria, whooping cough, and tetanus) vaccine. Since bacteria cause these diseases, she believes antibiotic therapy is a safe alternative to vaccination.

**43.** Explain to these concerned parents what a vaccine is and what it does in the body. Be sure to include the following:

- what is in a vaccine and how vaccination promotes immunity to various diseases
- the difference between antibiotics and vaccines in the prevention of bacterial diseases

**44.** State *one* advantage of using vaccinations in fighting bacterial diseases and *one* disadvantage of using antibiotics in fighting bacterial diseases.

### *Part C—Reading Comprehension*

*Base your answers to questions 45 to 47 on the information below and on your knowledge of biology.*

Avian flu virus H5N1 has been a major concern recently. Most humans have not been exposed to this strain of the virus, so they have not produced the necessary protective substances. A vaccine has been developed and is being made in large quantities. However, much more time is needed to manufacture enough vaccine to protect most of the human population of the world.

Most flu virus strains affect the upper respiratory tract, resulting in a runny nose and a sore throat. However, the H5N1 virus seems to go deeper into the lungs and causes severe pneumonia, which may be fatal for people infected by this virus.

So far, this virus has not been known to spread directly from one human to another. As long as H5N1 does not change to another strain that can be transferred from one human to another, a worldwide epidemic of the virus probably will not occur.

**45.** State *one* difference between the effect on the human body of the usual forms of flu virus and the effect of H5N1.

**46.** Identify the type of substance produced by the human body that protects against antigens such as the flu virus.

**47.** State the *one* change that could cause the H5N1 virus to turn into a worldwide flu epidemic.

# THEME IV

## Reproduction, Growth, and Development

# 12

# Mitosis and Cell Division

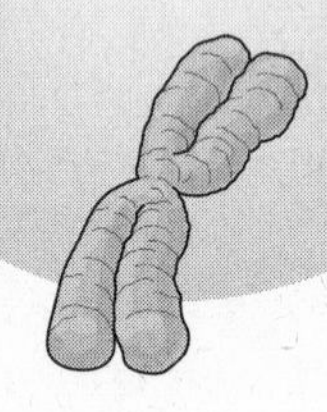

**Vocabulary**

| | | |
|---|---|---|
| asexual reproduction | cell division | deoxyribonucleic acid (DNA) |
| binary fission | chromosomes | mitosis |
| budding | cloning | replication |
| cancer | cytokinesis | reproduction |

### THE CONTINUITY OF LIFE

You have learned what it takes for an individual to maintain homeostasis and stay alive in a constantly changing environment. Although staying alive is important for every individual organism, it is not sufficient to maintain life on Earth. No individual organism lives forever. Every organism has a typical life span—the length of time between when its life begins and when it ends. The continuity of life requires **reproduction**, the ability of individuals within a species to produce more of their own kind. Individuals are members of populations. It is reproduction within populations that allows species to survive. It is reproduction that allows life on Earth to continue.

### THE LIFE OF A CELL

Every cell has a life of its own. This is as true for single-celled organisms as it is for each of the billions of cells that make up the bodies of plants and animals, including ourselves. Each cell has a beginning (with a period of growth), a middle stage, and then an ending—a process known as the *cell cycle*. (See Figure 12-1.)

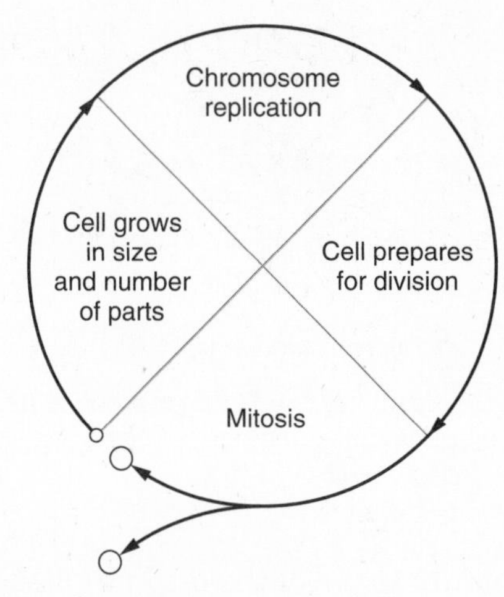


**Figure 12-1** The cell cycle is the series of events that occurs in the life of a cell—from its beginning to its ending (in mitosis or death).

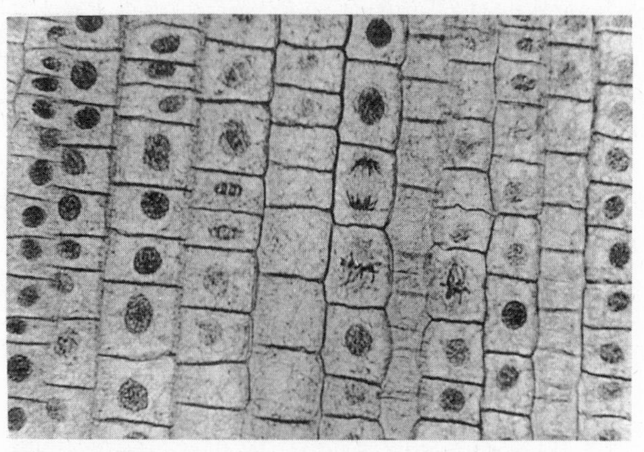

**Figure 12-2** Several plant cells can be seen here preparing for cell division: chromosomes have replicated and are dividing into two groups.

In the first stage, the cell begins to grow in size. Organic materials, such as amino acids and sugars, and inorganic materials, such as water, are moved into the cell. The cell increases in size by adding these materials to itself. The cell also increases the number of its parts. For example, its mitochondria divide in two to make more mitochondria. If it is a plant cell, the same thing happens to its chloroplasts.

During the next stage in the cycle, the cell stops getting larger. Now, the genetic material (the set of instructions received from the previous cell) duplicates—it makes an exact copy of itself. The genetic material is the building plan, similar in some ways to the set of blueprints used to build a house; it contains all the information about how the cell is to be built and how it functions. The genetic material is made up of the chemical called **DNA**, or **deoxyribonucleic acid**.

In reproducing cells, DNA is found in "packages" known as **chromosomes**. Bacterial cells may have just a single chromosome, a fruit fly has eight chromosomes in each body cell, a cabbage plant has 18, and a human has 46. The number of chromosomes in other organisms varies. The chromosome number is specific for each type of organism. And the exact chromosome number must be maintained for the species to continue. This means that as cells reproduce, the new cells must have the same number of chromosomes as the original cells.

The duplication of the cell's genetic material, during this middle stage in the cell's life cycle, is called **replication**. This is the most important stage in preparation for reproduction of the cell. Following this stage, some additional cell growth occurs. What is growing here is the material needed for the final stage in its life cycle, called **cell division**, because this is how a single cell reproduces. It divides into two new cells. (See Figure 12-2.)

## CELL DIVISION

During cell division, the genetic material must be equally divided. When a cell divides, it must send one copy of each of its chromosomes to each of the new cells. In addition, the cytoplasm and other cell parts must be divided between the two cells.

The division of the chromosomes occurs first. This division happens during a sequence of events called **mitosis**. During mitosis, the chromosomes of a cell are divided into two equal groupings. Following mitosis, the *cytoplasm* and contents of the cell divide in a process called **cytokinesis**. Afterward, each of the new cells has a complete set of chromosomes and organelles, just like the original cell. The two new cells are called *daughter cells*. The cell that they came from, which no longer exists, is called the *parent cell*. (See Figure 12-3.)

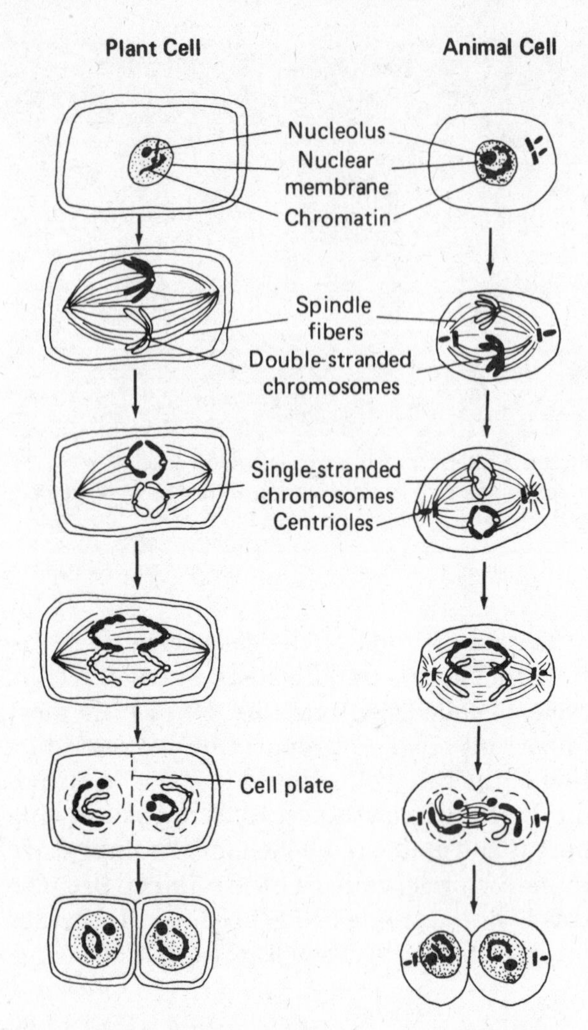


**Figure 12-3** Stages of mitosis leading to cell division.

## MAKING NEW INDIVIDUALS: ASEXUAL REPRODUCTION

Cell division produces two new daughter cells from one parent cell. The daughter cells are identical. They are also genetically identical to the parent cell. But have new individuals really been produced?

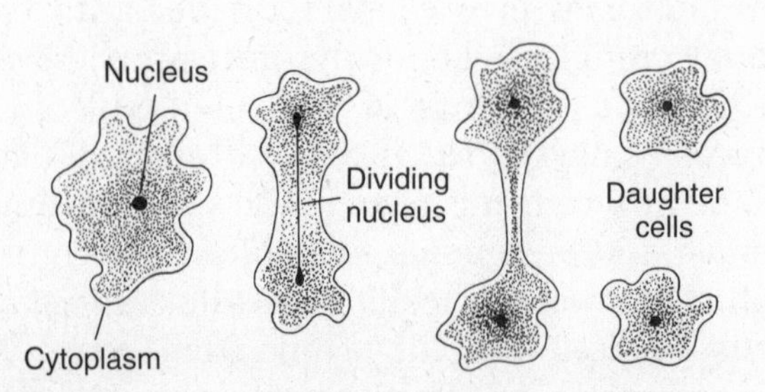


**Figure 12-4** Asexual reproduction, as seen in the ameba, requires only one parent to produce two new organisms.

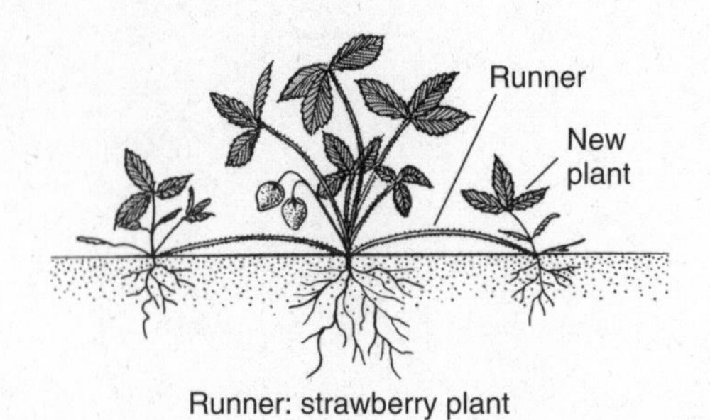


**Figure 12-5** Strawberry plants can reproduce asexually by means of runners; this method of producing identical offspring is called cloning.

The answer is *yes*, if the original parent was a single-celled organism. An ameba, through cell division, becomes two new, identical organisms. Reproduction in an ameba involves only one parent. Hence, this is called **asexual reproduction**. (For reproduction to be called *sexual*, it must involve two parents.) Single-celled organisms, such as the ameba, reproduce by a form of asexual reproduction known as **binary fission**. (See Figure 12-4.)

Plants carry out various types of asexual reproduction. In each type, a plant or a part of the plant reproduces itself through mitosis. As a result, the offspring are identical to the parent plant. For example, strawberry plants send out horizontal stems, called *runners*, above the ground. When these runners touch the surface of the soil at another spot, an entirely new, identical plant with roots and leaves begins to grow there. The production of identical genetic copies of a parent plant is called **cloning**. (See Figure 12-5.)

In some cases, animals can reproduce asexually, too. For example, if an arm of a sea star

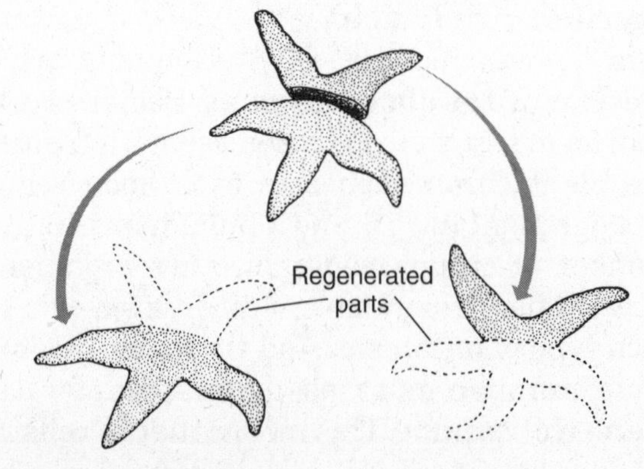


**Figure 12-6** Some animals, such as the sea star, can reproduce asexually by regrowing lost body parts when cut in two.

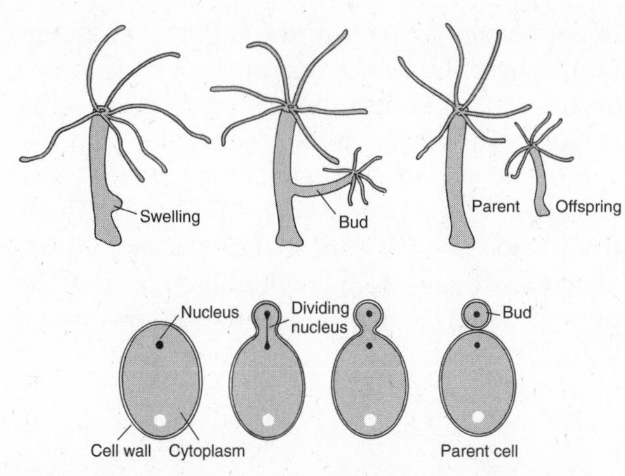


**Figure 12-7** The hydra (top) and yeasts (bottom) can reproduce asexually by budding. The offspring, or bud, is smaller than the parent cell or organism.

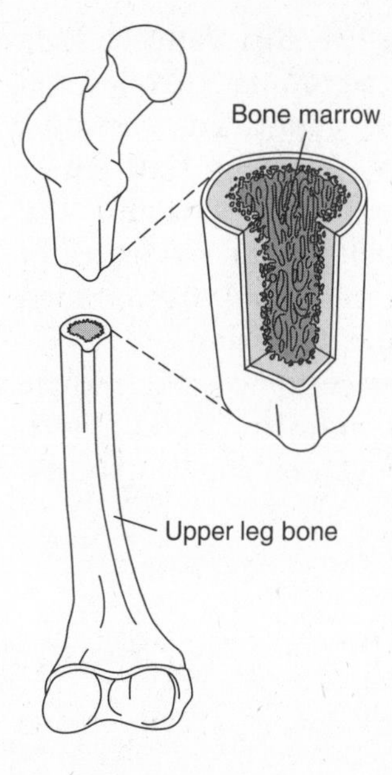


**Figure 12-8** The cells of different types of tissues divide at different rates, depending on their functions. Millions of red blood cells, which divide quickly, are made in the bone marrow each day. By contrast, the bone tissue cells divide much more slowly.

is broken off, the arm can sometimes grow into a whole new sea star. The sea star that lost the arm will regrow another one. This process is called *regeneration*. (See Figure 12-6.) Other organisms, such as yeasts, sponges, and hydra, can produce offspring by **budding**. During the process of budding, a new small individual begins to grow out of the side of the parent organism. The cells that form this new individual, or *bud*, result from mitotic cell division. The bud breaks free of the parent organism when it is large enough to live on its own. (See Figure 12-7.)

## THE RATE OF CELL DIVISION

When does a cell divide? How long does it take for one segment of cell division to begin and end? Do all types of cells divide at the same rate? The answers are very important to the process of growth and development in organisms.

Every multicellular organism is made up of various types of tissues and cells. For example, the human body contains blood, skin, muscle, bone, and nerve tissues, among other tissues. Controlling the rate at which cells of each particular kind of tissue divide is a necessary part of homeostasis. Red blood cells have a relatively short life span, and we need millions of them, so the cells that develop into red blood cells divide quickly. Bone cells, on the other hand, divide much more slowly. Skin cells normally take about 20 hours to complete one cell division; but their rate of division speeds up if you cut yourself. (See Figure 12-8.)

## CANCER: CELL DIVISION OUT OF CONTROL

The disease known as **cancer** results from uncontrolled cell division. Cancer cells do not seem to follow the rules or recognize the signals that control normal cell division. Uncontrolled cell growth can occur in many different types of cells. As a result, there are different types of cancer, such as skin cancer, breast cancer, prostate cancer, lung cancer, and many others. There does not seem to be a single cause for all types of cancer. (See Figure 12-9.)

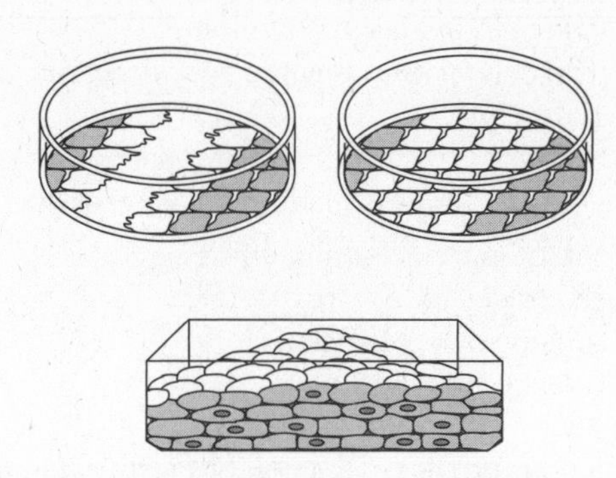

**Figure 12-9** Normal cells grown in a lab dish stop dividing when they touch each other (top). Cells that continue to divide after they touch each other exhibit the kind of uncontrolled cell growth seen in cancer (bottom).

Even though there are differences, it is quite certain that all types of uncontrolled cell division involve the cells' genetic instructions, which are made of DNA. Factors that cause cancer do so by damaging or changing the DNA. These factors include the exposure of cells to harmful substances, or *mutagens*, such as certain chemicals and radiation.

One of the most important ways to reduce the risk of cancer is to maintain good health habits. The immune system constantly attacks not only invading cells but also abnormal, cancerous cells from our own body. Much of the time, the immune system is successful. It destroys cancer cells before they can develop and cause problems. It is no surprise then that many people with AIDS actually die from some type of cancer. The patient's damaged immune system is not able to protect the person from cancerous cells. Therefore, a healthy immune system is one of the best protections against cancer.

## Chapter 12 Review

### *Part A—Multiple Choice*

**1.** An organism's typical life span is its

1 body length, from head to tail
2 time between birth and death
3 average age when it reproduces
4 normal population size

**2.** The survival of a species depends on

1 an environment that never changes
2 a continuously increasing life span
3 reproduction within populations
4 a limit of no more than two populations

**3.** During the first stage of the cell cycle, a cell

1 grows in size
2 divides in half
3 duplicates its genetic material
4 makes a copy of itself

**4.** What causes a cell to grow in size?

1 The cell takes in other cells.
2 The genetic material of the cell replicates.
3 The cell divides into two parts.
4 The cell takes in organic and inorganic materials.

**5.** The genetic material of the cell is most like

1 the blueprints for a building
2 the tracks for a train
3 an advertisement for a store
4 a fence for a house

**6.** In a reproducing cell, the DNA is found in the structures known as

1 lysosomes 3 vacuoles
2 chromosomes 4 ribosomes

**7.** The number of chromosomes

1 is specific for each type of organism
2 is the same for every species of organism
3 decreases from one generation to the next
4 increases from one generation to the next

**8.** Before cell division, the genetic material must undergo a process called

1 reduction
2 disintegration
3 replication
4 reproduction

**9.** During the process of mitosis, the chromosomes

1 are cut in half twice
2 are equally divided
3 form a circle in the cell
4 spread through the cell

**10.** The diagram below represents the chromosomes in a cell. Which of the following diagrams best illustrates the daughter cells that result from the normal division of this cell?

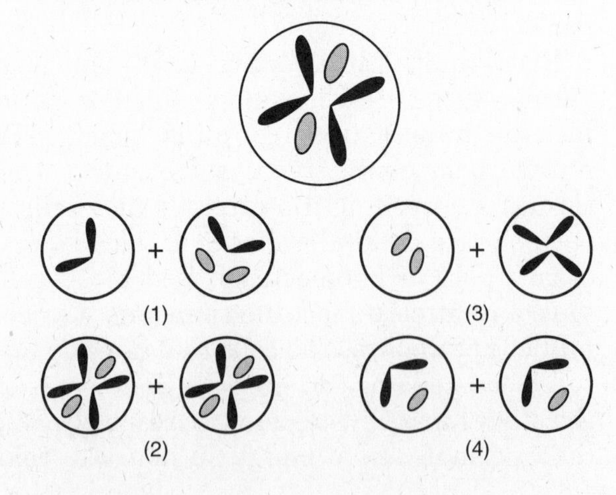

1 diagram 1
2 diagram 2
3 diagram 3
4 diagram 4

**11.** What happens *after* mitosis has occurred?

1 The cell grows in size again.
2 The genetic material replicates.
3 The genetic material forms a parent cell.
4 The cytoplasm of the cell divides in two.

**12.** Compared to the parent cell, each daughter cell that results from the normal mitotic division of the parent cell contains

1 the same number of chromosomes, but different genes from those of the parent cell
2 half the number of chromosomes, but different genes from those of the parent cell
3 the same number of chromosomes and identical genes to those of the parent cell
4 twice the number of chromosomes and identical genes to those of the parent cell

**13.** In asexual reproduction, the genetic material is supplied by

1 one daughter cell
2 one parent cell
3 two daughter cells
4 two parent cells

**14.** The diagram below represents a cell process. Which statement regarding this process is correct?

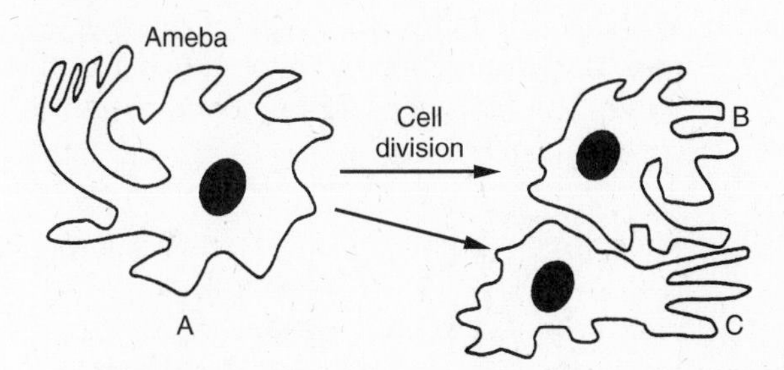


1 Cell B contains the same genetic information as cells A and C.
2 Cell A has DNA that is only 75 percent identical to cell B.
3 Cell C has DNA that is only 50 percent identical to cell B.
4 Cells A, B, and C each contain different genetic information.

**15.** The DNA of a plant produced by asexual reproduction would be

1 identical to that of the parent plant
2 similar, but not identical, to that of the parent plant
3 totally different from that of the parent plant
4 a combination of genetic information from several plants

**16.** A researcher determines that all the members of a certain population of plants on a lawn are genetically identical. The best explanation for this is that the plant

1 reproduces sexually, by cloning
2 reproduces sexually, by budding
3 reproduces asexually, by cloning
4 reproduces asexually, by budding

**17.** A new hydra can be produced from groups of cells that enlarge and stay attached to the parent hydra for a time before breaking off and becoming independent. This method of reproduction is called

1 sporulation
2 cloning by runners
3 binary fission
4 budding

**18.** One way to produce many genetically identical offspring is by

1 using radiation to change their genes
2 using chemicals to change their genes
3 cloning them, so they have the same genes
4 inserting a new DNA section into their genes

**19.** Which phrase does *not* describe cells cloned from a carrot?

1 they are genetically identical
2 they have the same DNA codes
3 they are reproduced sexually
4 they have identical chromosomes

**20.** Which statement about the rate of cell division is true?

1 All the cells of all organisms divide at the same rate.
2 The rate of cell division is related to a cell type's function.
3 All the cells within an organism divide at the same rate.
4 The rate of cell division is random in every organism.

**21.** Damage to a cell's DNA can cause cancer, which results from

1 a slower than normal cell division
2 a complete stop to all cell division
3 an uncontrolled type of cell division
4 no changes in the genetic instructions

### Part B—Analysis and Open Ended

**22.** Why does the survival of a species depend more on populations than on the life spans of individual organisms?

**23.** Briefly explain the main purpose of the genetic material in a cell.

**24.** Use the following terms to replace the definitions given within the boxes in the following flowchart: *Cell division; Cell growth; Mitosis; Replication.*

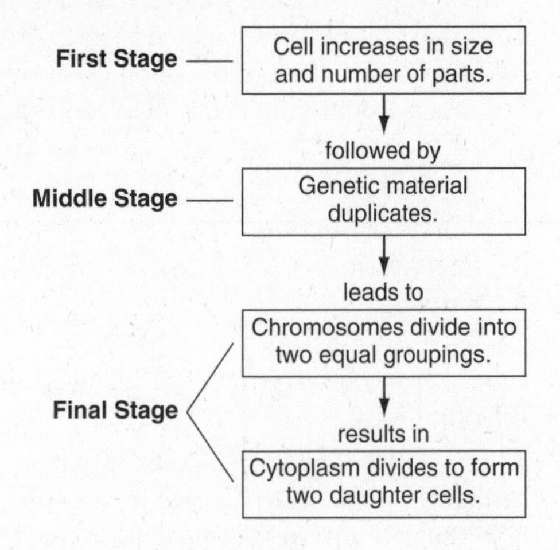


**25.** The best title for this concept map probably would be

1 The Functions of a Cell
2 The Life Cycle of a Cell
3 The Regeneration of a Cell
4 The Genetics of the Cell

**26.** How does the chromosome number of one species compare with that of another species?

**27.** Explain why the duplication of chromosomes is necessary for the process of cell division.

**28.** Why is cell division necessary for all living things? Your answer should include the following:

- why it is important for one-celled organisms
- why it is important for multicelled organisms
- *one* example of each type of organism discussed

**29.** During cell division, both the genetic material and the cytoplasm have to be equally divided. Which process occurs first? Which process is called mitosis?

*Refer to the following two diagrams to answer questions 30 and 31.*

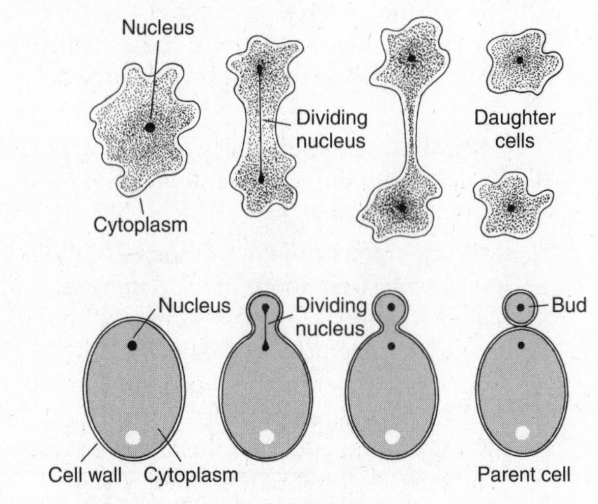


**30.** What types of asexual reproduction are illustrated in these two diagrams?

1 cloning and regrowing parts
2 binary fission and budding
3 budding and regrowing parts
4 cloning and binary fission

**31.** In what way is reproduction in the ameba (top) the same as reproduction in the yeast (bottom)? In what way is it different?

**32.** Which of the following diagrams represents asexual reproduction by mitosis? Explain why. (*Note:* The "n" stands for number of chromosomes in each cell.)

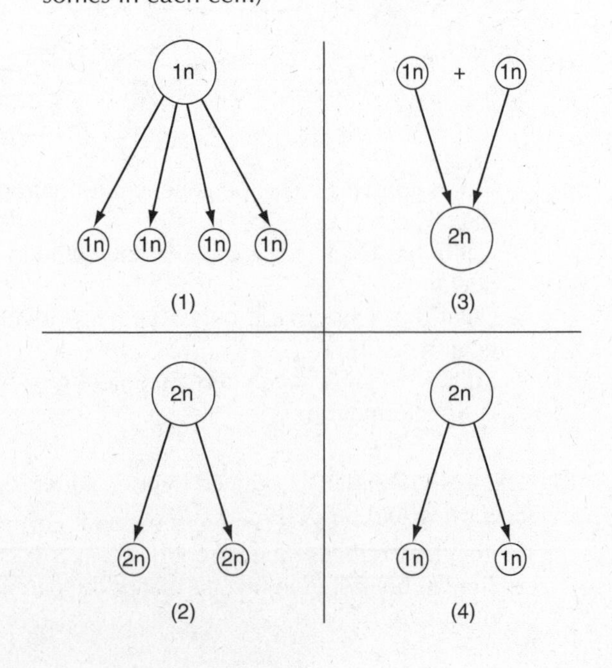

**33.** Does the rate of cell division differ from one tissue type to another within an organism? Why is this important for homeostasis? (Give *one* example.)

**34.** Explain why exposure to radiation and certain chemicals can cause uncontrolled cell division. What type of disease can this process lead to?

### *Part C—Reading Comprehension*

*Base your answers to questions 35 through 37 on the information below and on your knowledge of biology.*
Source: *Science News*, Vol. 162, p. 382.

By unleashing radio waves inside bone, researchers have stopped intractable pain in people with cancer that has spread to their skeletons.

Tumors that form inside bone when cancers spread can be especially painful. The new technique, called radio-frequency ablation, unleashes energy via a needle inserted into bone to reach the edge of the tumor. The radio waves create intense heat that kills nearby tumor cells within about 10 minutes, says study coauthor Matthew R. Callstrom of the Mayo Clinic in Rochester, Minn.

Targeting the surface where the tumor meets the bone seems critical, he says. "Our thought is that nerve fibers in that area—where tumor cells are eroding bone—are the pain generators," he says. Bone itself appears unaffected by the procedure.

The researchers treated 62 patients in whom conventional cancer therapy had failed. Of these, 59 reported significant pain relief, and 28 said they experienced total pain relief at some times, Callstrom says.

"We're not curing cancer with this treatment," he says. "But we're affecting the pain that patients have. The most important [concern] for all these patients is their quality of life."

**35.** Explain why researchers are using a new technique on cancer tumors inside bone.

**36.** State the process by which radio waves are being used to treat pain in cancer patients.

**37.** Why is targeting the areas where tumors meet the bone most important in this study?

# 13

# Meiosis and Sexual Reproduction

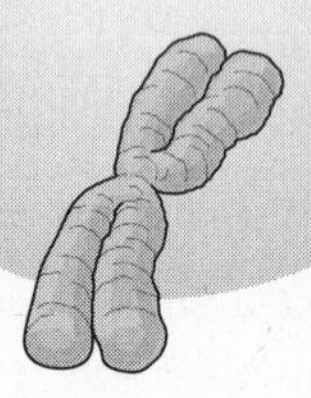

## Vocabulary

| | | |
|---|---|---|
| development | internal development | sex cells |
| egg cell | internal fertilization | sexual reproduction |
| embryo | meiosis | sperm cell |
| fertilization | placenta | zygote |
| gametes | recombination | |

## SEXUAL REPRODUCTION: IT TAKES TWO

For almost all animals, it takes two individuals to produce offspring: a male and a female. This is called **sexual reproduction**. Most plants use this method of reproduction to make more of their own kind, too. Sexual reproduction is very important in understanding the behavior and diversity of living things. It also plays a significant role in the process of evolution. To understand why this is so, we must look at the reproductive cells and study the chromosomes within them. (See Figure 13-1.)

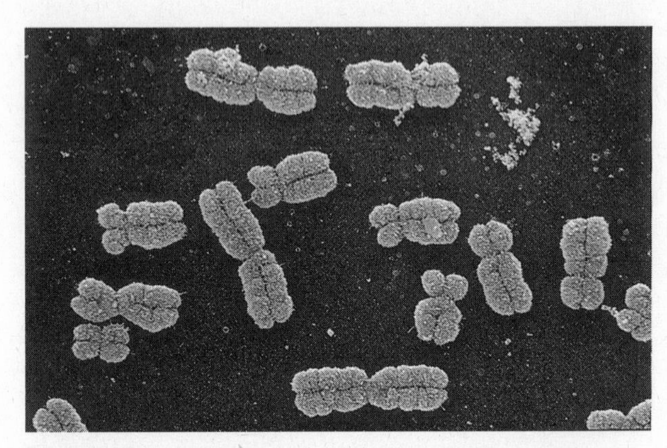

**Figure 13-1** A photograph of human chromosomes.

## IT'S ALL ABOUT CHROMOSOMES

Each of our cells contains chromosomes. The chromosomes contain the inherited information that has been passed along since the beginning of life on Earth. It is this information that determines an individual's characteristics. The chromosomes also contain the "know-how" that keeps our cells functioning correctly.

Why is sexual reproduction all about chromosomes? When a **sperm cell** and an **egg cell** unite during sexual reproduction, it is the nucleus from each cell that joins. And what does the nucleus contain? It contains the chromosomes. So, sexual reproduction is about the combining of chromosomes from two individuals, a male (the father) and a female (the mother).

Each human body cell contains 46 chromosomes. The first cell from which each of us came—the cell that resulted from the combination of a sperm cell and an egg cell—had 46 chromosomes. Every body cell now in you still has 46 chromosomes. The question is: How did a sperm cell and an egg cell combine to make a new cell with just 46 chromosomes?

There is only one way. Both the sperm and the egg must have had only 23 chromosomes each, half the amount of chromosomes from the normal number of 46. And indeed this is the case. A special type of cell division produces sperm and egg cells, each with that reduced number of chromosomes.

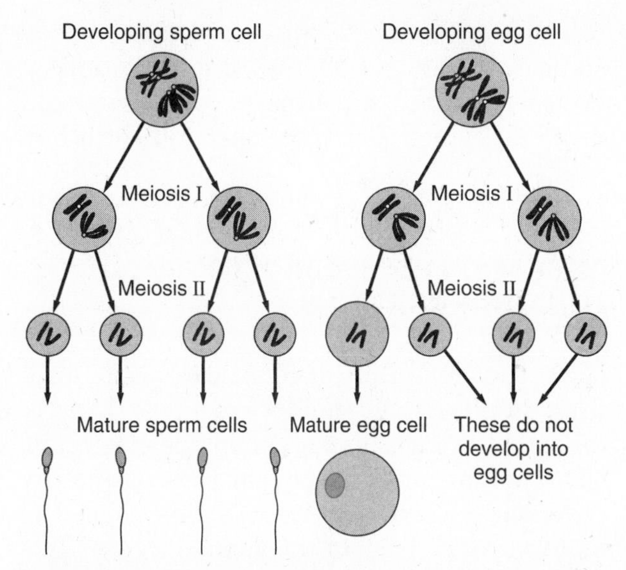


**Figure 13-3** As a result of meiosis, sperm cells and egg cells have half the normal number of chromosomes for their species.

## GAMETES

The sperm and egg cells, or **sex cells**, are also called **gametes**. In the process of sexual reproduction, the nuclei of the gametes join together. This fusion of the nuclei is called **fertilization**. The resulting cell, a fertilized egg cell, is called a **zygote**. (See Figure 13-2.) Each gamete, as we have said, has exactly half the normal number of chromosomes. The zygote and all body cells that come from the mitotic division of the zygote have two sets of chromosomes in them, one from each parent. Gametes are produced by a type of cell division that reduces the chromosome number by one-half, giving them just one set of chromosomes. Thus, when fertilization occurs, the normal number of chromosomes for the species is maintained. This special type of cell division is called **meiosis**.

## A CLOSER LOOK AT CHROMOSOMES

Our chromosomes exist in pairs. Essentially, we have two chromosomes of each type. And where does each of these two chromosomes come from? The answer is: one from each parent. Beginning with a normal body cell, which has the double set of chromosomes, gametes must be produced through meiotic cell division. Each gamete contains a single set of chromosomes. And it must be an exact set, meaning one and only one from each of the pairs of chromosomes. (See Figure 13-3.)

## MEIOSIS: REDUCING THE CHROMOSOME NUMBER

Mitosis and meiosis take place during cell division, and in some ways these two processes are similar. The chromosomes replicate before either process begins. However, the results of mitosis and meiosis are very different. When mitosis is completed, the chromosome number remains the same as in the original parent

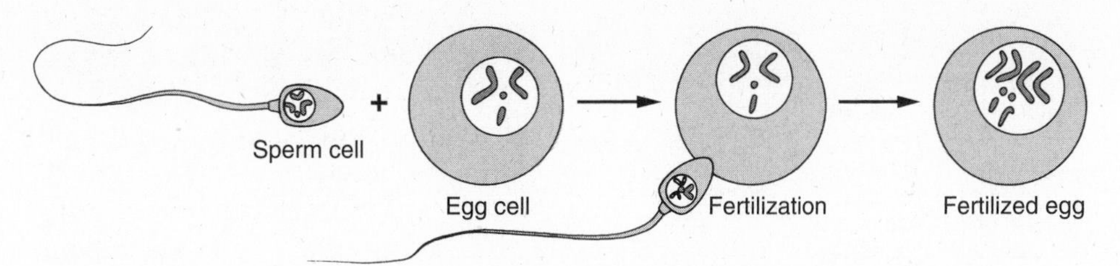


**Figure 13-2** Sexual reproduction involves the joining of chromosomes from a sperm cell and an egg cell in the process called fertilization.

cells. When meiosis is completed, the chromosome number is half the original number. Meiosis actually involves two separate cell divisions, which take place one after the other.

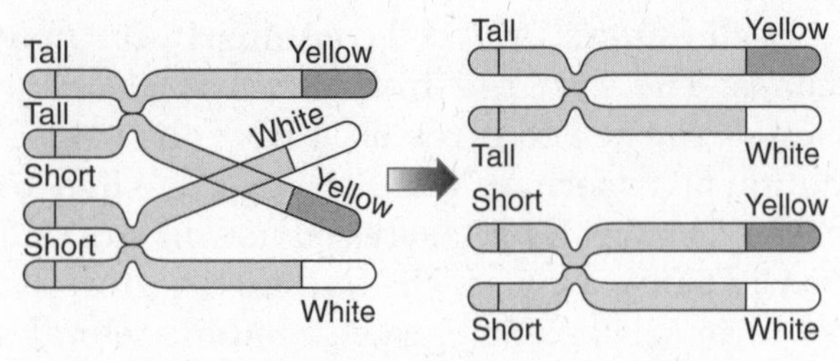


**Figure 13-4** During meiosis, genetic recombination occurs when chromosomes overlap and exchange pieces.

## MEIOSIS: THE SOURCE OF OUR DIFFERENCES

With the exception of identical twins, children in the same family are never exactly alike. Differences can occur in eye color, hair color, height, nose shape, ear size, and many other characteristics. Why is this so, if the children were born of the same parents? The explanation arises from one of the two important jobs of meiosis. The first job of meiosis, as we have said, is to maintain the normal species chromosome number by preparing gametes with single sets of chromosomes.

The second important job of meiosis is to increase the variation of traits in offspring by **recombination** of genes in the eggs and sperm. Genes may get exchanged between chromosomes during meiosis, in a process known as *crossing-over*. (See Figure 13-4.) Also, chromosomes may get resorted into new groupings. Because of this genetic recombination during meiosis, sexual reproduction results in offspring that are different from each other and from their parents. (See Figure 13-5.) This genetic variation is what natural selection acts on. A greater variety of characteristics in offspring increases the chances that some individuals will be better suited than others to survive in a particular place and time. As natural selection acts on the varied offspring in a population, generation after generation, the species evolves.

## UNUSUAL MEIOTIC EVENTS

The sorting of chromosomes that occurs in cell division (especially during meiosis) is a wonderfully complex sequence of events. However, it does not always proceed correctly. A gamete may have an extra chromosome because it receives both members of a pair of chromosomes, instead of only one. Or a gamete may be one chromosome short, having received neither member of a pair. If a gamete with such an abnormality fuses with another gamete, problems may occur. In most instances, the zygote fails to develop. However, in some cases, the zygote does develop into an individual with an abnormal chromosome number. One of the best-known examples of this is the disorder known as Down syndrome. In this case, a person has an extra copy of chromosome 21, resulting in a total of 47 chromosomes instead of the normal 46. The serious problems that may occur include mental and physical disabilities, greater risks of de-

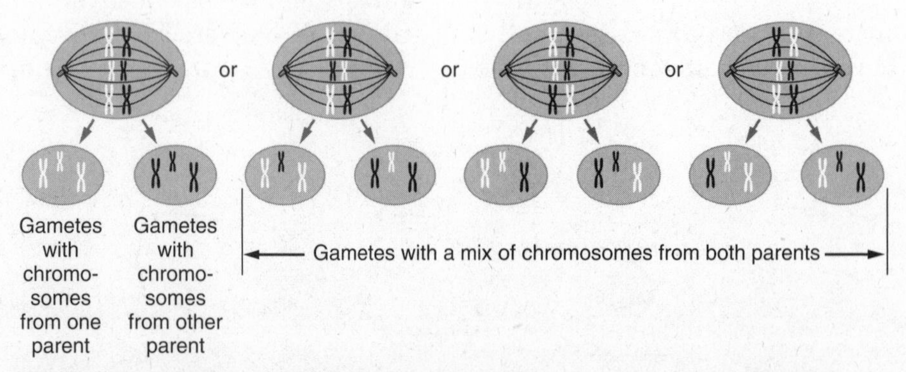


**Figure 13-5** Each time meiosis occurs, the chromosomes line up in a different arrangement, resulting in variability among offspring.

veloping leukemia and heart disease, and a shorter-than-normal life span.

## THE SEX LIFE OF FLOWERING PLANTS

Of the many types of plants on Earth, flowering plants are the group that has evolved most recently. Most types of plants reproduce sexually. As in animals, the male gamete (sperm cell) joins together with the female gamete (egg cell) to produce a zygote (the seed). The zygote then begins to grow into a new plant. In flowering plants, the place where the gametes join is very visible; it is often brightly colored, beautifully shaped, and sweet smelling. In other words, the place where sexual reproduction occurs in these plants *is* the flower. In fact, the parts that make up flowers include the sex organs of the plants: the ovary contains egg cells within its ovules and the anthers contain sperm cells within their pollen grains. The sperm cells fertilize the egg cells inside the ovary. (See Figure 13-6.)

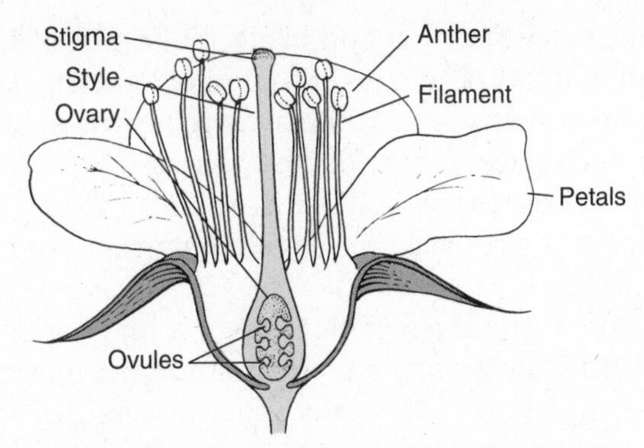


**Figure 13-6** In flowering plants, sexual reproduction takes place in the flowers, which contain the reproductive structures.

## SEXUAL REPRODUCTION: INTERNAL OR EXTERNAL?

Various methods of sexual reproduction occur in plant and animal species. Yet whatever the method, sexual reproduction always involves fertilization (the fusion of nuclei from two gametes) and **development**, the growth of the zygote into a new individual. One of the main differences in the types of reproduction involves the location of the events. Both fertilization and development may occur either inside or outside the bodies of the reproducing organisms.

For many aquatic plants, the sex cells meet in the open water; fertilization occurs and the zygote begins to develop. Many invertebrates simply release their gametes into the water, where fertilization and development then occur. Two groups of vertebrates, the fish and the amphibians, also reproduce in water. In most species of fish, the eggs and sperm are released directly into the water, where fertilization and development of the zygotes then occur. These events, which occur in the environment and not inside the organism, are known as *external fertilization* and *external development*. External fertilization and development are risky. Therefore, large numbers of eggs are released to increase the chances that some of the offspring will survive. This is natural selection at work. (See Figure 13-7.)

In the case of vertebrates that reproduce on land, the gametes still need moisture to meet

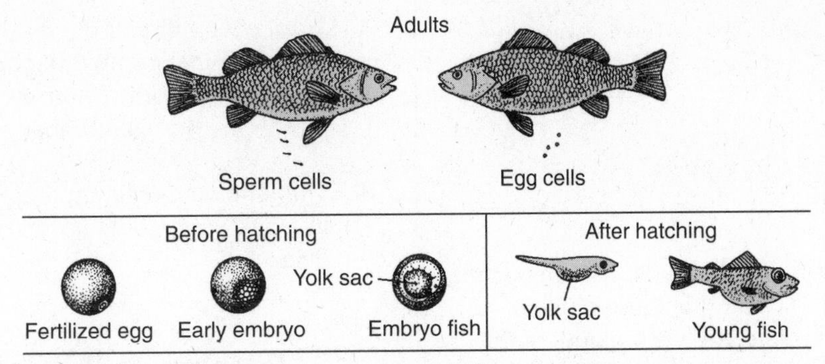


**Figure 13-7** In most species of fish, both fertilization and development are external; many eggs are released to ensure that some offspring will survive.

and fuse. Reptiles and birds make use of the fluids inside their bodies for fertilization. The male and the female must mate for the sperm to be deposited inside the female, a process known as **internal fertilization**. Then the zygote is prepared for development on land: a watertight membrane and protective shell form around the zygote; the egg is laid (usually within a nest); and development of the new organism occurs externally. Internal fertilization increases the chances of reproductive success and survival. Fewer eggs are produced, but there is some parental care to help protect the developing zygote. (See Figure 13-8.)

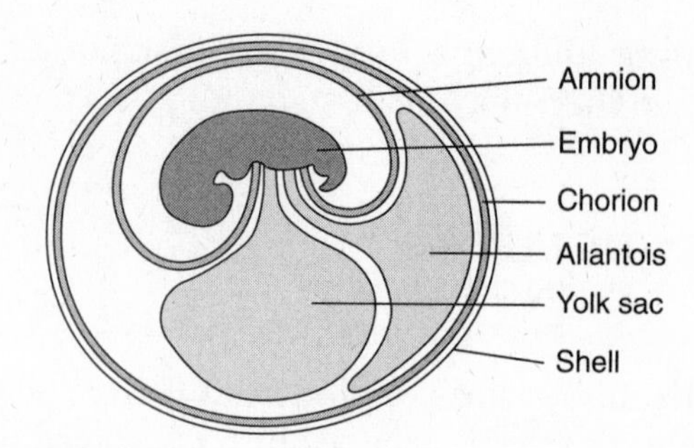


**Figure 13-8** In birds, fertilization is internal. The embryo is surrounded by a watertight membrane (the amnion) and then covered by a protective shell. The egg is laid and development is external.

One final pattern of sexual reproduction takes place mainly in mammals. Fertilization occurs internally, but the big difference from most other animal groups is that development of the zygote occurs within the female's body, too. Thus, mammals have **internal development**. The food for the developing **embryo** comes entirely from the body of the mother. A structure called the **placenta** has evolved to bring nutrients to the developing baby and to remove its wastes. (*Note:* The exceptions to this are the *marsupial* mammals, which complete their embryonic development within a protective pouch, and the egg-laying mammals.) After birth, continuing nourishment of the baby mammal occurs through its nursing on milk provided by the mother's mammary glands. Although the embryos are fewer in number, they have the most complete form of protection, since they develop within the body of the female and then receive more parental care after their birth.

## Chapter 13 Review

### Part A—Multiple Choice

**1.** During sexual reproduction, the chromosomes of

1 two separate individuals are combined together
2 one individual are transferred to another
3 one parent only are copied for its offspring
4 two separate individuals are split apart

**2.** If each human body cell has 46 chromosomes, how many were in the very first cell of your body?

1 23 3 92
2 46 4 100

**3.** Most cells in the body of a fruit fly contain eight chromosomes. How many of these chromosomes were contributed by each parent of the fruit fly?

1 8 3 2
2 16 4 4

**4.** Sperm cells of the Russian dwarf hamster contain 14 chromosomes. What is the total number of chromosomes that would be found in each cell of a normal, newly formed zygote of this species?

1 7 3 14
2 28 4 42

**5.** The gamete (sex cell) for any species should *always* contain

1 an even number of chromosomes
2 the normal number of chromosomes
3 twice the normal number of chromosomes
4 half the normal number of chromosomes

**6.** The following diagram represents some events in a cell undergoing normal meiotic cell division.

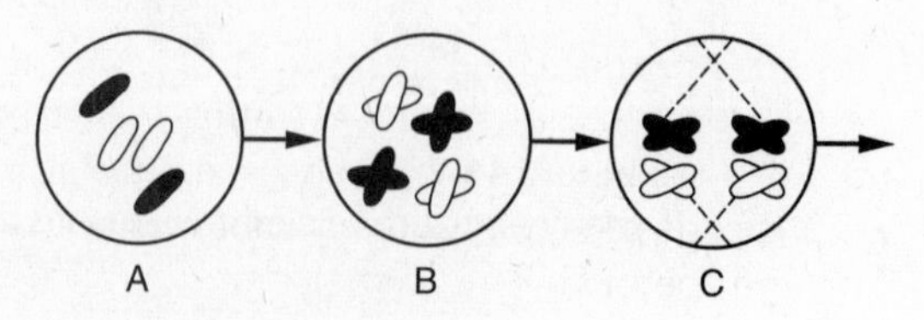

Which of the following diagrams most likely represents the next cell that would result from the process shown in the previous diagram?

(1)

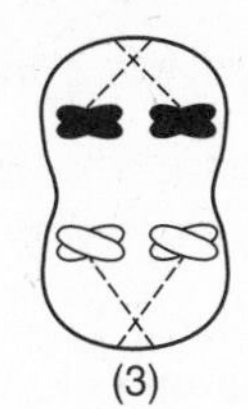

(3)

(2)

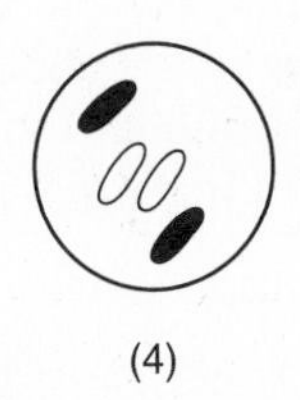

(4)

**7.** Compared to human cells resulting from mitotic cell division, human cells resulting from meiotic cell division should have

1 twice as many chromosomes
2 one-half as many chromosomes
3 the same number of chromosomes
4 one-quarter as many chromosomes

**8.** During fertilization, the parts of the sex cells that join are the

1 membranes
2 nuclei
3 ribosomes
4 vacuoles

**9.** Which of these is formed during fertilization?

1 an egg cell
2 a sperm cell
3 a zygote
4 a gamete

**10.** Most cells in the body of a fruit fly contain eight chromosomes. In some cells, only four chromosomes are present, a condition that is a direct result of

1 mitotic cell division
2 embryonic division
3 meiotic cell division
4 internal fertilization

**11.** Which statement best explains the significance of meiosis in the evolution of a species?

1 Meiosis produces egg cells and sperm cells that are completely alike.
2 Meiosis ensures the continuation of a species by asexual reproduction.
3 Meiosis produces equal numbers of egg cells and sperm cells in animals.
4 Meiosis results in genetic variation among the gametes that are produced.

**12.** Mitosis and meiosis are similar in that

1 the chromosomes are replicated before either process starts
2 the chromosome number is the same when each process is completed
3 two separate cell divisions occur during each process
4 each process combines genetic material from two individuals

**13.** Which diagram correctly represents part of the process of sperm formation in an organism that has a normal (species) chromosome number of eight?

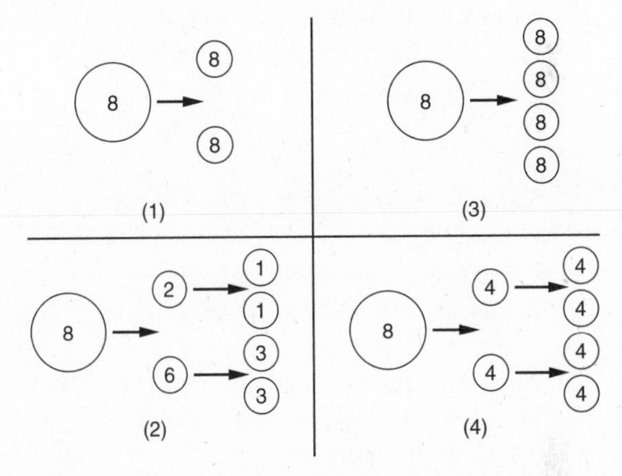


**14.** The great variety of possible gene combinations in a sexually reproducing species is due in part to the

1 sorting of genes as a result of gene replication
2 pairing of genes as a result of binary fission
3 sorting of genes as a result of meiosis
4 pairing of genes as a result of mitosis

**15.** During meiosis, recombining (gene exchange between chromosomes) may occur. Genetic recombination usually results in

1 overproduction of gametes
2 variation within the species
3 fertilization and development
4 formation of identical offspring

**16.** The following diagram shows a process that can occur during meiosis.

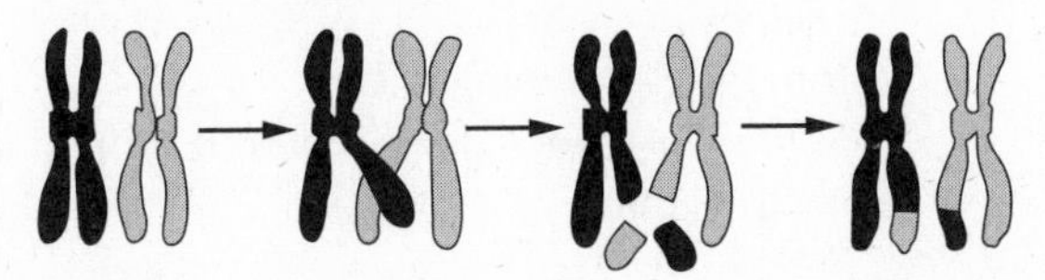

The most likely result of this process is

1 a new combination of inheritable traits that can appear in the offspring
2 a loss of genetic information that will produce a genetic disorder in the offspring
3 an inability to pass either of these chromosomes along to future offspring
4 an increase in the chromosome number of the organism in which this occurs

**17.** Mitosis produces new body cells and meiosis produces

1 new body cells, too
2 body cells and sex cells
3 sex cells, only
4 red blood cells

**18.** Which of the following is a characteristic found *only* in sexual (*not* asexual) reproduction?

1 cell division
2 cell growth
3 fertilization
4 chromosomes

**19.** Sexual reproduction in flowering plants occurs within the

1 roots
2 stems
3 leaves
4 flowers

**20.** An animal that has external fertilization will produce more eggs than an animal that has internal fertilization, because

1 the siblings help raise each other without any parental involvement
2 an animal can reproduce externally only once during its lifetime
3 it increases the chances that some of the offspring will survive
4 that way the parent animal can locate some eggs after they are fertilized

**21.** In terms of reproduction, how do mammals *differ* from most other animals?

1 The gametes are formed internally.
2 Fertilization takes place internally.
3 The zygote is formed externally.
4 The embryo develops internally.

**22.** What is the role of the placenta in the embryonic development of a mammal?

1 It forms a protective barrier around the developing baby.
2 It brings nutrients to and removes wastes from the developing baby.
3 It provides the location for fertilization of the egg to occur.
4 It provides a method of nourishing the baby after it is born.

### *Part B—Analysis and Open Ended*

**23.** Briefly state what chromosomes contain and what they determine in organisms.

**24.** The diagram below represents an incomplete process of meiosis in an animal's ovary. In your notebook, copy and complete the diagram by drawing in the chromosomes of cell A. Your drawing should show the usual result of meiosis in the formation of an egg cell.

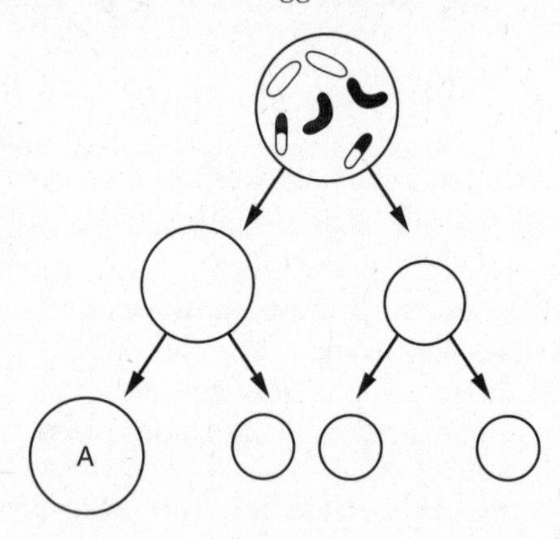


**25.** In what way is sexual reproduction basically all about chromosomes?

**26.** Use the words *gametes, zygote,* and *fertilization* in one sentence to explain sexual reproduction.

***Base your answers to questions 27 and 28 on the following diagram and on your knowledge of biology.***

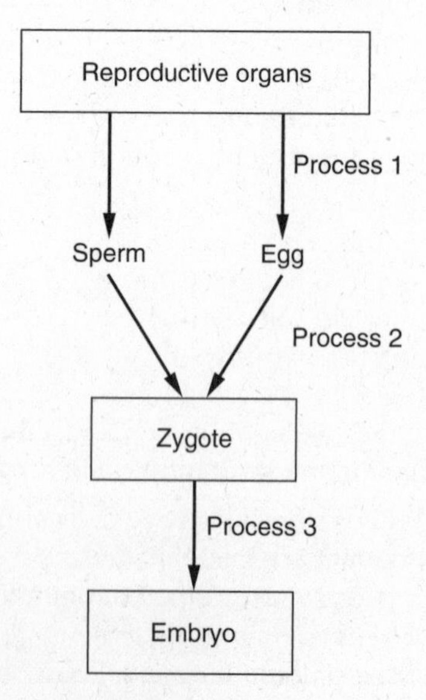

**27.** State why process 2 is necessary in sexual reproduction.

**28.** State *one* difference between the cells produced by process 1 and the cells produced by process 3.

**29.** Why are gametes essential to sexual reproduction, in terms of their chromosome number?

**30.** Explain why, in a mammal, a change in a gamete may contribute to evolution while a change in a body cell will not.

**31.** Although paramecia (single-celled organisms) usually reproduce asexually, some have developed a method by which they exchange genetic material with each other in a simple form of sexual reproduction. State *one* advantage this simple form of sexual reproduction would provide over asexual reproduction for the survival of these single-celled organisms.

*Refer to the diagram below to answer questions 32 to 34. (*Note: *The "n" stands for the number of chromosomes in each cell.)*

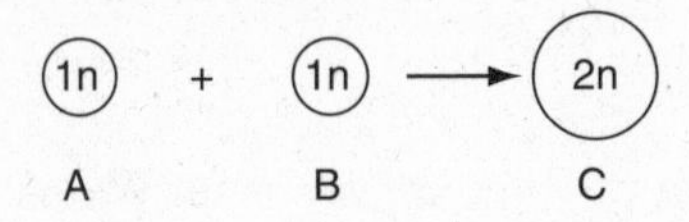


**32.** What reproductive process does the diagram represent?

1 gamete formation
2 cell division
3 fertilization
4 recombination

**33.** Which of the structures in the diagram represents a gamete?

1 A only
2 B only
3 A and B
4 C only

**34.** Which of the structures in the diagram represents a zygote?

1 A only
2 B only
3 C only
4 none of them

**35.** Which one of the following diagrams represents meiosis? (*Note:* The "n" stands for the number of chromosomes in each cell.)

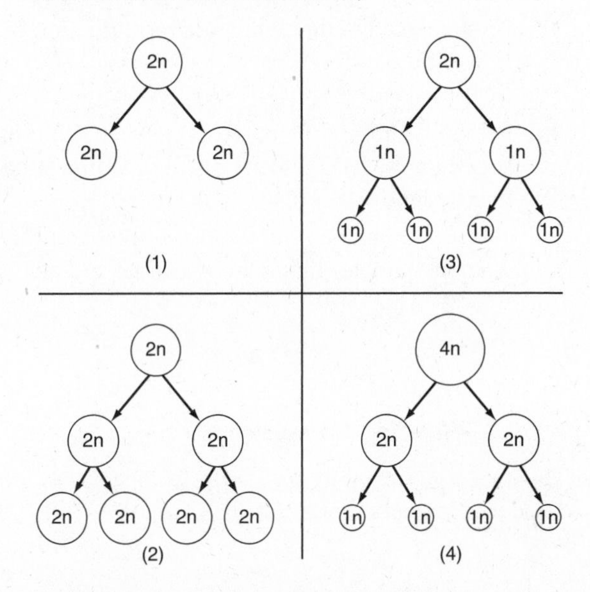


**36.** In terms of chromosome number, what is the main difference between the results of mitosis and meiosis?

**37.** Why are offspring of organisms that reproduce sexually *not* genetically identical to their parents?

**38.** Why is genetic variation (due to meiosis) important for the survival of offspring?

**39.** Compare asexual reproduction to sexual reproduction. In your comparison, be sure to include:

- which type of reproduction results in offspring that are usually genetically identical to the previous generation and explain why this occurs
- one other way these methods of reproduction differ

*Refer to the illustration below, which shows an important event that occurs during meiosis, to answer questions 40 and 41*

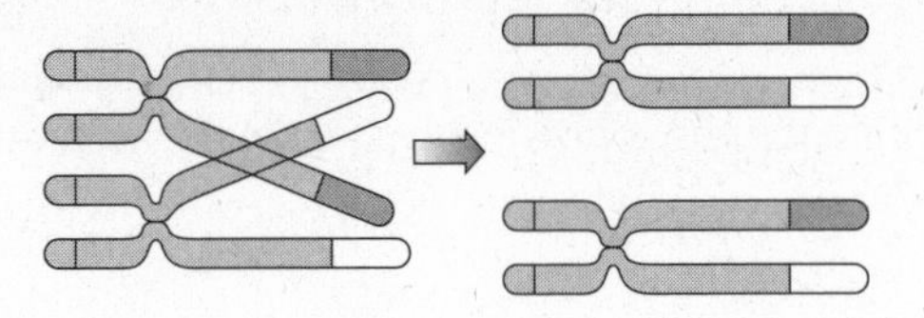

**40.** What is occurring during this process? How are the chromosomes at the end of the event different from those at the start?

**41.** Why is this process significant in terms of the offspring that are produced?

**42.** Briefly describe the process of reproduction in flowering plants.

**43.** Which characteristic of sexual reproduction has specifically favored the survival of animals that live on land?

1 fusion of gametes in the outside environment
2 male gametes that may be carried by the wind
3 fertilization within the body of the female
4 female gametes that develop within ovaries

**44.** Which process normally occurs at the placenta?

1 Oxygen diffuses from the fetus to the mother's blood.
2 Maternal blood is converted into fetal blood.
3 Nutrients are delivered to the fetus and wastes are removed.
4 Digestive enzymes pass from the mother's blood to the fetus.

### *Part C—Reading Comprehension*

*Base your answers to questions 45 through 47 on the information below and on your knowledge of biology.* Source: *Science News*, Vol. 162, pp. 189–190.

Biologists may have finally found what they call the "spark of life," a molecule in sperm that triggers a fertilized egg to begin developing.

Immediately after a sperm penetrates an egg, several waves of calcium flow out of the egg's stores of the ion. These calcium surges set off development of the fertilized egg. For more than a century, biologists have speculated that sperm must contain something that liberates this calcium. Several egg-activating factors have been proposed, but none has withstood scrutiny.

Because of its calcium-releasing role in some other cells, an enzyme called phospholipase C (PLC) was among the suspects. None of the known versions of PLC fits the bill as an egg activator, however.

Now, in the Aug. 15 *Development*, F. Anthony Lai of the University of Wales College of Medicine in Cardiff and his colleagues report the discovery of a new form of PLC that's present only in sperm. Moreover, when injected into an unfertilized egg, the enzyme stimulates calcium surges identical to those caused by sperm. This enzyme may provide a seemingly natural means of activating eggs in cloning or other forms of artificial reproduction, the scientists suggest.

Given the history of this issue, the role of the new PLC must be verified "10 times over," cautions Sergio Oehninger of the Jones Institute for Reproductive Medicine in Norfolk, Va.

**45.** Explain where the molecule called "spark of life" is found and what it does.

**46.** Describe the events that occur as soon as a sperm enters into an egg cell.

**47.** State the discovery that may explain why a fertilized egg begins developing.

# 14
# Human Reproduction

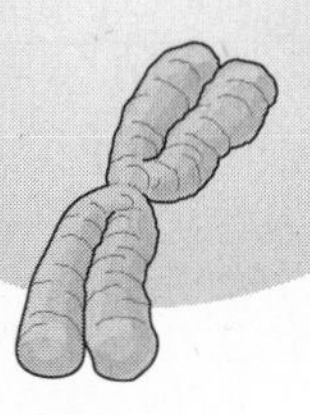

## Vocabulary

| | | |
|---|---|---|
| estrogen | pregnancy | testes |
| ovaries | progesterone | testosterone |
| ovulation | puberty | uterus |

## THE MALE REPRODUCTIVE SYSTEM

The male reproductive system in humans has three main functions: first, to produce the male gametes (sperm cells); second, to deposit the sperm cells it produces inside the female; and third, to provide a pathway for the removal of urine from the body.

The sperm cell formation occurs in the two **testes** (singular, *testis*). The formation of sperm requires a temperature that is a few degrees cooler than that of the rest of the body. This lower temperature occurs in the testes because they are not located within the body cavity. Instead, the testes are suspended in a sac called the *scrotum*. The scrotum is an adaptation that has evolved to increase the chances of producing healthy sperm. (See Figure 14-1.)

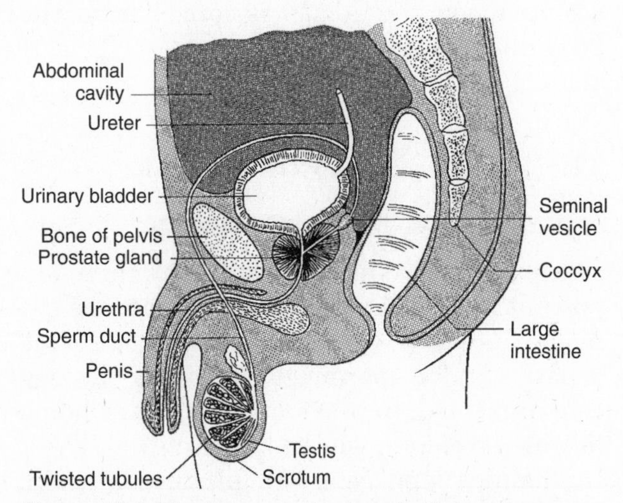


**Figure 14-1** The human male reproductive system.

Inside the testes are a great many tiny tubes, or *tubules*. As cells move through these tubules, they undergo the meiotic cell division that leads to formation of the gametes. Nowhere else in the male's body does meiotic cell division occur.

Sperm cells are highly specialized cells that are able to move. Each sperm cell's function is to try to deliver a single set of chromosomes from the male to an egg cell in the female. The structure of a mature sperm cell is well adapted to its function. Almost the entire head of the sperm is the nucleus, which holds the genetic information that is delivered to the egg. Attached to the head of the sperm is a long tail, called a *flagellum* (plural, *flagella*), which propels the cell along. Also present are large numbers of mitochondria that produce

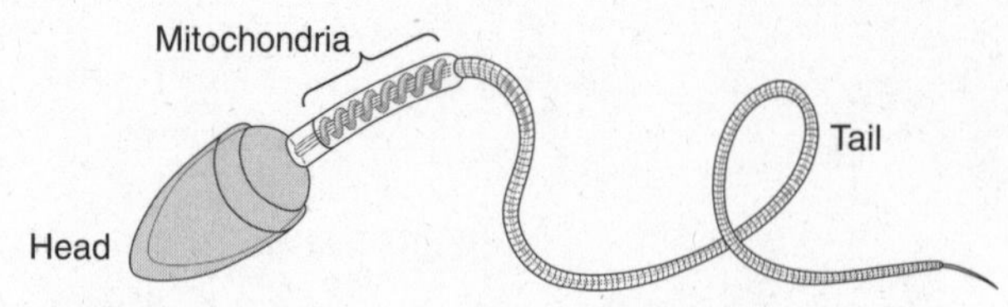


**Figure 14-2** The mature sperm cell is well adapted to deliver its single set of chromosomes to the egg cell in the female—the head contains the genetic information and the mitochondria supply the energy used by the tail to propel the sperm cell.

ATP, which supplies the energy that sperm use to propel themselves to the egg. (See Figure 14-2.)

After sperm move from the testes, a number of glands add fluids. The sperm and these fluids make up the *semen*. In fact, most of the semen consists of fructose, a sugar that provides an additional source of energy for the sperm.

The male reproductive system is adapted for internal fertilization. The penis is a structure that has evolved to deposit sperm safely within the female's reproductive tract. This occurs when the semen is forced from the body during ejaculation.

## THE FEMALE REPRODUCTIVE SYSTEM

The female reproductive system has three important functions as well: first, to produce the female gametes (egg cells); second, to provide a pathway for sperm cells to reach the egg cell; and third, to provide a temporary home for the developing embryo.

In females, egg cells are produced in the **ovaries**, a pair of reproductive organs. In an adult male, sperm production occurs all the time; approximately 30 million sperm cells are produced each day. In a female, nearly all potential eggs are already present when she is born. Throughout her reproductive life, a female releases only a few hundred of these eggs. Usually only a single egg matures and is released each month, packed with the nutrients needed to nourish an embryo right after fertilization. (See Figure 14-3.)

This development of egg cells occurs (in a sexually mature female) within the ovaries every month. One mature egg cell is released from one of the ovaries. This event is called **ovulation**. The egg cell gets swept into a long tubular structure called the *oviduct,* or fallopian tube, found next to each ovary. If fertilization occurs, the sperm usually joins the egg in the oviduct. The egg continues to move along the oviduct to the **uterus**, a pear-shaped organ with thick muscular walls. If the egg cell has been fertilized, the embryo becomes attached to the inside wall of the uterus and continues to develop. If fertilization does not occur, the egg cell breaks down within 24 hours of ovulation and is passed from the body. (See Figure 14-4.)

At the lower end of the uterus is the cervix, a narrow opening through which the sperm travel on their way to the egg cell. Connecting the cervix to the outside of the body is the vagina, which is made up of muscular tissue. It is into the vagina that sperm are ejaculated

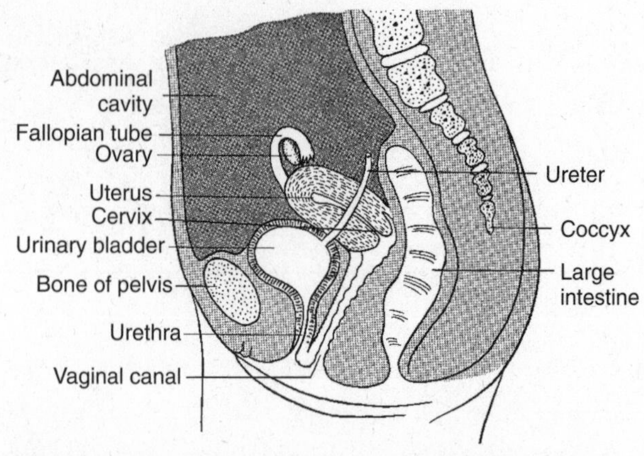


**Figure 14-3** The human female reproductive system (side view): egg cells are produced in the ovaries; sperm cells join the egg cells in the oviduct (fallopian tube); and the embryo develops in the uterus.

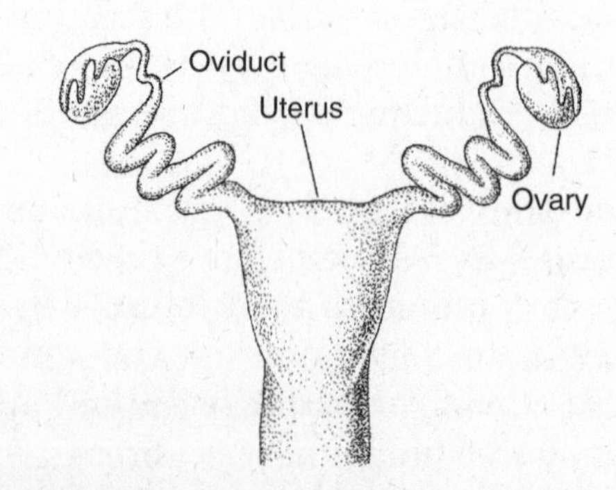


**Figure 14-4** Front view of the female reproductive system (simplified).

from the penis. Also, the vagina is the birth canal, through which the infant passes as it leaves the mother's body during birth.

Unlike in the male, the reproductive pathway in females is not combined with a pathway for excretion. Instead, the urine passes through an opening near the vagina.

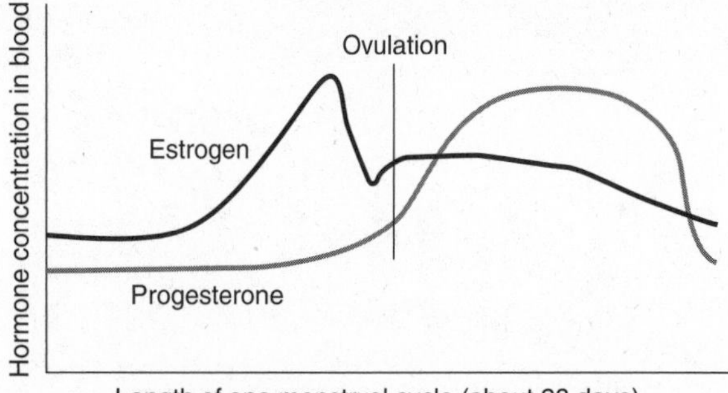


**Figure 14-5** The menstrual cycle and ovulation, which occur every month after puberty, are controlled by the release of hormones.

## HORMONES AND SEXUAL REPRODUCTION

During one's life, many changes and events occur in the body to make sexual reproduction possible. Hormones coordinate these changes. The main endocrine gland in charge of producing these hormones is the pituitary gland in the brain. The pituitary gland is controlled by the hypothalamus, a part of the brain.

The effects of a hormone depend not on the hormone itself but on its target tissue. The testes produce **testosterone**, the main male sex hormone. The effects of testosterone include the development of the male sex organs before and after birth. Around the age of 11, the level of testosterone suddenly increases in a boy's body. As a result, sperm production begins. This event is the beginning of **puberty**, or sexual maturation. During puberty, the penis and the testes begin to mature.

Testosterone also affects various other tissues in the male, causing the growth of facial and body hair, changes in body proportions, deepening of the voice, and other changes. These developments are called *secondary sex characteristics* because they are not directly related to sexual reproduction. In males, the level of testosterone in the body remains much the same for about 40 years after puberty, after which it gradually begins to decrease.

In females, the major sex hormones, **estrogen** and **progesterone**, are produced and released from the ovaries. The onset of puberty in females occurs somewhat earlier than in males. At about age 10, the levels of estrogen and progesterone increase dramatically, causing the uterus, vagina, and ovaries to mature. Secondary sex characteristics, such as the growth of body hair and breast (mammary gland) development, also are influenced by estrogen and progesterone. In addition, a monthly cycle of events known as the *menstrual cycle* begins. Remember, in males, sperm production occurs all the time after puberty. In females, the menstrual cycle occurs every month after puberty. Part of this cycle includes the release of an egg cell from the ovaries. (See Figure 14-5.)

In addition to ovulation, another critical function occurs during the menstrual cycle. The woman's body must be prepared in case fertilization occurs. Everything must be ready to nurture the developing embryo. So, during the first two weeks of the cycle, estrogen causes the lining of the uterus to thicken. There is also an increase in the amount of blood flow to that area.

During the second half of the cycle, after ovulation has occurred, progesterone, the pregnancy hormone, prepares the uterus for an embryo. If **pregnancy** occurs, the embryo becomes attached to the inner lining of the uterus. The *placenta* develops between the embryo and the uterus for the exchange of materials. The growing tissue then begins to release hormones to keep everything in the right condition. However, if fertilization does not occur, the continued preparations in the uterus are unnecessary. Toward the end of the four-week period of the menstrual cycle, the body senses that there is no embryo. The level of progesterone decreases; as a result, the uterine lining no longer remains intact, so it breaks down. The built-up tissue along with some blood and the unfertilized egg are released from the body. This flow of blood and tissue, called *menstruation*, lasts for about four days. Then the cycle begins again. (See Figure 14-6.)

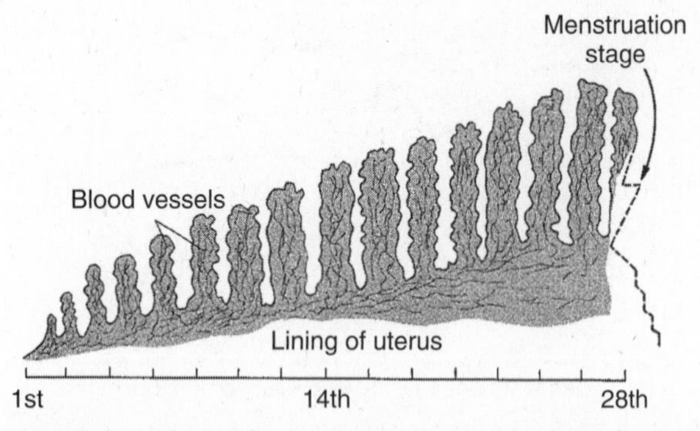


**Figure 14-6** Ovulation—the release of a mature egg from the ovary—occurs in the middle of the menstrual cycle, as hormones cause the lining of the uterus to thicken.

In women, the menstrual cycle continues for about 40 years from puberty. Between the ages of 45 and 55, the levels of hormones change; the menstrual cycle becomes less regular and eventually stops. This stage, called *menopause*, marks the point at which a female is no longer capable of reproducing. Menopause is a normal occurrence in all women. However, the effects of menopause vary widely from one woman to another. In men, by contrast, sperm production continues throughout life, although the number of healthy sperm likely declines with age.

## Chapter 14 Review

### Part A—Multiple Choice

**1.** The reproductive system of the human male produces gametes and

1 transfers gametes to the female for internal fertilization
2 releases hormones involved in external fertilization
3 produces enzymes that prevent fertilization
4 provides an area for fertilization of the gamete

**2.** The testes are adapted to produce

1 body cells involved in embryo formation
2 immature gametes that undergo mitosis only
3 sperm cells that may be involved in fertilization
4 gametes with large food supplies that nourish a developing embryo

**3.** The scrotum is located outside the body cavity, enabling healthy sperm to form because

1 blood does not flow to this region
2 the cells in the testes do not divide
3 its temperature is higher than that of the body
4 its temperature is lower than that of the body

**4.** In the male, meiotic cell division (that is, sperm cell formation) occurs within the

1 penis 3 testes
2 bladder 4 semen

**5.** The shape of a sperm cell can best be described as

1 an oval with four limbs
2 a head with a long tail
3 a tree with branches
4 a long, twisted tunnel

**6.** ATP is important to sperm cells because it

1 supplies the energy they need to move
2 enables the cells to replicate and divide
3 reduces their chromosome number
4 doubles their chromosome number

**7.** Most of the semen consists of

1 sperm 3 sugar
2 lipids 4 protein

**8.** The development of a human female's egg cells occurs within her

1 ovaries 3 cervix
2 oviduct 4 uterus

**9.** How does the production of male gametes differ from that of female gametes?

1 A male is born with all of his potential gametes, whereas a mature female produces them every day.
2 A female is born with all of her potential gametes, whereas a mature male produces them every day.
3 Female gametes are produced during fertilization, whereas male gametes are produced in advance.
4 Males produce one gamete at a time, whereas females produce millions each month.

**10.** The diagrams below represent cells that transport chromosomes. These cells are specialized for

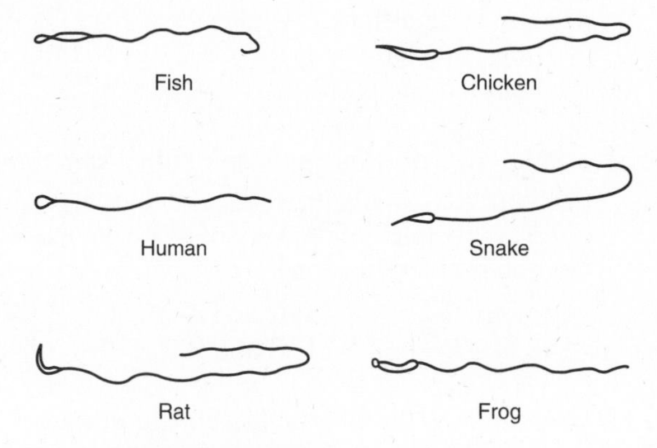


1 anaerobic respiration
2 sexual reproduction
3 chemical transmission
4 antibody production

**11.** Which statement does *not* correctly describe an adaptation of the human female reproductive system?

1 It produces gametes within the ovaries.
2 It provides for external fertilization of an egg.
3 It provides for internal development of the embryo.
4 It removes the wastes produced by the embryo.

**12.** Structures (in side view) in the human female are represented in the diagram below. Fertilization of the egg cell normally occurs within the part labeled

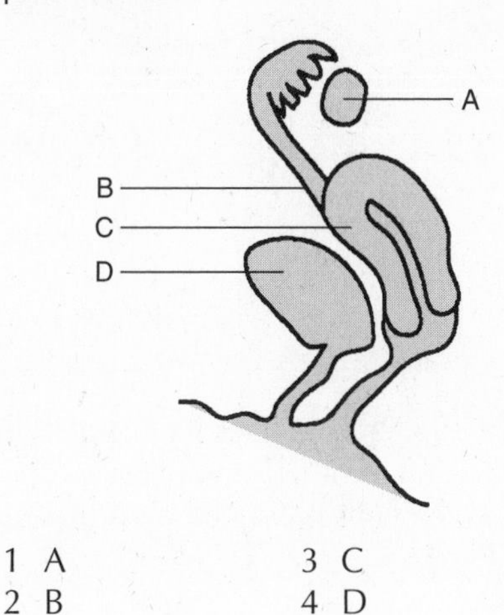


1 A 3 C
2 B 4 D

**13.** What happens if an egg cell is *not* fertilized after it is released?

1 It attaches to the wall of the uterus anyway.
2 It returns to the ovary to be released the next month.
3 It breaks down and is passed out of the female's body.
4 It remains in the oviduct until new sperm are introduced.

**14.** Regulation of sexual reproductive cycles of human males is related most directly to the presence of the hormone

1 estrogen 3 progesterone
2 testosterone 4 insulin

**15.** Estrogen and progesterone are examples of

1 organelles 3 hormones
2 tissues 4 enzymes

**16.** Secondary sex characteristics are traits that are

1 caused by sex hormones but are not directly related to reproduction
2 caused by the female sex hormones only
3 caused by the male sex hormones only
4 involved in reproduction, but only for the second child

**17.** What function does the placenta serve?

1 Oxygen diffuses from fetal blood to maternal blood.
2 Maternal blood is converted into fetal blood.
3 Materials are exchanged between fetal and maternal blood.
4 Digestive enzymes pass from maternal blood to fetal blood.

**18.** Why does the lining of the uterus thicken during the first half of the menstrual cycle?

1 to provide a place for sperm cells to attach
2 to produce a mature egg cell
3 to prepare to nurture an embryo
4 to rid the body of unfertilized egg cells

**19.** Why does menstruation occur?

1 to produce an egg in one of the ovaries
2 to release an egg from one of the ovaries
3 to allow an egg to be fertilized in the uterus
4 to remove an unfertilized egg from the uterus

**20.** If fertilization of the egg does *not* occur after ovulation, the

1 level of progesterone then increases
2 level of progesterone then decreases
3 level of testosterone then increases
4 level of estrogen then decreases

**21.** The following diagram represents a side view of the human male reproductive system.

If structure X were to be tied and cut off at the line, which change would occur in this system?

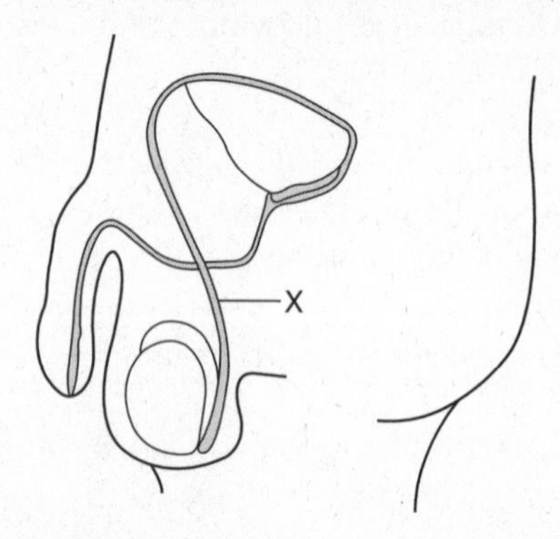


1 Sperm cells would no longer be produced.
2 Hormones would no longer be produced.
3 Sperm would be produced but no longer released from the body.
4 Urine would be produced but no longer released from the bladder.

**22.** After the onset of menopause, women normally
1 release more eggs than they did before
2 are no longer capable of reproducing
3 menstruate for longer periods of time
4 produce more estrogen and progesterone

### *Part B—Analysis and Open Ended*

**23.** Describe the *three* functions of the human male reproductive system.

**24.** In what way is the cell division that occurs in the testes different from that which occurs elsewhere in the male body?

*Base your answers to questions 25 and 26 on the diagram below, which represents the human male reproductive system.*

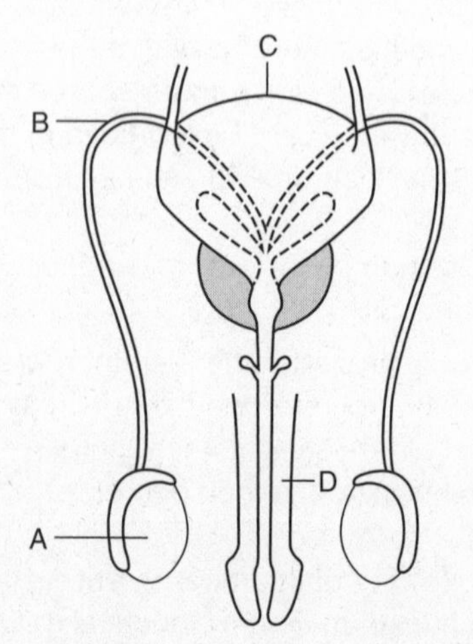


**25.** The hormone produced in structure A most directly brings about a change in the male's
1 blood sugar concentration
2 rate of digestion
3 physical characteristics
4 rate of respiration

**26.** Which pair of letters indicates both a structure that produces gametes and a structure that makes possible the delivery of those gametes for internal fertilization?
1 A and D
2 B and D
3 C and A
4 C and B

**27.** Explain how the structure of a mature sperm cell is related to its function. Your answer should mention the following features:

- the head of a sperm cell
- the tail of a sperm cell
- the mitochondria of sperm

**28.** List the *three* main functions of the female reproductive system. Include the specific structures involved in each of these functions.

**29.** What is the main difference between males and females in terms of the formation of their sex cells?

*Base your answers to questions 30 and 31 on the following diagram, which represents the human female reproductive system.*

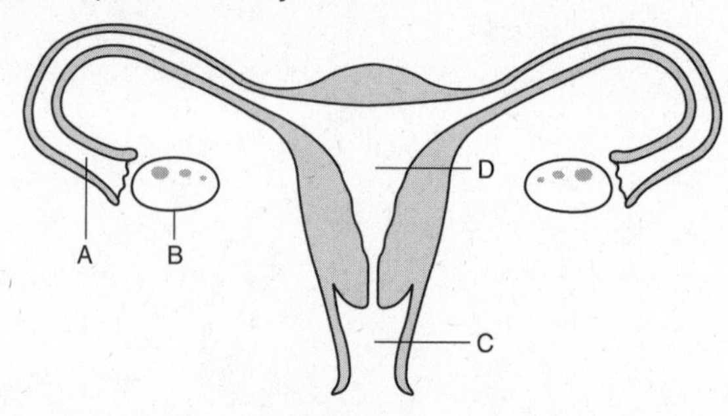


**30.** New inherited characteristics may appear in offspring as a result of new combinations of existing genes in the cells produced by structure
1 A
2 B
3 C
4 D

**31.** If pregnancy occurs, the embryo becomes attached to the inner lining of the structure labeled
1 A
2 B
3 C
4 D

**32.** Explain the meaning of the term *puberty*. How is the onset of puberty related to hormones and secondary sex characteristics?

*The diagrams below represent the reproductive organs of two individuals—one male and one female. The diagrams are followed by a list of sentences. For each phrase in questions 33 to 35, select the sentence from the list that best applies to that phrase.*

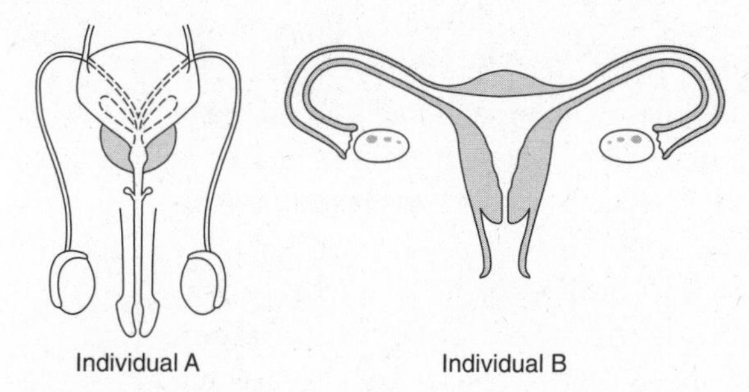


*Sentences:*

A The phrase is correct for both Individual A and Individual B.
B The phrase is *not* correct for either Individual A or Individual B.
C The phrase is correct for Individual A only.
D The phrase is correct for Individual B only.

**33.** Contains organs that produce gametes (sex cells).

1 A 3 C
2 B 4 D

**34.** Contains organs involved in internal fertilization.

1 A 3 C
2 B 4 D

**35.** Contains the structure in which a zygote develops.

1 A 3 C
2 B 4 D

**36.** Use the following terms to complete the boxes in the flowchart below: *facial hair; testosterone; wider hips; broader shoulders; estrogen; deeper voice; progesterone; ovaries and uterus mature; breasts develop; menstrual cycle begins; sperm production starts.*

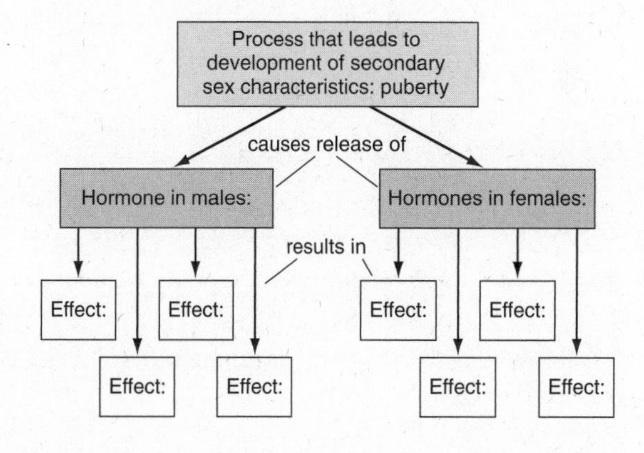


**37.** What is the main purpose of the menstrual cycle?

## *Part C—Reading Comprehension*

*Base your answers to questions 38 through 40 on the information below and on your knowledge of biology.*
Source: *Science News*, Vol. 163, p. 157.

> A woman's experiences in childbearing may presage her risk of heart disease, according to new research. Women who spontaneously lose one or more fetuses in pregnancy are about 50 percent more likely than other women to later suffer ischemic heart disease, in which constricted or obstructed blood vessels choke the flow of blood to the heart.
>
> The researchers reached these conclusions after analyzing data on all 129,290 women in Scotland who delivered their first live baby from 1981 through 1985. Additional data showed that those women who had had an early miscarriage in a previous pregnancy were more likely than other new mothers to have died from or been hospitalized with ischemic heart disease between 1981 and 1999.
>
> The loss of a fetus probably doesn't directly influence heart disease risk, Smith and his colleagues say in the Feb. 22 *British Medical Journal*. Rather, women with circulatory defects that predispose their blood vessels to become blocked face an elevated risk for both fetal loss and heart disease, the researchers hypothesize.

**38.** Explain what health risk in women may be connected to the loss of a fetus during pregnancy.

**39.** How was the research conducted that investigated the relation between miscarriages and heart disease?

**40.** Explain why there may be a connection between miscarriages and heart disease.

# 15
# Growth and Development

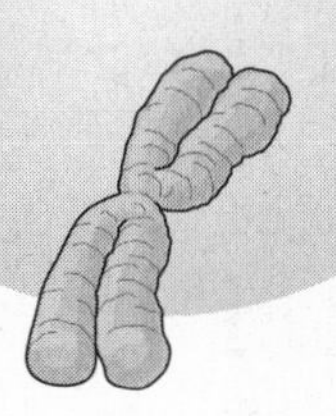

## Vocabulary

| | | |
|---|---|---|
| amnion | differentiation | gastrula |
| blastula | fetus | toxins |

## EMBRYONIC DEVELOPMENT: FROM THE BEGINNING

*Embryonic development* is the sequence of events that gradually changes a zygote into a functional organism. Most of the instructions that control this series of events are in the genetic material (chromosomes) of the zygote. In addition, the environment that surrounds a zygote can have profound effects on its development. The process of embryonic development is mostly the same for all animals, whether vertebrate or invertebrate. (See Figure 15-1.)

The beginning of embryonic development occurs as soon as the egg cell is fertilized. A zygote forms at the moment the cell membranes of the egg and the sperm join. This event makes it possible for the nuclei of the two cells to fuse.

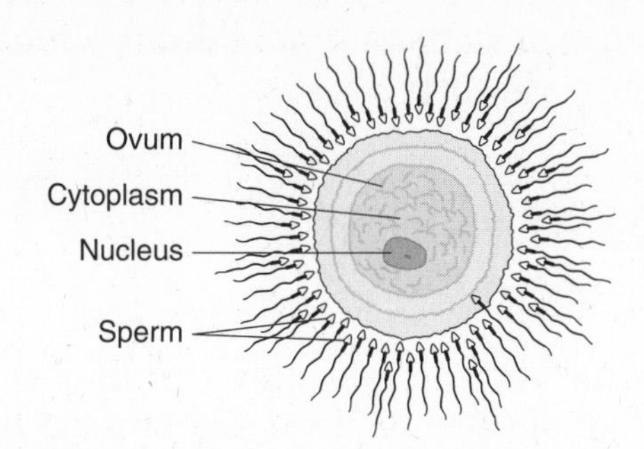


**Figure 15-1** Embryonic development begins after the egg is fertilized by a sperm.

The changes that begin to occur after fertilization, and which continue throughout life, are known as *development*. The most dramatic developmental changes—growth and differentiation—occur early in the life of an organism. Through *growth*, the organism becomes larger as its number of cells increases; **differentiation** occurs as these cells begin to develop their own specific structures and functions (due to controlled gene expression). We increase in size because our bodies are made up of many cells. However, we stay alive because our cells differentiate into more than 200 types, such as blood, skin, muscle, and bone cells.

All divisions of the zygote after fertilization are mitotic cell divisions. The number of chromosomes is maintained at each division. Therefore, all cells in the body still have the same complete set of chromosomes found in the fertilized egg cell. After the first series of mitotic cell divisions, the zygote becomes a hollow ball of cells, called a **blastula**. Although this ball of cells appears to have little organization, experiments have shown that each of the cells in a blastula already "knows"

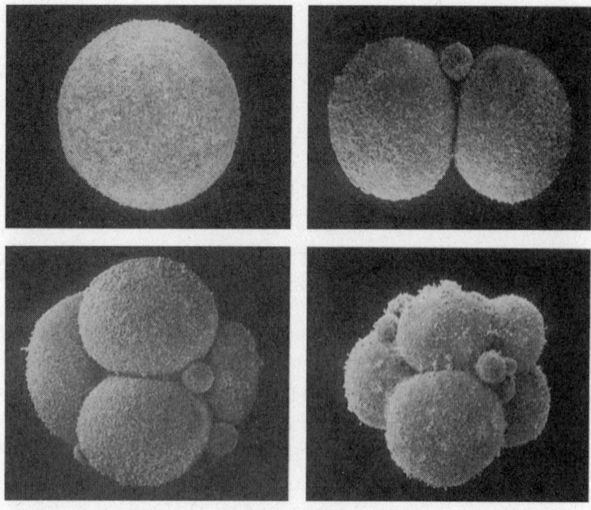

Figure 15-2 This series of scanning electron micrographs shows the mitotic cell division of a fertilized egg, or zygote, up to the eight-cell stage.

which part of the organism it will become. (See Figure 15-2.)

## DIFFERENTIATION OF CELLS

During the next stage of embryonic development, cells begin to move, changing position in a highly regulated fashion until a three-layered structure, called a **gastrula**, is formed. Each layer of the gastrula gives rise to particular body parts and body systems of the developing embryo. (See Figure 15-3.)

All eggs develop in a fluid environment. For animals that reproduce in the water, such as frogs, this presents no problem. However, for animals that reproduce on land, the need

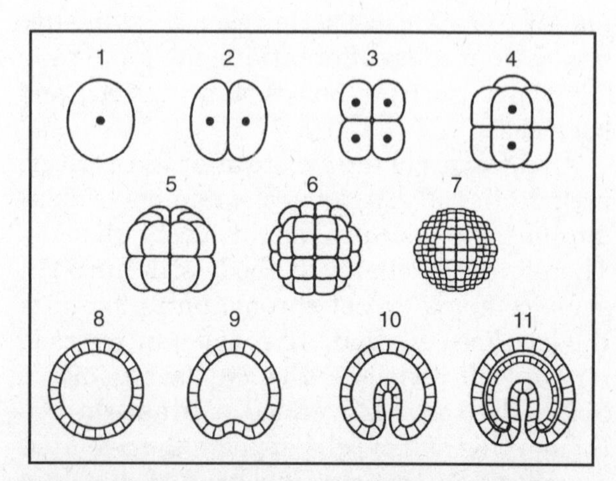


Figure 15-3 The three embryonic cell layers form in a regulated fashion from the zygote's hollow ball of cells. These cell layers give rise to all body parts and structures.

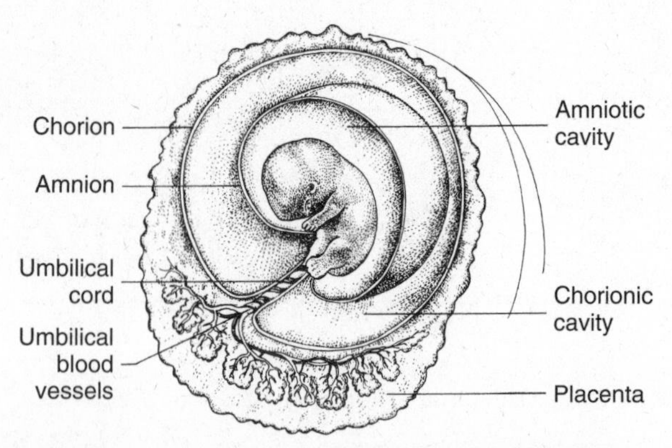


Figure 15-4 In mammals, the umbilical cord connects the placenta to the fetus. Nutrients and wastes are exchanged between the mother and fetus through the blood vessels of the umbilical cord and placenta.

for a watery environment poses an important problem. The solution to this problem evolved millions of years ago. A series of *membranes* forms around the developing embryo. The most important membrane, the **amnion**, has three main functions: it surrounds the embryo, protects the embryo, and holds in a fluid that cushions the embryo. Because of the amnion, the embryos of land animals develop in water just like those of aquatic animals do. (See Figure 15-4.)

At this point in development, the rest of the embryo's cells are still quite similar to each other. However, the process of cellular changes now begins. As the cells start to develop into muscle cells, skin cells, blood cells, or other tissue types, they begin to join to form organs. Cell differentiation is occurring. All of the cells contain the same genetic information, but each type of cell uses this information differently.

In most vertebrates, by the time birth occurs, all the major structures of the animal have been formed. The development that occurs after birth is mostly confined to an increase in size as the animal develops into its adult form. (See Figure 15-5.)

## THE DANGERS THAT THREATEN A FETUS

The **fetus** (an embryo after the first three months of development) is surrounded by a watery cushion, in which it is kept warm and nourished. It has its safety provided by the mother's womb, or *uterus*. Most infections that

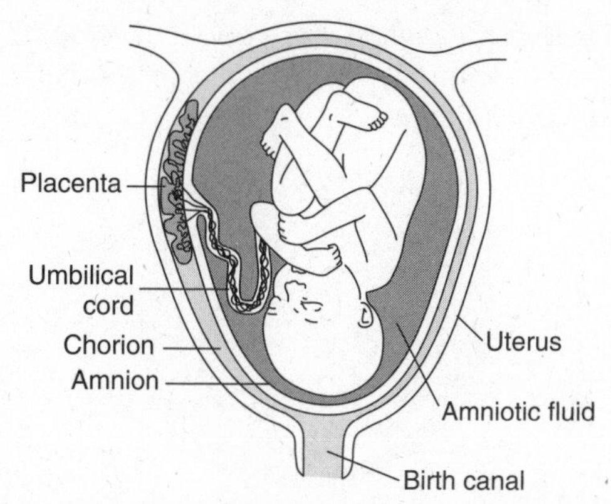


**Figure 15-5** At the final stage of development, the fetus grows in size until it is ready for birth.

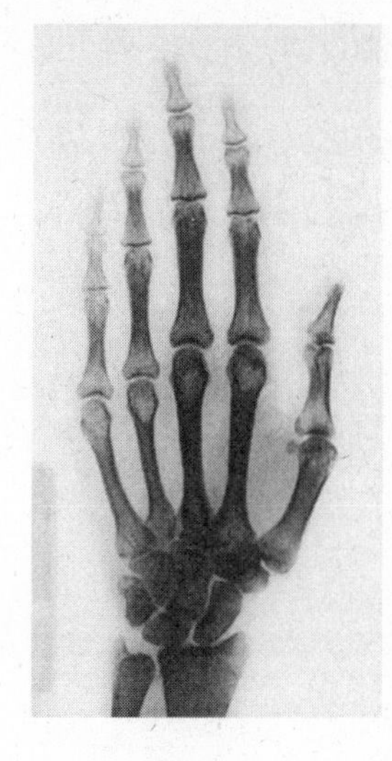

**Figure 15-6** During embryonic development, the cells are dividing and growing. Dividing cells are easily damaged by X rays, which can penetrate through soft tissues to show bones. For this reason, X rays are one of the things that can be harmful to a fetus and should be avoided during pregnancy.

may make the mother ill cannot cross over the placenta and into the fetus. However, the fetus is not entirely safe. Dangers can intrude into its small world.

Some forms of radiation can pass through the tissues of the mother and into the fetus. X rays, for example, can affect a fetus. (See Figure 15-6.) Cells are dividing, growing, and changing during embryonic development; and dividing cells are easily damaged by X rays. Some infectious microorganisms (such as HIV) within the mother can enter a fetus. Cigarette (tobacco) smoking by the mother, as well as chemicals, or **toxins**, taken in by her during pregnancy, can also harm a developing fetus. Use of heroin, LSD, cocaine, and alcohol can endanger a fetus, and many serious effects result. Babies can be born addicted to these drugs, and mental retardation can occur due to the alcohol that may pass from the mother's blood into the fetus.

## Chapter 15 Review

### *Part A—Multiple Choice*

**1.** What happens during embryonic development?

1 An egg cell is released from an ovary.
2 An egg cell is fertilized in the female.
3 A zygote changes into a functional organism.
4 A developed organism leaves its mother's body.

**2.** Which event does *not* occur between stages 2 and 11 in the process represented below?

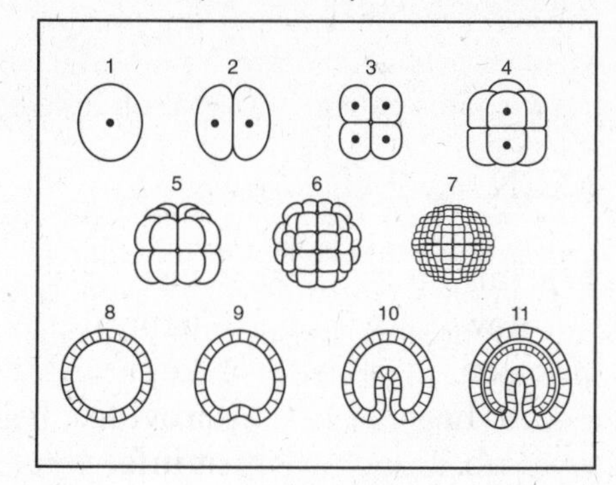


1 a decrease in individual cell size
2 development of embryonic layers
3 mitotic cell divisions
4 fertilization of the egg

**3.** Embryonic development begins when an egg cell is

1 produced in the ovary
2 released from the ovary
3 fertilized by the sperm cell
4 removed from the body

**4.** During the process of growth, the

1 number of cells in an organism increases
2 number of cells in an organism decreases
3 cells begin to develop specific structures
4 cells undergo mitosis and meiosis

**5.** The following diagram represents part of the human female reproductive system. Most embryonic development normally occurs within the structure labeled

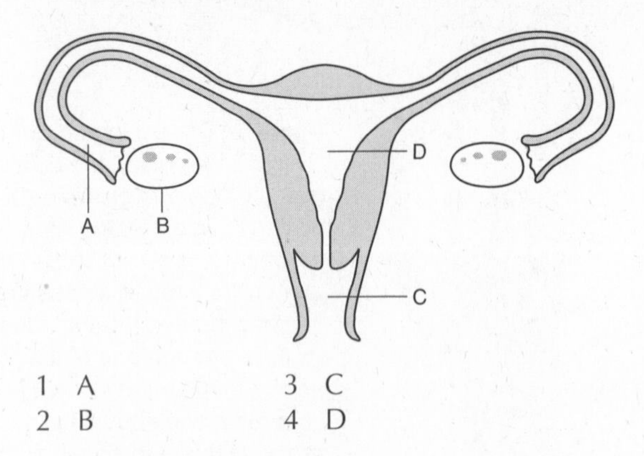


1 A
2 B
3 C
4 D

6. Compared with the number of chromosomes in a fertilized egg cell, each body cell of an adult organism has
   1 half as many chromosomes
   2 the same number of chromosomes
   3 twice as many chromosomes
   4 varying numbers, depending on function

7. Which phrase best describes the process represented in the diagram below?

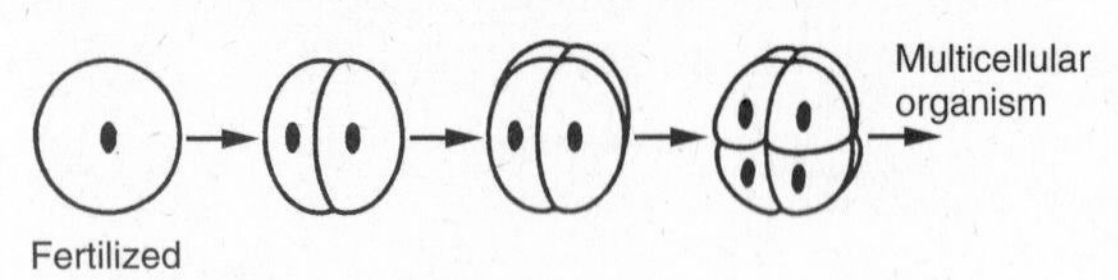


   1 a zygote dividing by mitosis
   2 a gamete dividing by mitosis
   3 a zygote dividing by meiosis
   4 a gamete dividing by meiosis

8. Cells develop into skin cells, muscle cells, bone cells, and so on, during the process of
   1 fertilization
   2 cell growth
   3 differentiation
   4 classification

9. In animals, the normal development of an embryo is dependent upon
   1 fertilization of a mature egg by many sperm cells
   2 production of body cells having half the number of chromosomes as the zygote
   3 production of new cells having twice the number of chromosomes as the zygote
   4 mitosis and the differentiation of cells after fertilization has occurred

10. The diagram below represents a developing bird embryo within an egg. What is the primary function of the egg?

   1 to serve as a food supply for wild predators
   2 to ensure survival of all the genes of that species
   3 to protect and nourish the developing bird embryo
   4 to give the parent birds freedom of movement

11. Heavy cigarette smoking and the use of alcohol throughout pregnancy usually increase the likelihood of the birth of
   1 identical twins
   2 a male baby
   3 a baby with a viral infection
   4 a baby with medical problems

12. Whether from frogs, snakes, cows, or humans, all fertilized eggs
   1 have half the chromosome number of body cells
   2 form a hard outer shell while developing
   3 develop within a fluid environment
   4 are ready to hatch within a few months

13. The role of the amnion membrane is to
   1 protect the embryo within a fluid
   2 provide a food source for the embryo
   3 control the differentiation of cells
   4 exchange gases and remove wastes

14. Cells undergo differentiation because they
   1 originate from different egg cells
   2 have very different genetic information
   3 each receive a different chromosome pair
   4 use the same genetic information differently

15. By the time most vertebrates are born,
   1 their three germ layers are still being formed
   2 all their major structures have been formed
   3 only half their major structures have been formed
   4 they no longer have to increase their body size

**16.** The term that describes an embryo in the uterus, after its first three months of development, is

1 placenta
2 zygote
3 fetus
4 newborn

**17.** During the last months of pregnancy, the brain of a human embryo undergoes a "growth spurt." Which action by the mother would most likely pose the greatest threat to the normal development of the fetus's nervous system at this time?

1 spraying pesticides in the garden
2 taking prescribed vitamins on a daily basis
3 maintaining a diet high in fiber and low in fat
4 not exercising anymore

**18.** When a pregnant woman ingests toxins such as alcohol and nicotine, the embryo is put at risk because these toxins can

1 diffuse from the mother's blood to the embryo's blood at the placenta
2 enter the embryo when it opens its mouth
3 transfer to the embryo through the mother's mammary glands
4 enter the uterus through the mother's navel

### *Part B—Analysis and Open Ended*

**19.** Some stages in the development of an organism are listed below. Which sequence represents the correct order of these stages?

(A) differentiation of cells into tissues
(B) fertilization of egg by sperm
(C) development of organs
(D) mitotic cell division of zygote

1 A→B→C→D
2 B→C→A→D
3 D→B→C→A
4 B→D→A→C

**20.** The sequence below represents some early events in the process of embryonic development. What factor helps regulate these cellular events?

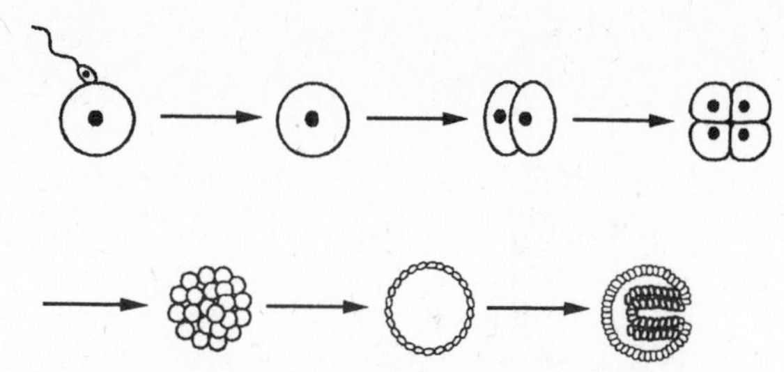

1 the presence of active genes in each cell
2 the removal of all enzymes from the cells
3 an increase in the number of genes in each cell
4 a decrease in mitotic activity in the cells

**21.** What are the two main factors that have an effect on embryonic development?

**22.** Distinguish between the terms *development, growth,* and *differentiation.* Which one of these terms includes the other two?

**23.** Arrange the following terms in the correct sequence in which they occur: *fertilization, growth, gamete formation, differentiation, zygote.*

**24.** The diagram below illustrates the formation of

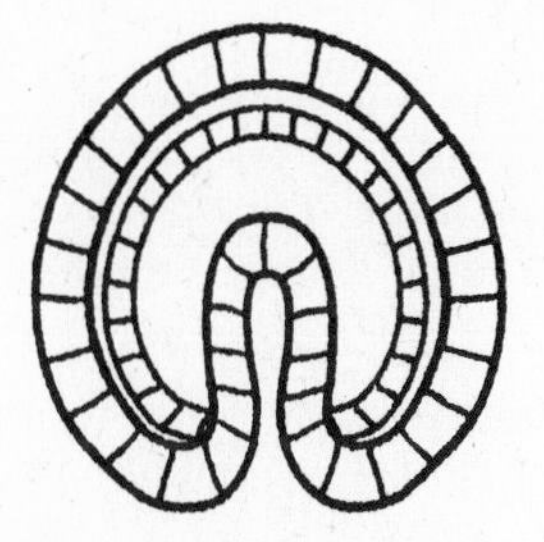

1 three mature egg cells
2 three separate organisms
3 a newly fertilized egg cell
4 three embryonic cell layers

**25.** How does the single-layered zygote change to give rise to all body parts?

**26.** Use the following terms to complete the definitions in the flowchart below: *three-layered structure; fetus; zygote; hollow ball of cells; differentiation.*

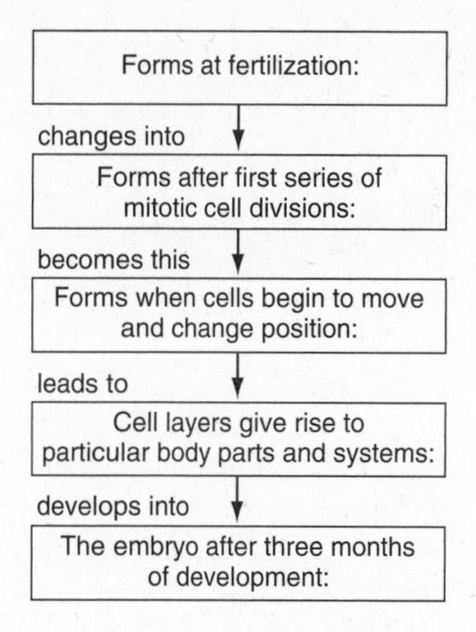

*Refer to the diagram below to answer questions 27 and 28.*

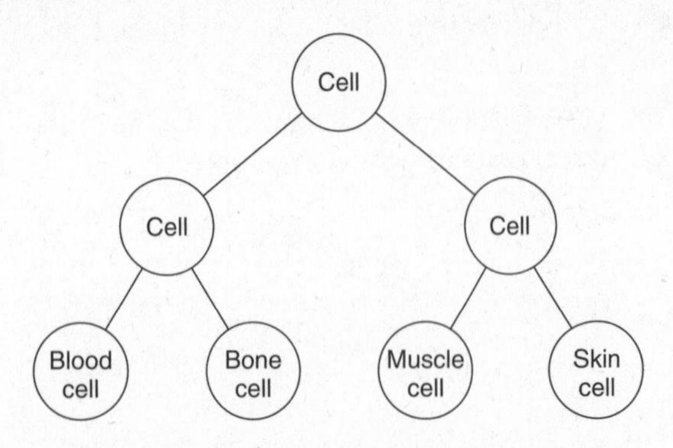


**27.** Which process is represented by the diagram?

1 fertilization
2 differentiation
3 meiosis
4 migration

**28.** Why is this process vital to the embryonic development of a fetus?

**29.** List the *three* functions of the amnion membrane. Explain why the amnion is so important for the embryonic development of land animals.

**30.** Why is the environment that surrounds an embryo important for its normal cell growth and development?

**31.** Use a specific example to identify *one* action taken by a mother that could have a negative effect on the embryonic development of her baby.

**32.** Which of the diagrams shown below represents development and growth in a fetus?

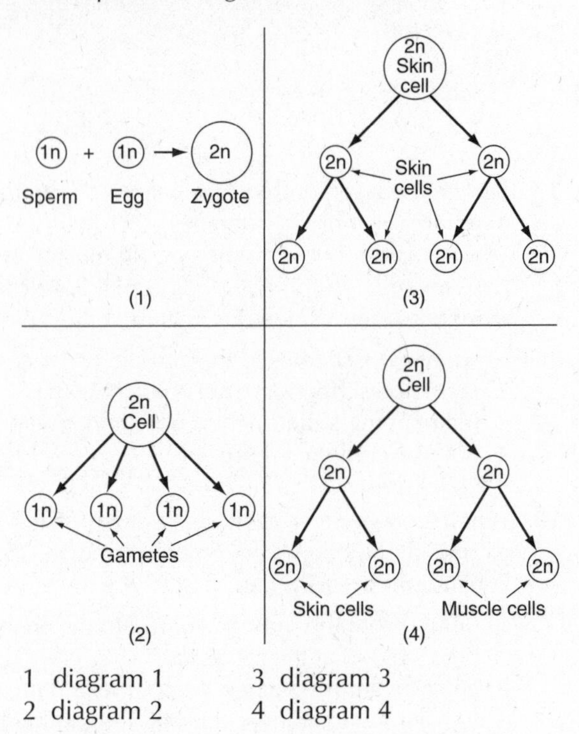


1 diagram 1
2 diagram 2
3 diagram 3
4 diagram 4

**33.** Identify *four* dangers that a fetus can encounter while inside the womb. Explain why they can be harmful to its development.

**34.** Write a brief essay that explains why someone who smokes cigarettes and has become pregnant should stop smoking cigarettes at that time.

### *Part C—Reading Comprehension*

*Base your answers to questions 35 through 37 on the information below and on your knowledge of biology.*
Source: *Science News*, Vol. 162, p. 301.

> The second year of life may be particularly memorable. Around the time of their first birthday, children make dramatic advances in remembering simple events for 4 months after witnessing them, a new study finds. This memory breakthrough depends on a proliferation of neural connections in memory-related brain structures known to develop as infants approach age 1, propose Harvard University psychologists Conor Liston and Jerome Kagan.
>
> The researchers recruited 12 babies and toddlers at each of three ages: 9 months, 17 months, and 24 months. Children watched an experimenter both perform and describe three action sequences. In one sequence, for example, the experimenter said "Clean-up time!" while wiping a table with a paper towel and then throwing the towel into a trash basket.
>
> Kids in the two older groups watched four demonstrations of each action sequence, and 9-month-olds saw six repetitions. After each presentation, the experimenter encouraged children to imitate what they had just seen.
>
> Four months later, the youngsters—then ages 13 months, 21 months, and 28 months—were asked to reenact each set of actions with the same materials after hearing the same verbal descriptions.
>
> The children now 28 months old correctly performed a majority of previously observed actions, usually in their original order, Liston and Kagan report in the Oct. 31 *Nature*. The 21-month-olds reenacted what they had seen almost as well as their older peers did. Far fewer signs of accurate recall appeared in 13-month-olds, the only participants who had been under 1 year of age during initial memory trials.

**35.** State the sudden change that occurs in a child's ability to remember as he or she passes the first birthday.

**36.** Explain what is thought to occur in the brain that results in the change in how infants remember.

**37.** Describe the design of the experiment that was used to conduct memory research on 1-year-olds and 2-year-olds.

# THEME V

## Genetics and Molecular Biology

# 16

# DNA Structure and Function

**Vocabulary**

linear sequence
mutation
nucleotides
template

### GENETIC MATERIAL: A JOB DESCRIPTION

During the 1940s and 1950s, several scientists conducted research to determine if it was the proteins or the DNA within a cell's chromosomes that contained the genetic information. After carrying out careful experiments and chemical analyses, they discovered that DNA (deoxyribonucleic acid) contains the information on which all life depends; that is, DNA *is* the genetic material. A substance that serves as the genetic material has the most significant job in the world: to carry on life itself. To carry out this job, the genetic material must do the following:

- It must be able to store information that can be passed on from one generation of cells to the next. It must be able to store enough information to make an organism like a tree or you.
- It must be able to make a copy of itself in order to pass this information from one generation to the next.
- It must be strong and stable so that it does not easily fall apart and perhaps cause harmful changes to its store of information.
- It must be able to *mutate*, or change, slightly from time to time. These changes allow a species to produce the variations

on which natural selection acts, which can lead to the evolution of new species.

We can now look at how a DNA molecule is built and how it functions to do these jobs.

## THE WORLD LEARNS OF THE DOUBLE HELIX

DNA is made up of smaller subunits. These subunits, or **nucleotides**, include four types of *bases*, which occur in two pairs. The amounts of bases adenine (A) and thymine (T) are always the same (A pairs with T). The amounts of bases guanine (G) and cytosine (C) are always the same (G pairs with C). In 1953, scientists James Watson and Francis Crick described the structure of DNA for the first time, as a *double helix*. (See Figure 16-1.)

To understand the double helix structure of DNA, picture a ladder that has been twisted, like a spiral staircase. The two sides of the ladder are parallel to each other, and the steps of the ladder link the two sides together. The sides of the ladder are the backbone of the DNA molecule (composed of sugar and phosphate molecules). Stretching between the two sides are the pairs of bases. The Watson-Crick model showed that the only possible way all the parts could fit was for each large adenine base to be matched opposite a smaller

Figure 16-1 Scientists James Watson (left) and Francis Crick (right) shown in 1953 with their model of part of a DNA molecule.

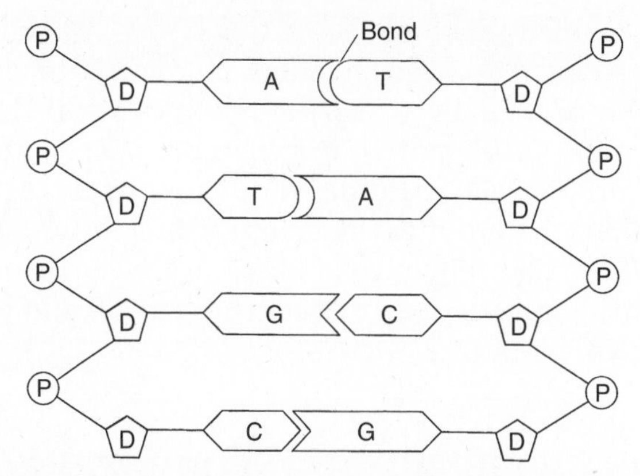


Figure 16-2 The structure of a DNA molecule—the nucleotide subunits include four types of bases (A, T, G, and C).

thymine base. Similarly, the large guanine had to be opposite a smaller cytosine. (See Figure 16-2.)

So, a molecule of DNA consists of two strands, opposite each other, connected by matching base pairs. If we look at one strand, we can describe it in terms of the order, or sequence, of its subunits. Because the subunits are in a long line, the order of the subunits is called a **linear sequence**. This linear sequence of nucleotides builds the DNA molecule, which may be very long. (Recall that long molecules such as DNA, which can contain thousands of nucleotide subunits in a sequence, are called *polymers*.) (See Figure 16-3.)

Imagine walking along a single strand of DNA. The bases in the subunits may occur in any order. The linear sequence on a short molecule of DNA might be A-T-T-G-A-C-C-G. Now imagine walking along the opposite strand, starting at the same place. Opposite the A in the first strand is a T. Because we know the sequence of bases in the first strand we automatically know the sequence of bases in the other strand. In this example, beginning with the T, the

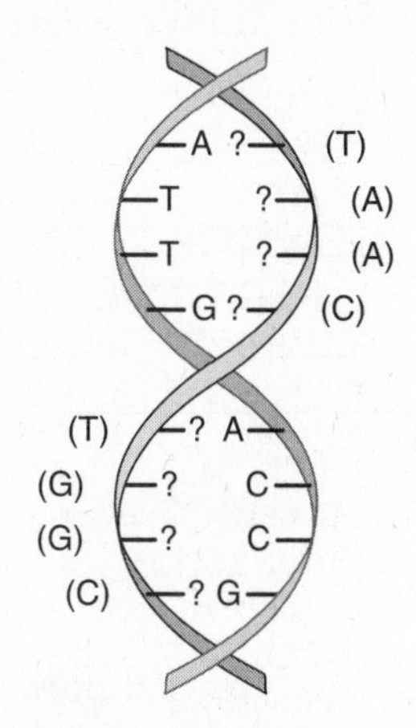


Figure 16-3 From the sequence of bases on one strand of DNA, we can determine the sequence on the opposite strand: A pairs with T, and C pairs with G.

sequence must be T-A-A-C-T-G-G-C. This is the key to how the DNA molecule copies itself. The process by which DNA copies itself depends on the matching base pairs in the subunits of each strand. What is so important about the order of the subunits in a strand of DNA? The sequence of bases in the subunits *is* the genetic information that the strand of DNA contains.

## DNA: A LIBRARY OF INFORMATION

In some ways, the bases in DNA are like the letters of an alphabet, only the DNA "letters" are chemical letters. Because there are only four letters (A, T, G, and C) in the DNA alphabet, scientists thought that DNA was too simple to contain the complex genetic information of life. But what is also significant about DNA is the sequence of the letters, not just the letters themselves. Using these four letters in long sequences, nature can create an almost unlimited variety of genetic messages.

When you realize that human DNA consists of three billion pairs of bases, you can begin to imagine how much information can be stored in the DNA of our cells. All the information for constructing our bodies—determining all of our characteristics or traits and keeping our bodies functioning—is stored in the linear sequences of bases in our DNA. The same is true for all other organisms on Earth. The evolutionary relationship between two organisms can be learned by comparing their DNA. The more similar their sequences of bases, the more recently the two organisms evolved from a common ancestor. (See Figure 16-4.)

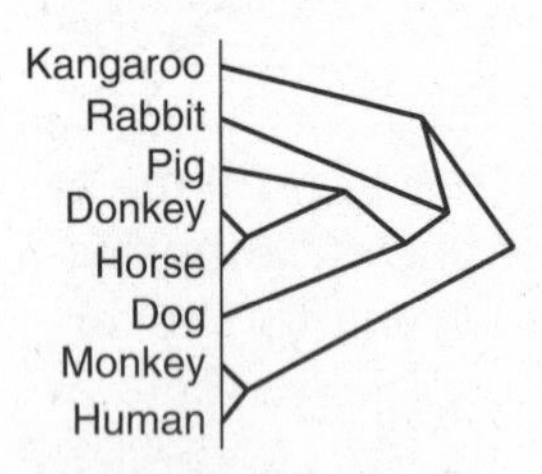


**Figure 16-4** Evolutionary relationships are confirmed by DNA closeness—the more similar their base sequences, the more recently two organisms evolved from a common ancestor. For example, the donkey and the horse are more closely related than are the pig and the horse.

To make use of the genetic information stored in DNA, organisms must change that information into proteins. Proteins are made up of amino acids, subunits that—like nucleotide bases—are joined in a linear sequence. The sequence of DNA subunits is used to direct the synthesis of proteins that have the correct sequence of amino acid subunits. In other words, through a chemical process, the order of the nucleotides determines the order of the amino acids in the proteins that are built.

## DNA REPLICATION: PASSING IT ON

To function as the genetic material, DNA has to be able to replicate, or make a copy of, itself. This process of DNA replication occurs during the middle of the cell cycle. What we already know about its structure is enough to explain how DNA replicates.

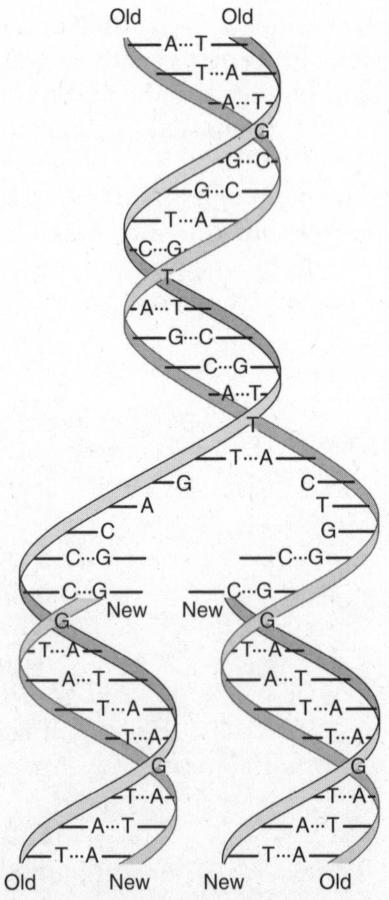


**Figure 16-5** During DNA replication, the double helix unwinds, the strands separate, and the new strands form opposite each of the original DNA strands.

To make a copy, you need an original, sometimes called a **template**. Because DNA is a double helix, it has templates built into it. To begin the process, the double helix unwinds. As with all metabolic activities, enzymes are needed for this process. Once the double-stranded molecule is untwisted, it begins to unzip, just like a zipper. Through the activity of an enzyme, the bonds between bases begin to break apart. (See Figure 16-5.)

As the bonds break, each strand of the DNA molecule becomes separate. Many free subunits float around in the cell. Specific enzymes, called *DNA polymerase*, match up these free subunits with the existing subunits in each DNA strand. Wherever a T is located on a strand, an A pairs to it; wherever a C is located, a G joins up, and so on. One by one, new subunits are joined to make a new strand opposite each old strand. The sequence of bases in the old strands determines the linear sequence of subunits in the new strands. When replication is complete, two double-stranded DNA molecules are formed. Each molecule is made up of one old strand joined to a newly synthesized strand. How do the two new DNA molecules compare to the original one? They are identical—DNA replication has occurred. (See Figure 16-6.)

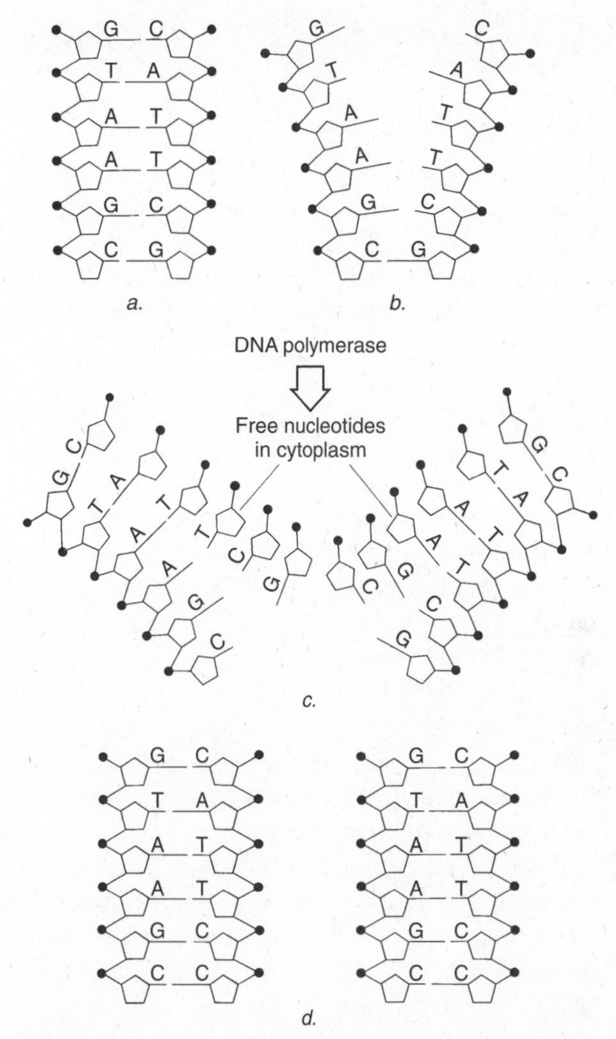


**Figure 16-6** Through the process of DNA replication, two identical double-stranded DNA molecules are formed.

## ERRORS IN DNA REPLICATION

Few things in life are perfect; this is true of DNA replication, too. The enzymes responsible for directing the correct pairing of subunits during DNA replication occasionally make mistakes. A nucleotide base may be left out. Or the wrong base may be matched up. Sometimes an extra one is added. These mistakes produce errors in the linear sequence in one strand of the DNA molecule. Such an error is called a genetic **mutation**. From what we know about the replication process, once an error occurs in a DNA strand, it may be copied again and again. Thus, a mutation in the genetic material of one cell can easily be passed on to future cells.

A mutation is simply a change. However, many changes in genetic material can be harmful and may make it impossible for future cells, or even the entire organism, to survive. Other mutations cause an unnoticeable change; rather than harming the organism, the mutation seems to produce no effect. Sometimes a mutation gives the organism a slight advantage that other similar organisms lack. Not only can mutations in DNA be good, but they are an important source of the genetic variation that is necessary for evolution to occur. In fact, much of the evolution of different life-forms on Earth has depended on the chance occurrence of these mutations. (Remember: Only mutations within the DNA of gametes can be passed along to offspring; mutations within the DNA of body cells cannot.)

# Chapter 16 Review

## Part A—Multiple Choice

1. Which is *not* a necessary characteristic of the genetic material?
   1 It must be able to make a copy of itself.
   2 It must be weak so that it can fall apart easily.
   3 It must be able to mutate from time to time.
   4 It must be able to store information.

2. If a set of instructions that determines all the characteristics of an organism is compared to a book, and a chromosome is compared to a chapter in the book, then what might be compared to a paragraph in the book?
   1 a starch molecule
   2 an amino acid
   3 a protein polymer
   4 a DNA molecule

3. A portion of a molecule is shown in the diagram below. Which statement best describes the main function of this type of molecule?

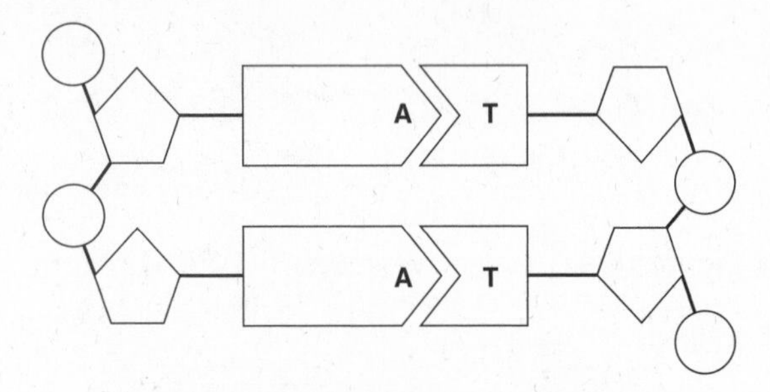


   1 It is a structural part of the cell wall.
   2 It determines what traits may be inherited.
   3 It stores energy for metabolic processes.
   4 It transports materials across the cell membrane.

4. The subunits of proteins are
   1 simple sugars
   2 phosphates
   3 amino acids
   4 enzymes

5. Watson and Crick contributed to the study of DNA by
   1 experimenting with pea plants
   2 recognizing that traits are inherited
   3 discovering the double helix structure of DNA
   4 mapping the entire human genome

6. The genetic code of a DNA molecule is determined by its specific sequence of
   1 ATP molecules
   2 carbohydrates
   3 sugar molecules
   4 nucleotide bases

7. The DNA molecule is formed from subunits arranged in a
   1 sequence with three kinds of bases
   2 circle with four kinds of bases
   3 sequence with four kinds of bases
   4 sequence with four kinds of acids

8. The base pairs in DNA are similar in arrangement to the
   1 sides of a ladder
   2 steps of a ladder
   3 railing of a staircase
   4 surface of a ramp

9. The order of the subunits in a strand of DNA is called a
   1 subunit sequence
   2 linear sequence
   3 strand sequence
   4 nucleotide sequence

10. If one strand of a DNA molecule is G-A-T-C-C-A-T, the sequence of the opposite strand is
    1 G-A-T-C-C-A-T
    2 C-T-A-G-G-T-A
    3 A-T-G-G-A-T-G
    4 T-A-C-C-T-A-G

11. The organization of bases in DNA can best be likened to the
    1 arrangement of letters in a word
    2 kinds of tools in a garage
    3 number of books in a library
    4 colors in a rainbow

12. When DNA separates into two strands, the DNA would most likely be directly involved in
    1 replication
    2 differentiation
    3 fertilization
    4 evolution

13. The sequence of subunits in a protein is most directly dependent upon the

1 region in the cell where enzymes are produced
2 type of cell in which starch is found
3 DNA in the chromosomes in a cell
4 kinds of materials in the cell membrane

**14.** In the diagram below, strands I and II represent sections of a DNA molecule. Strand II would normally include (top to bottom)

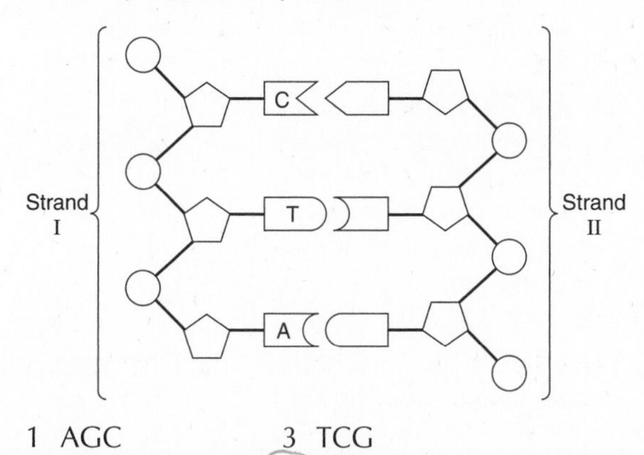


1 AGC
2 TAC
3 TCG
4 GAT

**15.** During the first step in the replication of DNA, the

1 double helix unwinds
2 base template is created
3 subunits of DNA form pairs
4 double helix rewinds itself

**16.** What causes the base pairs of DNA to break apart?

1 a mutation during replication
2 the activity of an enzyme
3 the production of new bases
4 the introduction of a fifth base

**17.** After DNA replication, the new DNA molecules are

1 the reverse of the original
2 the mirror image of the original
3 identical to the original
4 totally different from the original

**18.** Which statement is true regarding an alteration or change in DNA?

1 It is always referred to as a mutation.
2 It is always passed on to the offspring.
3 It is always advantageous to an individual.
4 It is always detected by chromatography.

**19.** A mutation occurs in a cell. Which sequence best represents the order of events for this mutation to affect traits expressed by the cell?

1 amino acids joining in sequence→a change in the sequence of DNA bases→appearance of characteristic
2 a change in the sequence of DNA bases→amino acids joining in sequence→appearance of new characteristic
3 appearance of new characteristic→amino acids joining in sequence→a change in the sequence of DNA bases
4 a change in the sequence of DNA bases→appearance of new characteristic→amino acids joining in sequence

**20.** A mutation is considered positive when it

1 makes it hard for the organism to survive
2 has absolutely no effect on the organism
3 changes the organism in an undetectable way
4 provides a sudden advantage that aids survival

### *Part B—Analysis and Open Ended*

**21.** What four qualities must the genetic material have in order to do its job?

**22.** List the four bases of the DNA nucleotides and tell which bases pair together.

**23.** Explain the basic structure of DNA as described by Watson and Crick.

**24.** Why did scientists once think that DNA was too simple to contain the genetic information of living things? Explain why their reason was not correct.

**25.** Molecule 1 represents a section of hereditary information, and molecule 2 represents part of the substance that is determined by the information in molecule 1. What will most likely happen if there is a change in the first three subunits on the upper strand of molecule 1 shown below?

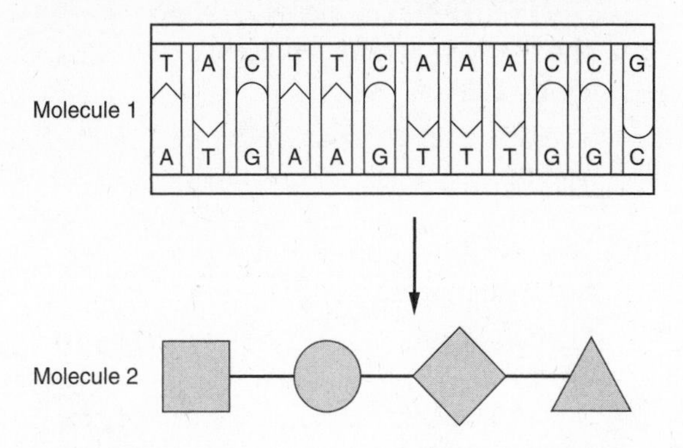

1 The remaining subunits in molecule 1 will also change.
2 Molecule 1 will split apart, triggering an immune response.
3 A portion of molecule 2 may be formed differently.
4 Molecule 2 may form two strands rather than one.

**26.** In an experiment, DNA from dead pathogenic bacteria was transferred into living bacteria that were, normally, not pathogenic. These altered bacteria were then injected into healthy mice. The mice died of the same disease caused by the original pathogens. Based on this information, which statement would be a valid conclusion?

1 DNA is present only in living organisms.
2 DNA functions only in the original organism from which it comes.
3 DNA changes the organism receiving an injection into another organism.
4 DNA from a dead organism can become active in another organism.

**27.** Briefly explain how the genetic information is arranged within a DNA molecule.

**28.** You see a photograph of a famous man and his teenaged son. You notice that they look very much alike, and that they even wear similar eyeglasses. What conclusion can you draw from this observation?

1 The DNA present in their body cells is identical.
2 Their percentage of having the same proteins is high.
3 The base sequences of their genes are all identical.
4 The mutation rate is the same in their body cells.

*Refer to the figure at right to answer questions 29 to 31.*

**29.** The diagram at right represents a molecule of

1 ATP
2 RNA
3 DNA
4 FSH

**30.** The structures labeled G, C, T, and A all represent

1 acids
2 sugars
3 bases
4 phosphates

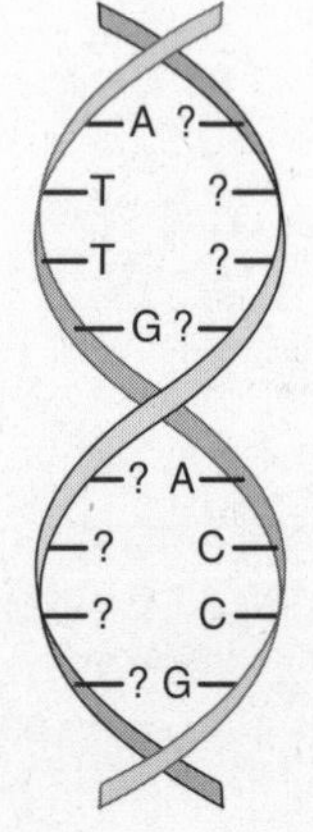


**31.** Starting from the top of the diagram, what would be the letters of the missing units on the matching strand?

**32.** Complete the analogy: Nucleotide bases are to DNA as amino acids are to

1 sugars
2 proteins
3 lipids
4 nucleic acids

**33.** Use data from the diagram at right to explain why DNA nucleotide sequencing is important to the study of evolution.

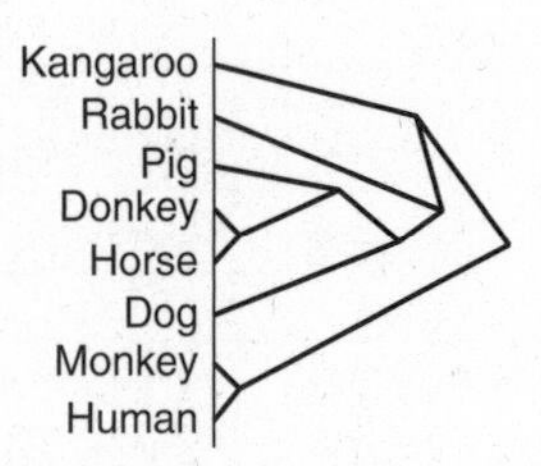


**34.** How do the nucleotides of the DNA molecule allow it to replicate?

**35.** Briefly describe the process of DNA replication. Your answer should include the following terms (but not necessarily in this order):

- template
- enzymes
- subunits

*Base your answers to questions 36 and 37 on the following chart, which provides information about heredity, and on your knowledge of biology.*

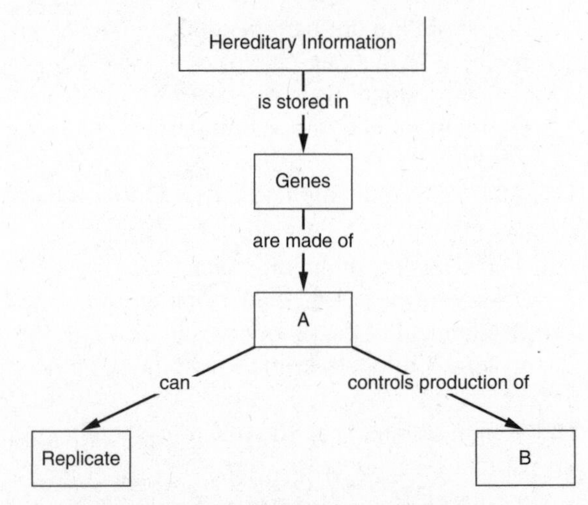


**36.** The molecule that is represented by box A serves as a template. Identify the type of molecule and explain how it is a template.

**37.** Which types of molecules are represented by box B?

1 bases
2 proteins
3 lipids
4 sugars

**38.** How would you explain to someone who has never heard of DNA why it is such an important molecule?

**39.** Mutations can be helpful to organisms; yet people fear the effects of substances that can cause mutations. Explain how mutations can be both helpful and harmful.

***Part C—Reading Comprehension***

*Base your answers to questions 40 through 42 on the passage below and on your knowledge of biology.*

When making movies about dinosaurs, film producers have sometimes used ordinary lizards and enlarged their images thousands of times. We all know, however, that while they may look like dinosaurs and be related to dinosaurs, modern lizards are not actually dinosaurs.

Recently, some scientists have developed a hypothesis that challenges this view. These scientists suggest that some dinosaurs were actually the same species as some modern lizards that had grown to unbelievable sizes. They think that such growth might be due to a special type of DNA called *repetitive DNA*, often referred to as "junk" DNA because scientists do not understand its functions.

The scientists studied pumpkins that can reach sizes of nearly 1000 pounds and found them to contain large amounts of repetitive DNA. Other pumpkins that grow to only a few pounds in weight have very little of this kind of DNA. In addition, cells that reproduce uncontrollably have almost always been found to contain large amounts of repetitive DNA.

**40.** State *one* reason why scientists formerly thought of repetitive DNA as "junk."

**41.** Which kind of cells would most likely contain large amounts of repetitive DNA? Why?

1 red blood cells
2 cancer cells
3 nerve cells
4 skin cells

**42.** Which fact best supports the hypothesis that large amounts of repetitive DNA are responsible for increased sizes of organisms?

1 Lizards look very much like little dinosaurs.
2 Modern lizards may be related to dinosaurs.
3 Large pumpkins contain a lot of repetitive DNA.
4 Another term for repetitive DNA is "junk" DNA.

# 17
# Genes and Gene Action

**Vocabulary**

codon
gene
gene expression
genetic code
ribosomes

## GENES AND PROTEINS

Now that it has been shown that DNA is what makes up the genetic material, it is time to look more closely at *genes*. What is a gene? A **gene** is a specific segment along a DNA molecule, made up of linear sequences of nucleotide subunits. It is, in effect, a package of information that tells a cell how to make proteins. Likewise, proteins are polymers made up of linear sequences of amino acid subunits. As you learned already, there are 20 different types of amino acids. The order in which the amino acids are joined determines which protein is made. Every different protein has a unique sequence of amino acids. This sequence determines the shape of a protein molecule. It is the shape of the protein that allows the molecule to do its work in the cell.

Cells have to use the linear sequence of subunits in DNA to build a linear sequence of amino acids for a protein. Yet, in all cells except for bacteria, DNA is stored in the nucleus, and protein synthesis occurs outside the nuclear membrane, at the **ribosomes**. These small organelles are distributed throughout the cytoplasm. How does the genetic information contained in DNA within the nucleus get out to the ribosomes? A third type of molecule, called *ribonucleic acid*, or *RNA*, works as a helper to transmit the information. That is, the genetic information flows from the DNA to the RNA to a protein. (See Figure 17-1.)

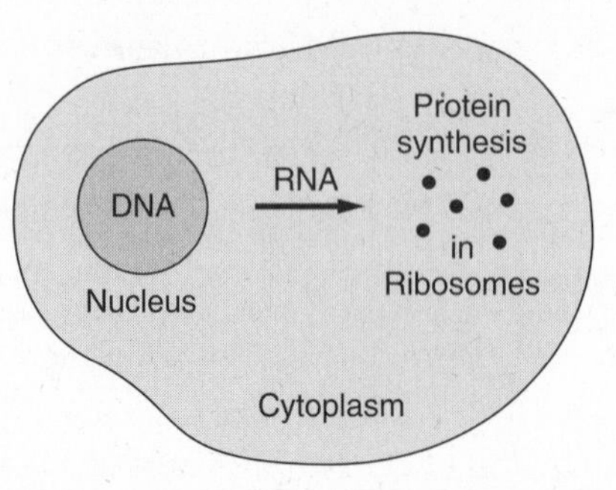


**Figure 17-1** The flow of genetic information in a cell—from DNA in the nucleus to RNA to amino acids at the ribosomes.

## FROM DNA TO RNA

Each gene is a portion of a chromosome, in effect a portion of the DNA chain. An RNA molecule called *messenger RNA* (mRNA) does the job of moving the information in the linear base sequence out to the ribosomes. DNA is copied into RNA in a process called *transcription*, which is similar to DNA replication.

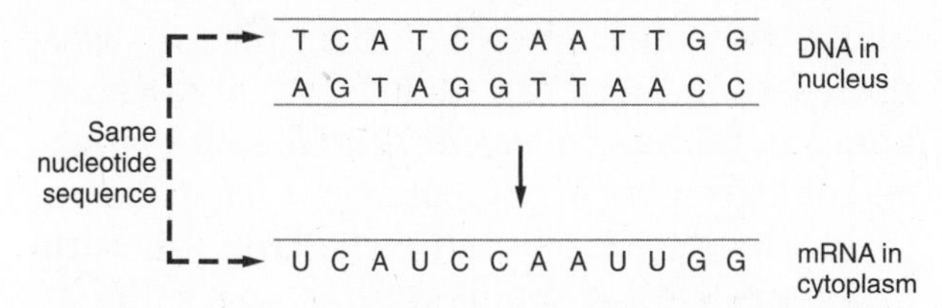


**Figure 17-2** The DNA sequence is copied into messenger RNA, which goes out to the ribosomes in the cytoplasm. *Note:* In RNA, the base uracil (U) substitutes for the DNA base thymine (T).

First the DNA double helix opens up where a particular gene is located. Special enzymes begin to match up RNA subunits with the correct DNA subunits. The new RNA molecule has the same base sequence as one strand of the original DNA (except that in RNA the base *uracil* substitutes for the DNA base thymine). This RNA molecule then goes out of the nucleus, through pores in the nuclear membrane, to ribosomes in the cytoplasm. (See Figure 17-2.)

## FROM RNA TO PROTEIN

So far, the genetic information, stored as a base sequence, has moved from the nucleus to the cytoplasm by using RNA. Another problem remains: how to use the nucleotide base sequence in the RNA to build a protein with the correct sequence of amino acids. This problem involves a change of "language," from the base-sequence language of RNA into the amino-acid language of proteins. This process is called *translation*, and it occurs at the ribosomes.

Built into every living cell in the world is a **genetic code**. It is called the *triplet code*. Each different combination of three bases (hence *triplet* code) makes up a word, called a **codon**. Each codon represents a specific amino acid. Each of the 20 amino acids has at least one codon, and most have more than one. This genetic code is universal; in other words, all organisms on Earth use the same genetic code. For example, the codon GCA stands for the amino acid alanine in all life-forms, from bacteria to trees to humans. This similarity among living things is good evidence that all organisms evolved from a common ancestral life-form in Earth's distant past. (See Figure 17-3.)

## MUTATIONS: A CLOSER LOOK

In Chapter 16, a *mutation* was defined as a change in the base sequence of a DNA molecule. The possible effects of a mutation can now be explained in terms of what you know about protein synthesis.

| First Position | Second Position: U | Second Position: C | Second Position: A | Second Position: G | Third Position |
|---|---|---|---|---|---|
| U | UUU, UUC } Phe<br>UUA, UUG } Leu | UCU, UCC, UCA, UCG } Ser | UAU, UAC } Tyr<br>UAA *Stop*<br>UAG *Stop* | UGU, UGC } Cys<br>UGA *Stop*<br>UGG Trp | U<br>C<br>A<br>G |
| C | CUU, CUC, CUA, CUG } Leu | CCU, CCC, CCA, CCG } Pro | CAU, CAC } His<br>CAA, CAG } Gln | CGU, CGC, CGA, CGG } Arg | U<br>C<br>A<br>G |
| A | AUU, AUC, AUA } Ile<br>AUG Met | ACU, ACC, ACA, ACG } Thr | AAU, AAC } Asn<br>AAA, AAG } Lys | AGU, AGC } Ser<br>AGA, AGG } Arg | U<br>C<br>A<br>G |
| G | GUU, GUC, GUA, GUG } Val | GCU, GCC, GCA, GCG } Ala | GAU, GAC } Asp<br>GAA, GAG } Glu | GGU, GGC, GGA, GGG } Gly | U<br>C<br>A<br>G |

**Figure 17-3** The amino acid triplet codes. Note that most amino acids are represented by more than one codon.

The order of bases in DNA determines the order of amino acids in proteins. In certain cases, a mutation in one subunit will change the triplet code, which in turn may make a change in an amino acid. If this change occurs in a body cell, then all other cells in the organism's body that reproduced (through mitosis) from that cell will have the same change. It is more important, however, if the mutation occurs in the DNA of a gamete. If that gamete fuses with another gamete in sexual reproduction, then the mutation will be inherited. The change in the DNA will be passed on to succeeding generations. The new organism will have the mutation, as will all offspring of that organism. This will be an inherited condition. If the mutation is harmful, the individual and its offspring may have a genetic disease.

## GENE EXPRESSION AND CELL DIFFERENTIATION

Chromosomes contain extremely long DNA molecules. Many genes are stretched out along these molecules. For example, it is estimated that there are about 25,000 different genes in human cells. After fertilization, every cell of a growing organism arises from the mitotic cell division of other cells. Through mitosis, every cell in our body has the same 46 chromosomes with the same DNA as the original fertilized egg cell.

You learned in Chapter 15 that there are different types of cells in our bodies. We have skin cells, muscle cells, bone cells, nerve cells, blood cells, and so on. If all these cells have the same DNA, why are they so different from each other? The answer is that only certain genes are used in certain cells. This gives the cell its own structure, enzymes, functions, and physical characteristics. A muscle cell contracts, a nerve cell transmits an impulse, and a skin cell helps form a flat, protective layer.

The use of specific information from a gene is called **gene expression**. It turns out that genes are actually few and far between on the DNA, with large sections of DNA in between that do not code for traits. While the noncoding DNA does not make proteins, it does

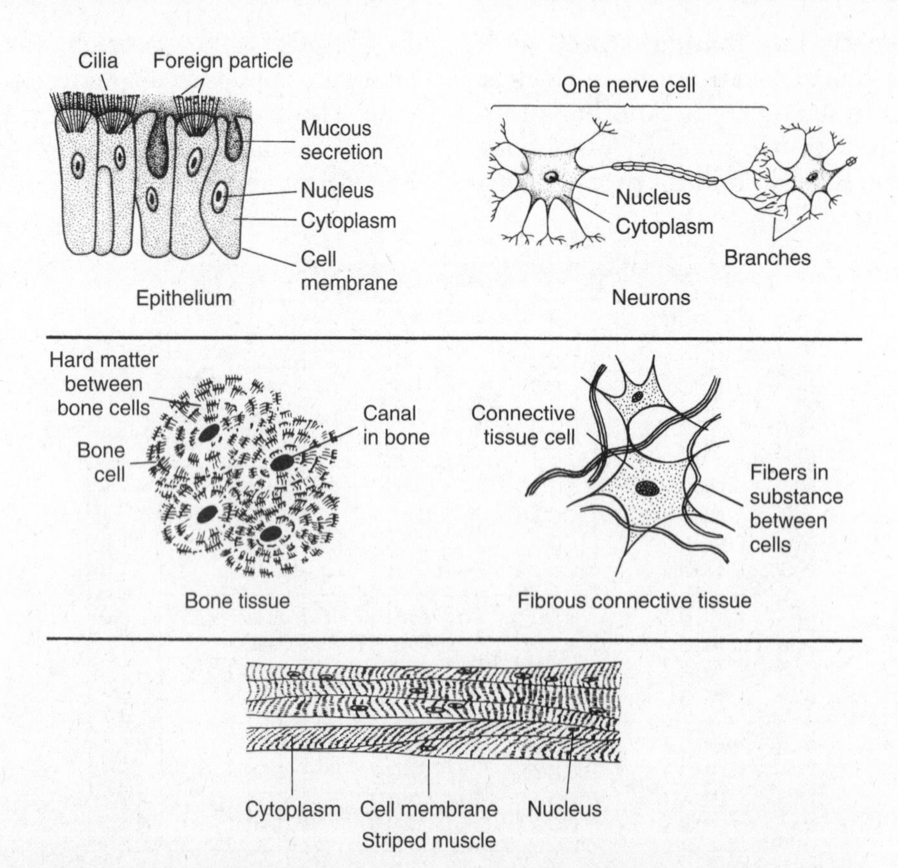


**Figure 17-4** Many different types of cells make up the human body. This cell differentiation results from differences in gene expression—only some genes are "turned on" to make the specific proteins needed for each cell type.

have a great deal to do with gene expression and regulation. Proteins are synthesized only from genes that are being expressed, or "turned on." All other genes in the cell are kept silent, or "turned off." Sometimes a particular trait is controlled by the expression of more than one gene. The process by which special types of cells are formed through controlled gene expression is called *cell differentiation*. This is an essential process of life. Without cell differentiation, we could not survive, because our bodies would be made up of only one type of cell. While the exact process is not known for certain, it is thought that environmental factors—both outside and inside each cell—influence gene expression. (See Figure 17-4.)

## Chapter 17 Review

### Part A—Multiple Choice

**1.** Genes can best be described as

1 directions for making DNA
2 directions for making proteins
3 subunits of proteins
4 molecules that transfer information out of the nucleus

**2.** Which path correctly describes the flow of information in cells?

1 DNA→RNA→protein
2 protein→RNA→DNA
3 protein→DNA→RNA
4 RNA→DNA→protein

**3.** The kinds of genes that an organism has are determined by the

1 type of amino acids in its cells
2 size of simple sugar molecules in its organs
3 sequence of the subunits A, T, C, and G in its DNA
4 shape of the protein molecules in its organelles

**4.** A change in the order of DNA bases that code for a respiratory protein will most likely cause

1 the production of a starch that has a similar function
2 a change in the sequence of amino acids determined by the gene
3 the digestion of the altered gene by enzymes
4 the release of antibodies by certain cells to correct the error

**5.** The role of messenger RNA is to

1 prevent mutations during DNA replications
2 match ribose-containing subunits to subunits of DNA
3 move the information in a base sequence out to the ribosomes
4 translate the base sequence at the ribosomes

**6.** RNA receives information from DNA by

1 binding with a double helix as a third strand
2 matching with subunits of a single strand of DNA
3 making an exact copy of the DNA molecule
4 accepting proteins through pores in the nuclear membrane

**7.** What happens at the ribosome?

1 The DNA strands separate.
2 RNA matches up with DNA strands.
3 Genetic information is mutated.
4 RNA is translated into amino acids.

**8.** The diagram below represents a process that occurs within a cell in the human pancreas. This process is known as

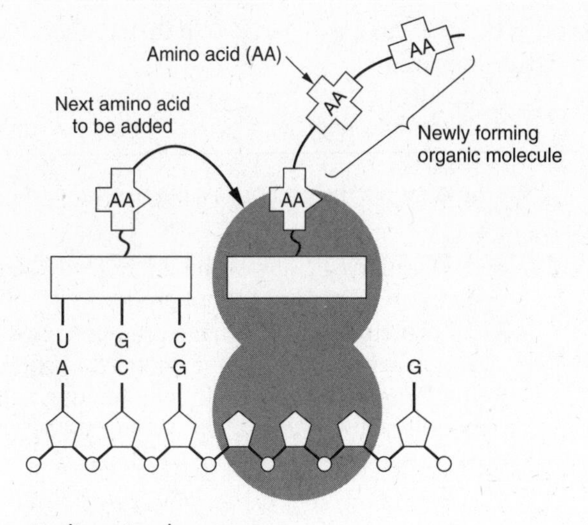


1 digestion by enzymes
2 energy production
3 protein synthesis
4 replication of DNA

**9.** How many bases make up a codon?

1 one
2 two
3 three
4 four

**10.** What does a codon represent?

1 a specific amino acid
2 a specific base
3 an RNA molecule
4 an enzyme

**11.** The genetic code is

1 different for every organism
2 the same for all organisms
3 constantly changing
4 impossible to identify

**12.** The sequence of amino acids in a protein is determined by the

1 speed at which translation occurs
2 size of the cell involved
3 number of ribosomes in a cell
4 order of bases in the DNA

**13.** The diagram below provides some information concerning proteins. Which phrase does the letter A represent?

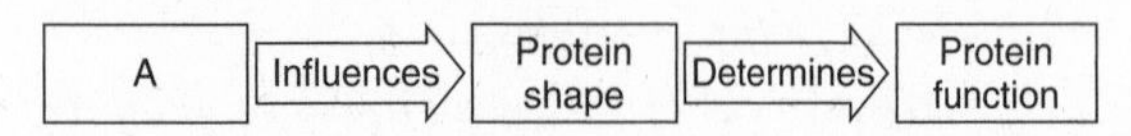


1 Sequence of amino acids
2 Sequence of starch molecules
3 Sequence of simple sugars
4 Sequence of ATP molecules

**14.** A mutation is always inherited if it

1 occurs in a gamete used in sexual reproduction
2 occurs in a cell that undergoes mitosis
3 gives the organism a better chance for survival
4 endangers the organism's chance for survival

**15.** People with cystic fibrosis inherit defective genetic information and cannot produce normal CFTR proteins. Scientists have used gene therapy to insert normal DNA segments that code for the missing CFTR protein into the lung cells of people with cystic fibrosis. Which statement does *not* describe a result of this therapy?

1 Altered lung cells can produce the normal CFTR protein.
2 The normal CFTR gene may be expressed in altered lung cells.
3 Altered lung cells can divide to produce other lung cells with the normal CFTR gene.
4 Offspring of someone with altered lung cells will inherit the normal CFTR gene.

**16.** About how many genes are contained on the 46 chromosomes in each human body cell?

1 5000
2 10,000
3 25,000
4 55,000

**17.** The cells that make up a person's skin have some functions that are different from those of the cells that make up the person's liver. This is because

1 all types of cells have a common ancestor
2 environment and past history have no influence on cell function
3 different cell types have completely different genetic material
4 different cell types use different parts of the genetic instructions

**18.** Gene expression means that

1 different genes are found in different cells
2 genes are passed through the nuclear membrane
3 only some genes are turned on in each type of cell
4 some cells have genes and some cells do not

**19.** Scientific studies have shown that identical twins who were separated at birth and raised in different homes may vary in height, weight, and intelligence. The most probable explanation for these differences is that

1 the original genes of each twin increased in number as they developed
2 the environments in which they were raised were different enough to affect the expression of their genes
3 one twin received genes only from the mother while the other twin received genes only from the father
4 the environments in which they were raised were different enough to change the genetic makeup of both individuals

**20.** After a series of cell divisions, an embryo develops different types of body cells, such as muscle cells, nerve cells, and blood cells. This development occurs because

1 the genetic code changes as the cells divide
2 different genetic instructions are created to meet the needs of new types of cells

3 different segments of the genetic instructions are used to produce different cell types
4 some sections of the genetic material are lost as a result of fertilization

### *Part B—Analysis and Open Ended*

**21.** What is so important about the order in which amino acids are joined?

**22.** The diagram below shows two different structures, 1 and 2, that are present in many single-celled organisms. Structure 1 contains protein A, but not protein B; and structure 2 contains protein B, but not protein A. Which statement is correct concerning protein A and protein B?

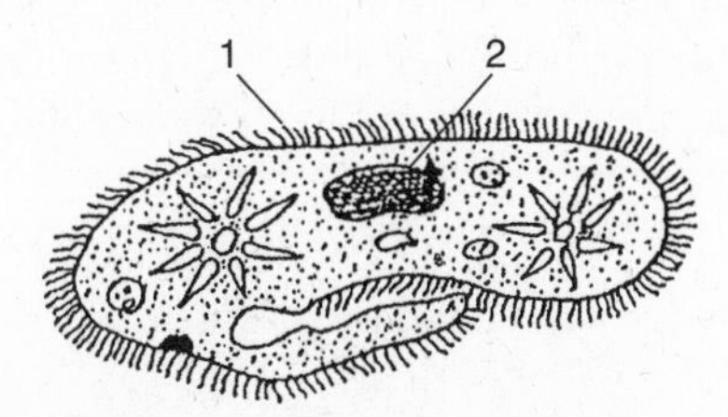


1 Proteins A and B have different functions and different amino acid chains.
2 Proteins A and B have the same function but a different sequence of bases (A, C, T, and G).
3 Proteins A and B have different functions but the same amino acid chains.
4 Proteins A and B have the same function and the same sequence of bases (A, C, T, and G).

**23.** The letters in the diagram at right represent genes on a particular chromosome. Gene B contains the code for an enzyme that cannot be synthesized unless gene A is also active. Which statement best explains why this can occur?

A
B
C
D
E

1 Some traits are controlled by the expression of more than one gene.
2 All the genes on a chromosome act to produce a single trait.
3 Genes are made up of double-stranded segments of DNA.
4 The first gene on each chromosome controls all the other genes.

**24.** Hemoglobin is a complex protein molecule found in red blood cells. Hemoglobin with the normal amino acid sequence can effectively carry oxygen to body cells. In the disorder known as sickle-cell anemia, one amino acid is substituted for another in the hemoglobin. One characteristic of this disorder is poor distribution of oxygen to the body cells. Explain how one change in the amino acid sequence of this protein could cause the results described.

**25.** In what way are the structures of DNA and protein similar?

**26.** Why is the location of DNA within the nucleus a potential problem for protein synthesis? How does RNA solve this problem for the cell?

**27.** Arrange the following structures from the largest to the smallest: DNA molecule; chromosome; nucleus; gene.

**28.** Explain how the process of copying DNA into messenger RNA is similar to DNA replication. Your answer should include the following terms:

- double helix
- enzymes
- subunits
- base sequence

**29.** Why must the bases be grouped into triplets in order to represent amino acids?

*Base your answers to questions 30 through 32 on the table below, which provides the DNA codes for several amino acids.*

| Amino Acid | DNA Code Sequence |
|---|---|
| Cysteine | ACA or ACG |
| Tryptophan | ACC |
| Valine | CAA or CAC or CAG or CAT |
| Proline | GGA or GGC or GGG or GGT |
| Asparagine | TTA or TTG |
| Methionine | TAC |

**30.** A certain DNA strand has the following base sequence: TACACACAAACGGGG. What is the sequence of amino acids that would be synthesized from this code (if it is read from left to right)?

**31.** Suppose the DNA sequence undergoes the following change: TACACACAAACGGGG→ TACACCCAAACGGGG. How would the sequence of amino acids be changed as a result of this mutation?

**32.** The original DNA sequence undergoes the following change: TACACACAAACGGGG→TACACACAAACGGGT. Why does this mutation produce *no change* in the function of the final molecule that is synthesized from this code?

*Refer to the flowchart below to answer questions 33 and 34.*

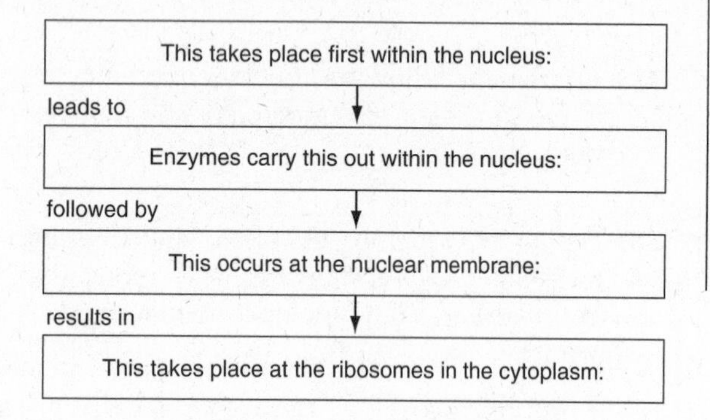


**33.** Use the following phrases to complete the steps listed in the flowchart boxes: a new RNA molecule moves out through pores; RNA base sequences translate into amino acid sequences; the DNA double helix opens up and unwinds; RNA subunits match up with DNA subunits.

**34.** The *best* title for this flowchart probably would be

1 How RNA Is Made
2 How DNA Is Made
3 How Proteins Are Made
4 How Ribosomes Are Made

**35.** How does RNA allow for translation from the genetic base sequences to the amino acid sequences?

**36.** Explain why the genetic code is referred to as "universal." What is the evolutionary significance of this fact?

**37.** In what way do ribosomes help in the process of translation?

**38.** How might a mutation in a DNA molecule result in a different protein?

**39.** Suppose the two cells in the diagram below come from the same body. How is gene expression related to cell differentiation? Do they have the same proteins? Do they have the same DNA?

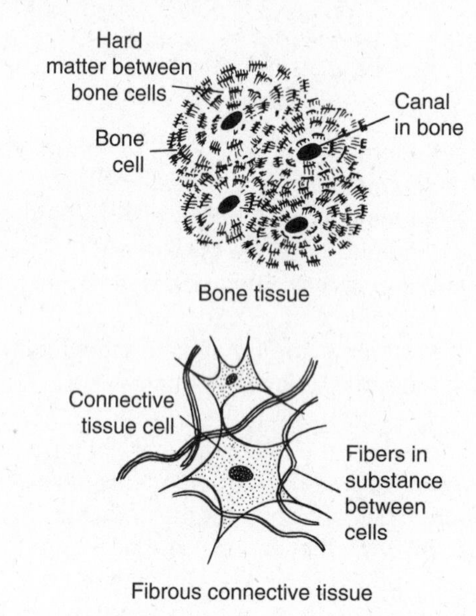


### *Part C—Reading Comprehension*

*Base your answers to questions 40 through 42 on the information below and on your knowledge of biology.* Source: *Science News*, Vol. 163, p. 54.

> An unusual study of the brains of women and girls who had received transplants of bone marrow from men indicates that marrow cells can transform into nerve cells. Researchers found that each female brain had nerve cells containing a Y chromosome, presumably derived from the transplanted bone marrow.
>
> Over the past several years, numerous research groups have reported that bone marrow, the source of a person's blood cells, can transform into cells of the skin, muscle, heart, liver, and even brain. These lab and animal studies have

raised hopes that bone marrow or cells derived from it could repair hearts, cure neurological disorders, and treat many other medical conditions.

Some investigators, however, have challenged the bone-marrow results. The stakes are high because of the politicized debate over whether adult stem cells, such as those in bone marrow, are as promising a therapeutic tool as stem cells derived from embryos are.

In an upcoming *Proceedings of the National Academy of Sciences*, Eva Mezey of the National Institute of Neurological Disorders and Stroke in Bethesda, Md., and her colleagues report their analysis of the brain tissue of two girls and two women. Each had received a bone-marrow transplant from a male donor in a futile attempt to treat her illness. Mezey's group exposed brain-tissue samples from the four females to a marker that attaches to a DNA sequence unique to a male's Y chromosome. The investigators also applied antibodies specific to nerve cells.

In each case, Mezey and her colleagues identified a small number of nerve cells with Y chromosomes. For example, one girl studied had received a bone-marrow transplant when she was 9 months old and died less than a year later. When researchers examined 182,000 of her brain cells, they found Y chromosomes in 519—and 19 of those males cells also displayed nerve cell markers.

Another research team's unpublished findings mirror Mezey's study. Last year, Martin Körbling of the University of Texas M.D. Anderson Cancer Center in Houston and his colleagues employed the same Y chromosome-based strategy to discover bone-marrow-derived skin, gut, and liver cells in a half-dozen women who had received marrow transplants before dying. Now, Körbling tells *Science News*, "we have data showing similar results in midbrain and cortex tissue."

Diane Krause of Yale University notes that her research team and many others are vigorously studying the mechanisms by which bone-marrow cells may transform into cells other than blood cells. Unless researchers can enhance the pace of this natural cellular makeover, the phenomenon is unlikely to be of much medical use, both she and Mezey caution.

**40.** State the basic research question about transplanted bone-marrow cells that was being discussed in this report.

**41.** Explain why brain tissue from two girls and two women was being analyzed by the researchers in Bethesda, Maryland.

**42.** State the conclusions made by the researchers about the transplanted bone-marrow cells in these four individuals.

# 18 Patterns of Inheritance

## Vocabulary

alleles
genetic cross
inheritance
selective breeding

## GREGOR MENDEL: THE FOUNDER OF GENETICS

How characteristics are passed from parents to offspring was a question that puzzled people for thousands of years. A set of experiments completed almost 150 years ago by Gregor Mendel helped our understanding of **inheritance**, that is, how traits are passed from one generation to the next. Mendel, an Austrian monk, conducted hundreds of experiments using thousands of pea plants. By applying careful mathematical analysis to his work—something that had rarely been done in biology before—Mendel discovered much about the way heredity works. The study of inheritance really began with Mendel; he can rightly be called the "founder of the science of genetics." (See Figure 18-1.)

Mendel discovered that hereditary information is passed from parents to offspring in individual units that he called *factors*, which we now call genes. He also discovered that the factors were passed on in specific, predictable patterns from both parents. These patterns of inheritance are described in Mendel's laws.

Mendel identified specific characteristics, or *traits*, in his pea plants. The characteristics he identified appeared in two different, contrasting versions. For example, some of the pea plants Mendel worked with had purple flowers; other plants had white flowers. Some of the plants produced yellow peas; others pro-

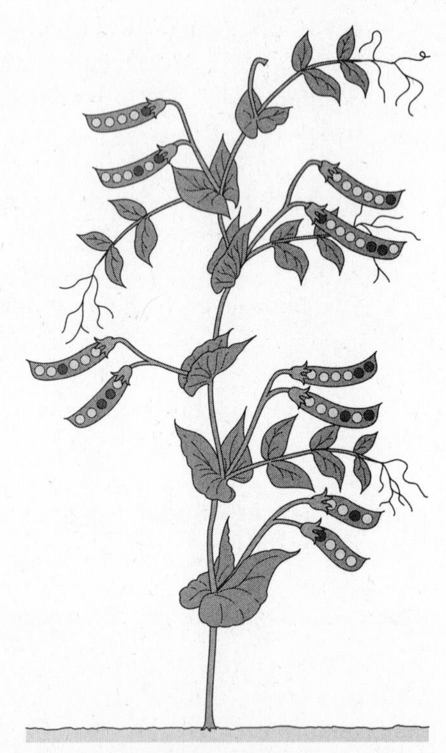

**Figure 18-1** Gregor Mendel investigated the inheritance of traits from one generation to the next by carefully studying characteristics of the common pea plant.

duced green peas. The shape of the peas varied, too; they were either smooth and round, or wrinkled. In addition, the pea plants could be either tall or short. The characteristics that Mendel studied occurred as contrasting traits that were easy to observe. He would breed two plants that had contrasting traits to better understand how they were inherited. This type of experiment, in which two organisms with different traits are bred, is called a **genetic cross**.

## MENDEL'S IDEAS: THE LAWS OF INHERITANCE

Mendel proposed several ideas that explained his results. These ideas were correct, even though Mendel knew nothing about chromosomes, genes, or DNA. Since that time, scientists have been able to confirm Mendel's ideas and combine them with what has been learned about genetics.

Mendel's first main idea was that each characteristic, or trait, exists in two versions. These two forms of a gene are called **alleles**. For example, in pea plants, there are two alleles for the gene for flower color: one allele for purple flowers and one allele for white flowers. Genes exist at specific locations on chromosomes. So, one chromosome may have an allele for purple at the location for flower color, while another chromosome may have an allele for white flowers at the same spot. The DNA at these locations consists of a sequence of subunits. At one allele, the DNA subunits code for proteins that result in the color purple. At the other allele, a different sequence of DNA subunits codes for proteins that make the color white.

Mendel's second main idea was that for each characteristic, an individual inherits two alleles (that is, versions), one from each parent. We now know that all offspring—plant or animal—that result from sexual reproduction have a double set of chromosomes made up of chromosome pairs. Each pair of chromosomes consists of one chromosome from the mother and one chromosome from the father. Since corresponding genes (such as for flower color) occur at the same place on each chromosome in a pair, any particular gene exists twice in each cell. Thus, every cell has two alleles for each gene. (See Figure 18-2.)

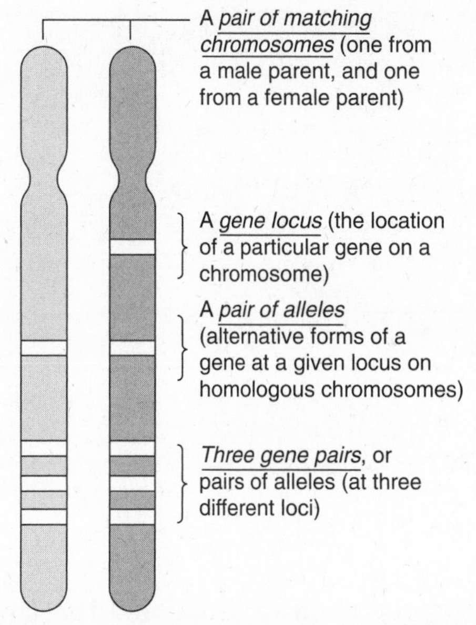


**Figure 18-2** The gene for one trait, such as flower color, exists at the same place on each member of a pair of matching chromosomes. Each gene has two versions, or alleles, of that trait, for example, either purple or white for flowers.

Mendel next explained the results obtained by crossing plants that show the two different alleles. The one that appears in the offspring is the *dominant* allele. The one that does not appear in the offspring is the *recessive* allele. For example, the allele for purple flower color is dominant and the allele for white flower color is recessive. Thus, in a cross between a pure plant with purple flowers and a pure plant with white flowers, only purple flowers will show up in the offspring.

A dominant allele usually codes for an enzyme that works, for example, to make the purple color. A recessive allele usually codes for a form of the enzyme that does not work. As long as one allele for purple flowers is present in a plant, some of the enzyme will be produced to make the purple color.

## GENE EXPRESSION

You already know that the genes of an organism determine its characteristics. Yet the environment in which an organism lives can also affect the way its genes are expressed. The fur coloration pattern of Siamese cats shows this

Figure 18-3 The expression of the gene that codes for dark-colored fur in Siamese cats depends on temperature—the parts of the body that have a lower (cooler) temperature allow that gene to be expressed. The rest of the body, which is warmer, has light fur.

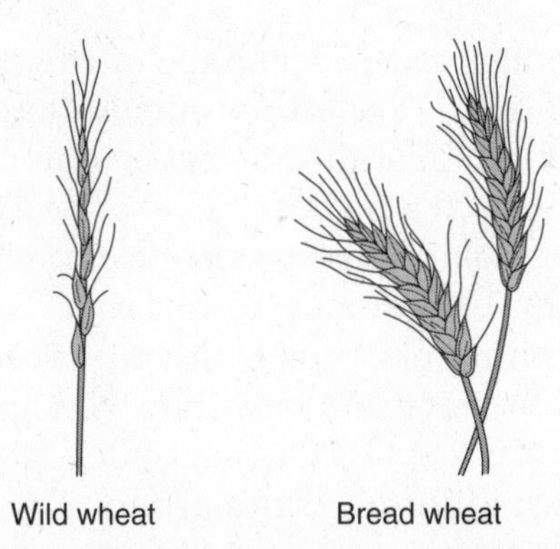


Figure 18-4 As a result of centuries of selective breeding, the small kernels of wild wheat have been transformed into the large kernels of bread wheat, which is a more useful crop for people.

type of interaction of genes and the environment. Siamese cats have a gene that codes for an enzyme that produces dark fur. However, this enzyme works only at cool temperatures. Most of a cat's body is too warm for this enzyme to work. But you can easily identify the cooler areas where the enzyme does work—the dark ears, paws, face, and tail that are typical of a Siamese cat. Even though all parts of a cat's body have the same combination of genes, the way that the genes are expressed differs from one part to another because of different environmental influences. Thus, the environment frequently affects the final appearance of an organism. (See Figure 18-3.)

## PLANT AND ANIMAL BREEDING

For centuries, people have chosen to breed plants and animals that had desirable traits. This is called **selective breeding**. By allowing only those organisms to reproduce, traits such as size, shape, and color have been altered over time.

The breeding of plants and animals has been greatly helped by the discoveries of Mendel and later geneticists. Often, breeders can identify the exact traits in which they are interested and then prepare the best genetic crosses. For example, plant breeders have produced new crops that are resistant to disease, can live in new environments, grow more plentifully, and look and taste better. (See Figure 18-4.)

People have bred animals for many different purposes, too. For example, sheep have been bred to produce more wool of better quality, turkeys have been bred to produce more white meat, chickens to lay larger eggs, cows for more milk, and pigs to produce meat that contains less fat.

# Chapter 18 Review

### *Part A—Multiple Choice*

**1.** Mendel studied inheritance patterns in

1 pink roses
2 fruit flies
3 Siamese cats
4 pea plants

**2.** Mendel is credited with

1 discovering the structure of DNA
2 beginning the science of genetics
3 recognizing the role of RNA
4 beginning animal-breeding programs

**3.** An allele is a

1 version of a gene
2 specialized enzyme
3 subunit of DNA
4 three-base code

**4.** One of Mendel's ideas was that

1 alleles are responsible for mitosis
2 alleles can cause mutations
3 there are two alleles for each trait
4 each gene exists as only one allele

**5.** In the plants Mendel studied, one allele produces purple flowers while the other produces white flowers because

1 it depends on how the sunlight reflects off the flowers' petals
2 the DNA subunits at those alleles code for two different proteins
3 the DNA subunits at those alleles code for different carbohydrates
4 it depends on how hot or cold the environment is when they bloom

**6.** According to Mendel, for each trait inherited, offspring receive

1 just one allele per cell
2 one allele from each parent
3 two alleles from each parent
4 several pairs of alleles

**7.** How are an organism's traits related to the environment?

1 An organism inherits different genes depending on the environment.
2 Genetic information is never affected by the environment.
3 The environment can affect the expression of genetic traits.
4 The environment affects genetic traits only in wild organisms.

**8.** In a particular variety of corn, the kernels turn red when exposed to sunlight. In the absence of sunlight, the kernels remain yellow. Based on this information, it can be concluded that the color of these corn kernels is due to

1 a different type of DNA that is produced when sunlight is present
2 the effect of sunlight on the number of chromosomes inherited
3 a different species of corn that is produced only in sunlight
4 the effect of the environment on gene expression in the corn

**9.** In Siamese cats, the fur on the ears, paws, tail, and face (that is, the extremities) is usually black or brown, while the rest of the body fur is almost white. If a Siamese cat stays indoors, where it is warm, it may grow fur that is almost white on its extremities. In contrast, if a Siamese cat mostly stays outside, where it is cold, it will grow fur that is quite dark on its extremities. The best explanation for these changes in fur color is that

1 an environmental factor influences the expression of this inherited trait
2 skin cells that produce pigments have a higher mutation rate than other cells
3 the location of pigment-producing cells determines the DNA code of the genes
4 the alleles for fur color are mutated by interactions with the environment

**10.** The diagram below represents the change in a sprouting onion bulb when sunlight is present and then when sunlight is no longer present. Which statement best explains this change?

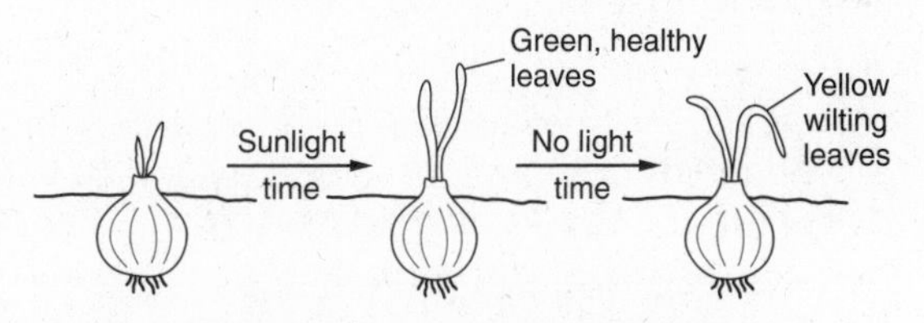


1 Plants need carbon dioxide to survive.
2 Plants produce hormones for growth.
3 Environmental conditions never affect genetic traits.
4 The environment can influence the expression of traits.

**11.** Fruit flies with the curly-wing trait will develop straight wings if kept at a temperature of 16°C during development and curly wings if kept at 25°C. The best explanation for this change in the shape of wings is that the

1 genes for curly wings and for straight wings are found on different chromosomes
2 outside environment affects the expression of the genes for this trait
3 type of gene present in the fruit fly is dependent on environmental temperature
4 lower outside temperature always produces the same genetic mutation

**12.** To produce large tomatoes that are resistant to cracking and splitting, some seed companies use the pollen from one variety of tomato plant to fertilize a different variety of tomato plant. This process is an example of

1 selective breeding
2 direct harvesting
3 DNA sequencing
4 plant cloning

**13.** Mendel experimented by carrying out selective breeding and

1 natural selection
2 mathematical analysis
3 molecular selection
4 animal husbandry

**14.** Research applications of the basic principles of genetics have contributed greatly to the rapid production of new varieties of plants and animals. Which activity is an example of such an application?

1 testing new chemical fertilizers on food crops
2 developing new irrigation methods to conserve water
3 selective breeding of crops that show a resistance to disease
4 using natural predators to control insect pests

**15.** Which process has been used by farmers for hundreds of years to develop new animal varieties?

1 genetic cloning
2 DNA splicing
3 genetic engineering
4 selective breeding

**16.** When humans first domesticated dogs, there was very little physical diversity in the species. Today there are many varieties, such as the German shepherd and the Boston terrier. This increase in diversity is most closely associated with

1 cloning of selected body cells
2 years of mitotic cell division
3 selective breeding for desirable traits
4 environmental influences on inherited traits

### *Part B—Analysis and Open Ended*

**17.** Even though Mendel did not know about genes, he can be called the "founder" of the science of genetics. Why?

**18.** In one sentence, tell what Mendel noticed about the "factors" that were passed from parents to offspring among pea plants.

*Refer to the figure below to answer questions 19 and 20.*

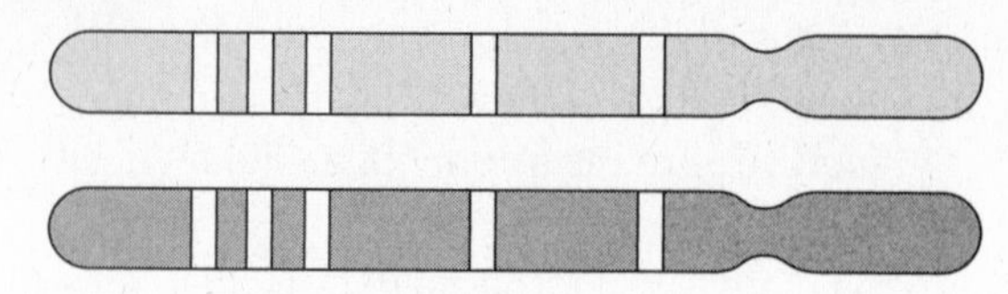

**19.** The two main structures represent

1 the pods of two different pea plants
2 one pair of genes from a pea plant
3 one pair of matching chromosomes
4 five mutations in the genes of a trait

**20.** The figure illustrates that

1 every chromosome has five separate genes on it
2 Mendel's pea pods always grew in matching pairs
3 several pairs of alleles are needed to determine every trait
4 alleles for a trait are at the same spot on matching chromosomes

**21.** How are Mendel's ideas about "factors" explained by what we now know about genes and chromosomes?

**22.** Explain how fur color in Siamese cats demonstrates an important fact about the expression of genes. In what way might this be an adaptive feature?

*Refer to the following figures and paragraph to answer question 23.*

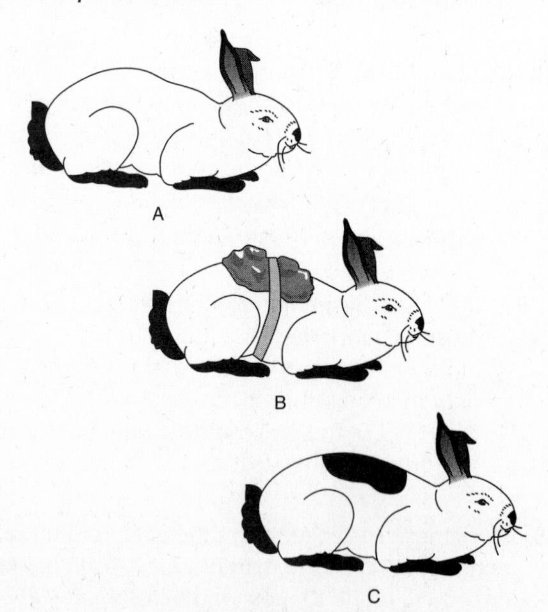

The normal color pattern of a Himalayan rabbit's fur is shown in Figure A. In Figure B, the rabbit is shown with the fur shaved from an area on its back and an ice pack applied to that area. Figure C shows the same rabbit after new fur has grown in the shaved area.

**23.** The best explanation for this change in the rabbit's color pattern is that the low temperature of the ice pack

1 caused a genetic mutation in the fur
2 deleted the gene for white-colored fur
3 allowed the gene for dark fur to be expressed
4 had no impact; it was due to a change in diet

**24.** List five examples in which humans have altered the characteristics of plants or animals by preparing particular genetic crosses.

**25.** For many years, people have used a variety of techniques to influence the genetic makeup of organisms. These techniques have led to the production of new varieties of organisms that possess characteristics that are useful to humans. Identify *one* technique presently being used to alter the genetic makeup of an organism, and explain how people can benefit from this change. Your answer must include:

- the name of the technique used to alter the genetic makeup
- a brief description of what is involved in this technique
- *one* specific example of how this technique has been used
- how humans have benefited from the production of this new variety of organism

**26.** When people carry out specific genetic crosses (such as those done by Mendel), are they working with organisms that reproduce sexually or asexually? Explain.

### *Part C—Reading Comprehension*

*Base your answers to questions 27 through 29 on the information below and on your knowledge of biology.* Source: *Science News*, Vol. 163, p. 350.

> Canadian researchers have demonstrated that, in principle, they can engineer genetically modified (GM) crops to be incapable of breeding with conventional crops or wild relatives. The new approach could help contain the unintended spread of artificial traits, which is a major source of public concern about GM crops.
>
> Although several alternative strategies for such containment exist, "there is no perfect solution," says Johann P. Schernthaner of Agriculture and Agri-Food Canada in Ottawa.
>
> To engineer a crop that would theoretically require no intervention by farmers to keep it reproductively contained, Schernthaner and his colleagues inserted into some tobacco plants a genetic element called a seed-lethal trait. That element prevents the plants' seeds from germinating under any circumstances.
>
> To enable the GM plants to reproduce amongst themselves, the researchers then inserted another artificial trait that represses the seed-lethal construct. GM plants with both traits develop and reproduce normally, the researchers report in an upcoming *Proceedings of the National Academy of Sciences*.
>
> To create an inherently containable GM crop, the researchers suggest that a different seed-lethal construct be placed on each member of a pair of chromosomes, so the constructs wouldn't be inherited together. Each chromosome would also receive a repressor trait that inactivates the seed-lethal construct on the other chromosome. That way, unintended crossings of the GM crop with related plants should produce nonviable seeds because they'd contain only one of the two engineered chromosomes.
>
> The new approach faces several potential problems, says Henry Daniell of the University of Central Florida in Orlando. For example, it's still possible that with certain chromosomal rearrangements, both the seed-lethal and the repressor traits could spread to a non-GM plant.

**27.** State the major concern that exists about growing genetically modified (GM) crops with artificial traits in open fields.

**28.** Explain the purpose of the first seed trait that was inserted into the tobacco plants' genes.

**29.** Describe the changes made to other chromosomes on the GM plants that allow them to reproduce among themselves.

# 19
# Human Genetics

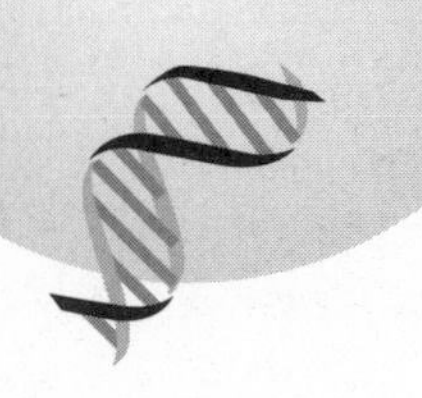

## Vocabulary

autosomes
carrier
karyotype
linkage
polygenic inheritance
sex chromosomes
sex-linked

## GENETIC COMPLEXITIES: BEYOND THE PEA PLANTS

Mendel's experiments provided the foundation for the study of inheritance. Much of Mendel's success was due to the traits he studied. The pea plants he worked with were either tall or short, and their seeds were either yellow or green, smooth or wrinkled, and so on. But are people either tall or short? Are there only two skin colors? Do people have only blond or black hair? Of course not. In humans, and in most organisms, almost all traits are not as clearly defined as the traits Mendel studied in pea plants. Yet Mendel was correct in his explanations.

The traits Mendel picked to study were, luckily for him, each determined by single genes. The height of pea plants is due to a single gene that occurs in just two different versions, or alleles. Human height, however, is a different story. Height in humans is determined by several genes. Pieces of DNA on different chromosomes code for the proteins that affect human height. A trait that is determined by several genes is said to follow a pattern of **polygenic inheritance**. If a large group of people were arranged according to height, those with average height would be the most numerous. There would be fewer extremely short and extremely tall people. Grouping individuals in this way produces a bell-shaped curve. (See Figure 19-1.)

**Figure 19-1** When a group of men are arranged by height, a bell-shaped curve is produced, which shows variation within a population.

There are other traits that can be determined by more than one gene. For example, the many variations in human skin color show that multiple genes are involved in this trait, too.

## LINKED TRAITS

In some cases, the genes for one type of trait are inherited along with the genes for another particular type of trait. Such traits are said to exhibit **linkage**, because they are located on the same chromosome. For example, why do most people with red hair also have freckles? The reason is that the genes for red hair and for freckles are on the same chromosome. These genes are linked, so they tend to be inherited together. Now we will examine more closely the types of chromosomes found in human body cells.

## IS IT A BOY OR A GIRL?

Scientists can take a photograph of the chromosomes in a human cell by using a camera attached to a microscope. The chromosomes are paired up and then numbered from largest to smallest. The picture that results is called a **karyotype**. If we examine a human karyotype, we find 22 perfectly matched pairs of chromosomes. These chromosome pairs, numbered from 1 to 22, are called **autosomes**. This accounts for 44 of the 46 chromosomes found in the cell. However, the last two chromosomes do not always make a matched pair. It depends on whether the cell came from a male or a female.

If the cell came from a male, the last two chromosomes will not be an identical match. Cells from males contain an X and a much smaller Y chromosome. If the cell came from a female, the last two chromosomes will both be X chromosomes, a matched pair. Differences in the last pair of chromosomes—whether a person has an X and a Y chromosome or two X chromosomes—determine the person's sex. Therefore, these last two chromosomes are called the **sex chromosomes**. (See Figure 19-2.)

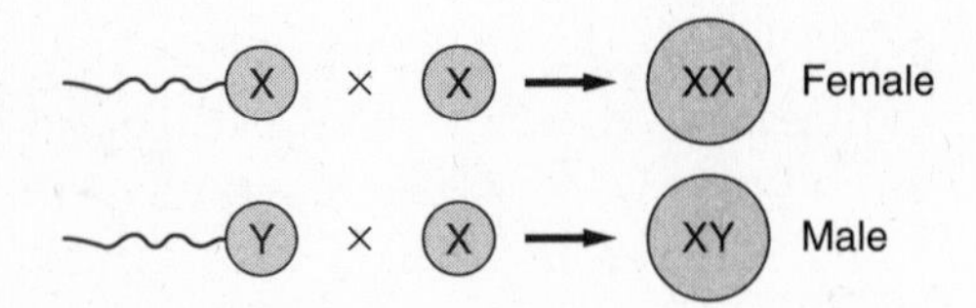


**Figure 19-2** The X and Y chromosomes are called the sex chromosomes. The sex of the offspring is determined by the sperm cell.

Egg cells are produced in the mother via meiosis. Her cells have 22 pairs of autosomes and one pair of X's (her sex chromosomes), for a total of 46 chromosomes. Through meiosis, every egg cell gets one of each of the 22 autosomes and one X chromosome. Now consider sperm cells. Every cell in the father's body has 22 pairs of autosomes and an X and a Y chromosome. During meiosis, 50 percent of his sperm cells get 22 autosomes and an X chromosome, and 50 percent get 22 autosomes and a Y chromosome. Egg cells have only X chromosomes in addition to the autosomes. So who determines the sex of the child? The father does. If a sperm with a Y chromosome fertilizes the egg, the zygote is XY, and a male develops. If a sperm with an X chromosome fertilizes the egg, the zygote is XX, and a female develops.

## SEX-LINKED TRAITS AND GENETIC DISORDERS

Genetic disorders occur in some infants. These disorders are not caused by infectious microorganisms. Instead, genetic disorders result from inborn errors that are caused by defects in genes; that is, gene mutations that are inherited from parents. Red-green color blindness (an inability to distinguish the colors red and green) occurs most often in males. Hemophilia, a disease in which blood does not clot properly, also occurs most often in males. Duchenne muscular dystrophy, a disease that slowly destroys muscles, occurs primarily in males, too. What accounts for this pattern?

These traits are all **sex-linked**. They occur when a particular allele with a defective portion of DNA is present on the X chromosome. The much smaller Y chromosome lacks alleles for these three traits. A female who has a recessive defective allele on one X chromosome but not on the other X chromosome is called a **carrier**. These females carry the genetic disorder, but they do not actively have the condition it causes (since one X chromosome has the "healthy" allele). However, they can pass the condition on to their offspring. (A female will have the condition only if the defective allele is on both of her X chromosomes.)

## OTHER GENETIC DISORDERS

Genetic disorders can result from an abnormal number of chromosomes. As described in Chapter 13, unusual events can happen during meiosis as the paired chromosomes separate. Abnormal chromosome numbers cause disorders that can be detected prior to or at birth. For example, Down syndrome is due to an extra chromosome number 21. (See Figure 19-3.)

Sickle-cell anemia is a disorder that is due to a single gene defect, not an abnormal chromosome number. A mutation in the DNA base sequence of the gene for hemoglobin causes this disease, which reduces the ability of red blood cells to carry oxygen. This is a recessive trait, so a person will have the disease if they inherit two copies of the defective allele; that is, one from each parent. (See Figure 19-4.)

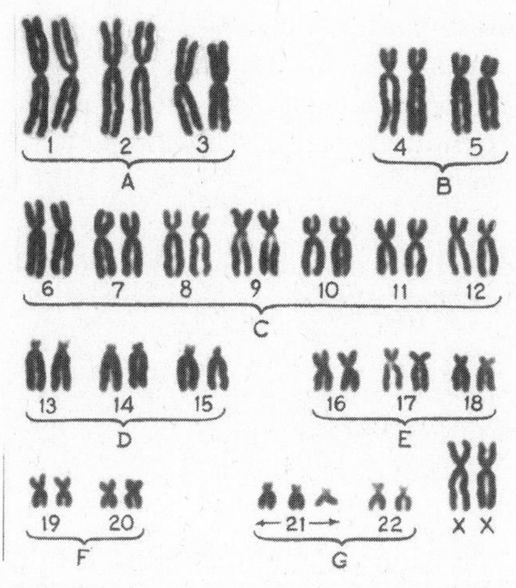


**Figure 19-3** The karyotype of a person with Down syndrome shows that the disorder is caused by inheritance of three copies of chromosome 21.

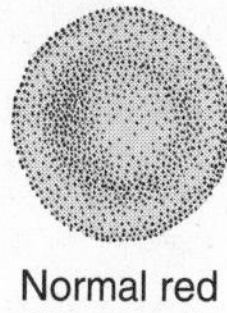


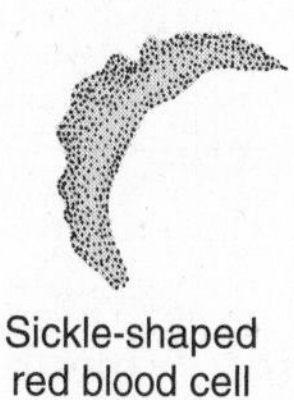


**Figure 19-4** The sickle-shaped red blood cell is characteristic of sickle-cell anemia—a genetic disorder caused by a mutation in the base sequence of the gene for hemoglobin.

Phenylketonuria is one of the most studied genetic disorders. Like sickle-cell anemia, it is a recessive trait; two copies of the allele must be inherited to cause the disorder. The allele for this condition prevents a newborn from producing the enzyme that breaks down the amino acid phenylalanine. As a result, phenylalanine builds up in the baby's blood, which interferes with the development of the brain, causing mental retardation. Fortunately, some genetic disorders can be treated. Routine tests for this disease are now done on newborns. With such early detection, the baby's diet can be changed to prevent the disease's effects from developing, and the baby can lead a normal life.

## FAMILY HISTORY AND GENETIC DISORDERS

Now that the human genome has been decoded, new and powerful information-processing tools are being used to locate genes that cause various human genetic disorders. As we learn more about these disorders, we realize that there are risks in having children. To assess the risks, people must have information about their family's medical history. A man and a woman who want to become parents can go to a trained genetics counselor for help in determining their risks of having a child with a genetic disorder. A genetics counselor prepares a chart showing the occurrence of any genetic disorders in past generations of the couple's families. Such a chart showing a person's family history for a particular trait can be analyzed, and patterns of inheritance can be determined. This information helps prospective parents make informed decisions about having children.

It is now is possible to learn if a baby has certain genetic disorders before birth. The presence of some genetic disorders can be determined by *prenatal* (meaning "before birth") tests. Biochemical tests can be done to show the presence of a genetic disorder. More often today, the DNA of a fetus is studied directly to see if something is abnormal in its genes (for example, by examining its karyotype). Other prenatal tests are also possible. Sound waves (ultrasound technology) can be used to make images of the fetus (while it remains safely in its mother's uterus) in order to see if it has any physical abnormalities. (See Figure 19-5.)

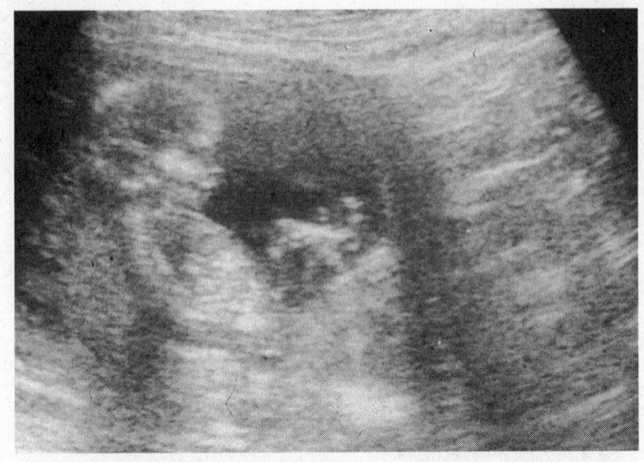

**Figure 19-5** This ultrasonograph shows the outline of a developing fetus. (*Note:* The fetus is facing the center of the image.) This prenatal test is used to see if the fetus has any physical abnormalities.

## Chapter 19 Review

### Part A—Multiple Choice

**1.** The traits Mendel studied were particularly useful because they were

1 inherited on one chromosome
2 determined by single genes
3 carried by just a few genes
4 carried by several genes

**2.** Unlike height in pea plants, human height is determined by

1 a single gene
2 several genes
3 two alleles
4 proteins instead of genes

**3.** A bell-shaped curve for height indicates that most people are

1 short
2 tall
3 either very short or very tall
4 average height

**4.** People with red hair also have freckles because these traits are

1 very common in people
2 found in all chromosomes
3 on the same chromosome
4 on two linked chromosomes

**5.** The term *karyotype* refers to a

1 group of similar alleles
2 photograph of chromosome pairs
3 cross between two plants or animals
4 pair of traits that are linked

**6.** Autosomes differ from sex chromosomes in that

1 autosomes do not occur in pairs
2 there are 22 sex chromosomes
3 autosomes are always perfectly matched
4 sex chromosomes are always perfectly matched

**7.** In humans, a cell that has an X chromosome and a Y chromosome

1 comes from a male
2 comes from a female
3 causes Down syndrome
4 is missing a chromosome

**8.** The following diagram represents the organization of genetic information within a cell nucleus. The circle labeled Z most likely represents the

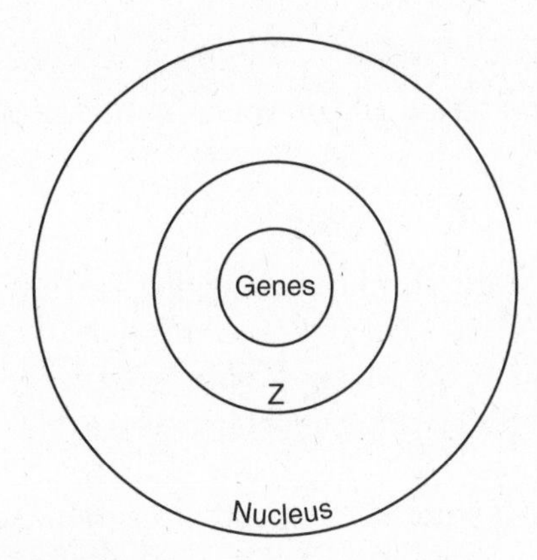


1 amino acids
2 vacuoles
3 chromosomes
4 molecular bases

**9.** In humans, a zygote will develop into a male if

1 a sperm with an X chromosome has fertilized an egg cell
2 a sperm with a Y chromosome has fertilized an egg cell
3 an egg with a Y chromosome has been fertilized by any sperm cell
4 the egg cell and sperm cell both have Y chromosomes

**10.** Couples may consult with a genetics counselor to

1 map the genes for their future children
2 determine the exact genes that their children will have
3 determine patterns of inheritance and risks of disorders
4 correct all genetic disorders before their child is born

**11.** Inborn genetic disorders can result from all of the following *except*

1 an abnormal chromosome number
2 a single gene defect
3 an infectious microorganism
4 a mutation in a DNA base sequence

### *Part B—Analysis and Open Ended*

**12.** How did Mendel's choice of pea plants contribute to his scientific success?

**13.** Why do traits such as human height show a very wide range of variations?

**14.** Explain why red hair and freckles are most often inherited together.

**15.** The sex chromosomes are identified for each cell in the figure below. Copy the figure and correctly label each cell using the following terms: *Sperm cell; Egg cell; Male zygote; Female zygote.*

X × X → XX

Y × X → XY

**16.** Discuss the importance of the sex chromosomes. Your answer should explain the following:

- In what way are the two sex chromosomes different from the autosomes?
- How do chromosomal differences determine whether a person is male or female?
- In what way is the father responsible for determining the sex of a child?

**17.** How can a chart that shows a family's medical history be used to assist a couple that is thinking about having children?

**18.** Sickle-cell anemia is due to a single gene defect that affects hemoglobin. Use your knowledge of genetics to explain how one defect in a DNA base sequence could upset production of this protein.

**19.** Why is early detection important for treating the disease phenylketonuria?

**20.** Refer to Figure 19-5 on page 156 to answer this question. The image of the fetus was formed by a diagnostic method that uses

1 X rays
2 light rays
3 sound waves
4 chemical tests

**21.** Why is important for a couple that has a family history of a genetic disorder to seek a prenatal diagnosis?

**22.** How can sound waves be used for prenatal diagnosis? What can they show?

### *Part C—Reading Comprehension*

*Base your answers to questions 23 through 25 on the information below and on your knowledge of biology.* Source: *Science News*, Vol. 162, p. 254.

> Rumors of the human Y chromosome's eventual death may be exaggerated. David Page of the Whitehead Institute for Biomedical Research in Cambridge, Mass., and his colleagues have identified a means by which the Y chromosome may forestall, or at least delay, the gradual degradation that some biologists argue will ultimately delete it from the human genome.
>
> In the Feb. 28 *Nature*, two Australian scientists summarized recent research showing that the human Y chromosome gradually accumulates mutations that deactivate its few genes. Non-sex chromosome pairs have a chance to replace mutation-ridden sections by swapping DNA with each other. But most of the Y is unable to exchange DNA with its partner, the X chromosome. "At the present rate of decay, the Y chromosome will self-destruct in around 10 million years," the two researchers said.
>
> Page's group isn't so sure. These researchers have sequenced all the DNA on the Y [chromosome] and discovered that many of its genes have neighboring doubles. A gene and its copy often form DNA palindromes, in which a gene's DNA sequence is followed by nearly the same sequence, but in reverse. (Palindromes in literature are words or phrases that read the same forward or backward, such as, "Was it a rat I saw?").
>
> Further scrutiny of the Y palindromes led the researchers to conclude that each gene and its backward copy are so similar in sequence that they must regularly exchange DNA in a novel form of the recombination observed in non-sex chromosomes.
>
> "There is, in fact, intense recombination within the arms of the palindrome," says Page. "It really changes the way we think about the Y chromosome." He speculates that the palindromes enable the human Y chromosome to constantly repair mutations, staving off its own degradation.
>
> "It's a great story," comments Eric Green of the National Human Genome Research Institute in Bethesda, Md. He suggests that researchers should look for such palindromes in sex chromosomes of other animals.

**23.** Explain the conclusion that was reached by two Australian scientists about the human Y chromosome and the deactivation of its genes.

**24.** State what Page and his research team have discovered next to many of the genes on the human Y chromosome.

**25.** Explain the importance of recombination between sections of DNA on the human Y chromosome.

# 20
# Biotechnology

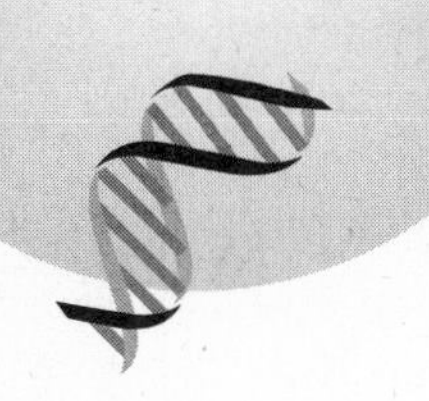

**Vocabulary**

| | | |
|---|---|---|
| biotechnology | recombinant DNA | vectors |
| genetic engineering | restriction enzyme | |

## RECOMBINANT DNA TECHNOLOGY: A BRIEF DESCRIPTION

A revolution in biology began with the discovery of the structure and function of DNA, the molecule of life. This revolution has increased in importance through advances in *recombinant DNA technology*.

We know that genes are made of DNA and that they determine the characteristics of every organism on Earth. Now scientists have learned how to identify and find individual genes. Once found, these pieces of DNA can be removed and put together, or *recombined*, with other pieces of DNA. The genes can then be moved from one cell into another. The methods for doing this make up **recombinant DNA** technology.

For thousands of years, people have selectively bred certain characteristics in plants and animals to produce different crops and breeds. Now, recombinant DNA technology makes it possible to put "new" genes into organisms—that is, to actually change the genetic makeup of plants and animals.

Through recombinant DNA technology—also called **biotechnology** or **genetic engineering**—human genes can be inserted into the genetic material of bacteria. These altered bacteria then become tiny "factories" that produce human proteins. Many other types of genes can be inserted into the genetic material of bacteria and other organisms, creating what are known as *transgenic organisms*. For example, agricultural scientists improve crops by inserting genes that make them disease-resistant, and improve livestock by inserting genes that make them grow faster or produce more milk. Perhaps most important, through biotechnology, human *gene therapy* may be used to treat some genetic disorders. Although progress has been slow, someday it may be possible to remove defective genes that cause certain disorders. In their place, healthy genes may be inserted that will prevent people from developing the disorder.

## BASIC TOOLS OF RECOMBINANT DNA TECHNOLOGY

As you read in Chapter 17, scientists now estimate that humans have about 25,000 different genes in each cell. Although this figure is much lower than the 100,000 genes that had been estimated years earlier, it is still a very large number of genes. From these genes, after considerable processing, come a much larger number of different proteins in our cells. Scientists wanted to know where the genes for specific proteins were located in the DNA. They wanted to be able to move the genes from one organism and place them into another. The task seemed hopelessly complex until the 1970s, when restriction enzymes were discovered. A **restriction enzyme** is a molecule that recognizes a small sequence of base pairs within a DNA strand. Whenever it finds this specific sequence, the restriction enzyme cuts the DNA. The place where the cut occurs is called a *restriction site*. (See Figure 20-1.)

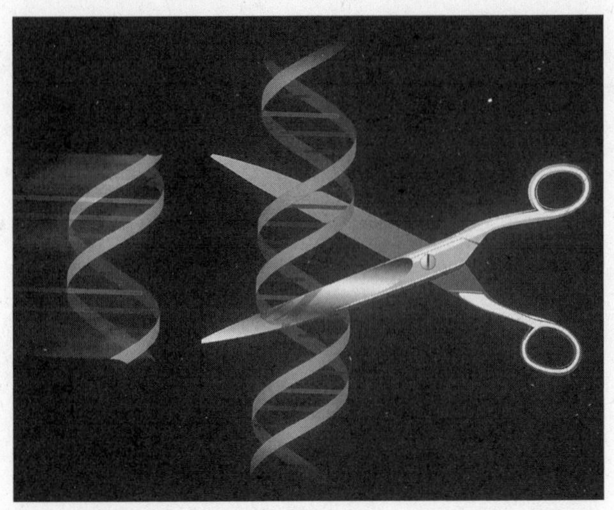

**Figure 20-1** In this piece of computer artwork, the scissors represent the restriction enzyme that is used to cut a piece of DNA. Another piece of DNA, perhaps the gene to produce human insulin, would be inserted where the cut is made.

For a scientist, restriction enzymes are very powerful tools. With them, DNA can be cut at precise locations. In addition, the same restriction enzyme can be used to cut DNA from two completely different organisms, such as a frog and a bacterium. Then, the pieces of DNA from one organism can be inserted, or *spliced*, into the DNA of another organism. Remember, those "pieces of DNA" *are* the genes themselves. (See Figure 20-2.)

Yet, restriction enzymes are not enough. Scientists also need a way to move pieces of DNA from the cell of one organism to that of another organism, where the foreign DNA can replicate. So scientists developed a way to use special molecules, called **vectors**, which can move pieces of DNA from one organism to another. The vector is usually a small circular

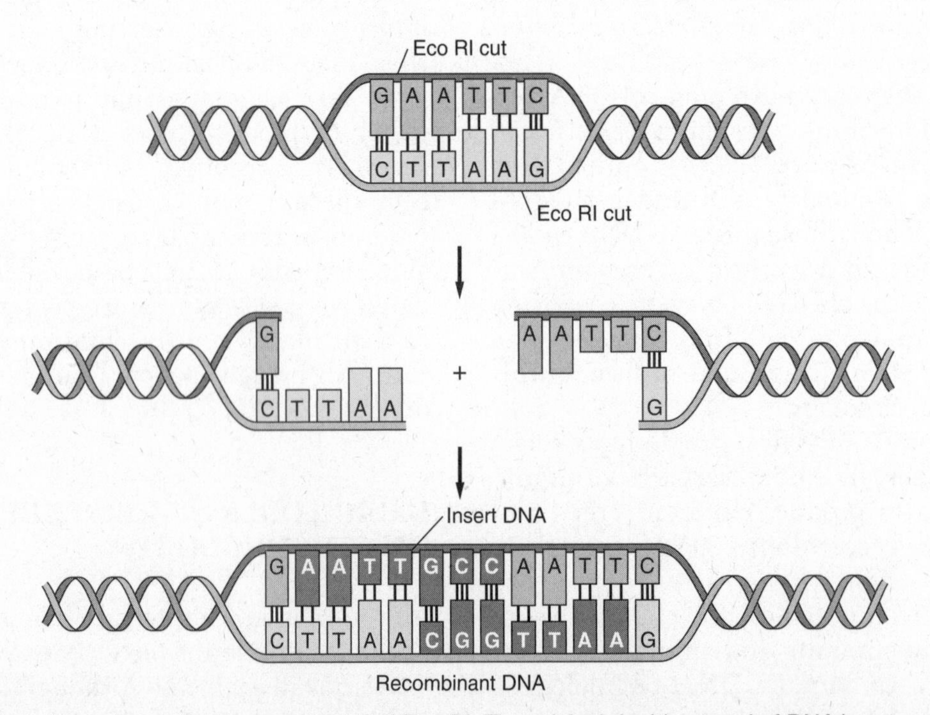


**Figure 20-2** The action of a restriction enzyme called Eco RI: The original double strand of DNA is cut, and another piece of DNA that has been cut (perhaps from a different organism) is inserted to form recombinant DNA.

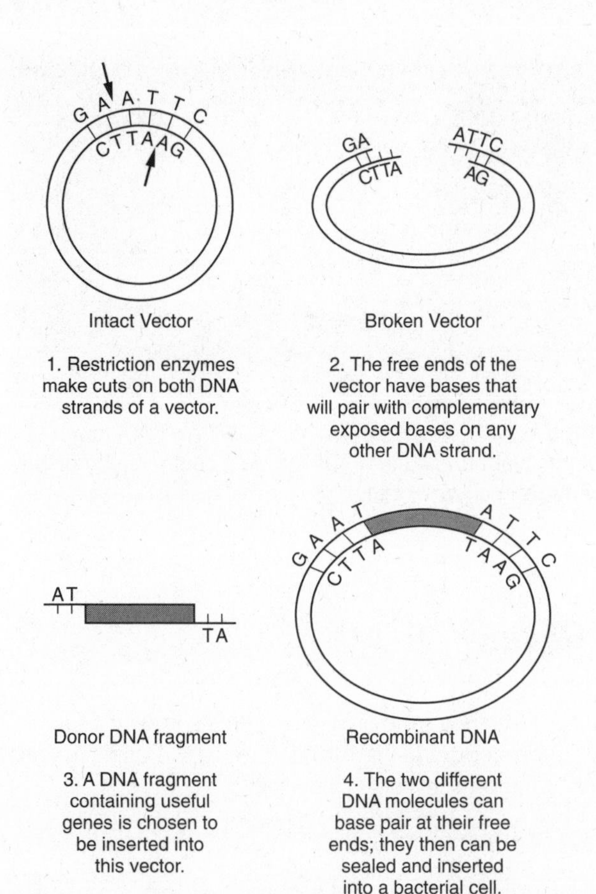


**Figure 20-3** Scientists use vectors, such as circular pieces of bacterial DNA, to insert and move genes from one organism to another. The vector is usually placed within a bacterial cell, because it can reproduce quickly and make many more copies of the recombinant DNA.

piece of bacterial DNA, called a *plasmid*, or a virus. (See Figure 20-3.)

Bacterial cells are used most often to receive the piece of DNA from a vector. Bacteria reproduce quickly (and asexually). Since the altered genes are passed on to every cell that develops (through mitosis), soon there are thousands of bacterial cells that contain the new DNA. In other words, once the piece of new DNA is in the reproducing bacteria, the amount of it increases as the number of bacteria reproduce and increase. This is one way to make much larger quantities of the recombinant DNA.

There is another reason why the bacteria are encouraged to reproduce. The new DNA in the bacteria is coding for the production of a useful plant protein or animal protein. The more bacterial cells there are, the more protein that is produced. Bacterial cells are not useful for some purposes. Instead, eukaryotic cells are used to receive the piece of DNA. These cells may be from yeast, plant, or animal (including human) cells that are grown or cultured in the laboratory.

## USES OF RECOMBINANT DNA TECHNOLOGY

Medicine is benefiting from the use of biotechnology. To date, more than 600 human genetic disorders have been diagnosed with the use of recombinant DNA technology. Often, individuals can be diagnosed with a disease even before they show any symptoms. This is because the gene that causes the disease can be identified in their DNA. This identification can even be performed on a fetus in the mother's womb.

Some medicines are now being produced in bacteria through biotechnology. For example, genetically engineered bacteria use a human gene to produce insulin. It is, therefore, pure human insulin. (See Figure 20-4.) A type of

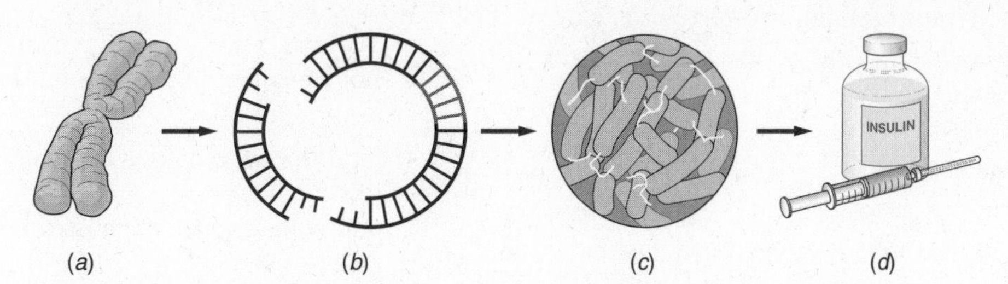


**Figure 20-4** The insulin gene from a human chromosome *(a)* is inserted into a vector *(b)*, a circular piece of bacterial DNA. The bacteria that receive this vector then contain the gene to produce insulin *(c)*, which is collected for use by people who have diabetes *(d)*.

biotechnology called *DNA fingerprinting* is now used as evidence in legal cases to determine the guilt or innocence of individuals in criminal trials. (See Figure 20-5.) Biotechnology has been enlisted to help clean up the environment, too. For example, genes are inserted into bacteria to give them the ability to remove hazardous substances from the environment. As mentioned before, farm animals are receiving genes from other animals to make them grow faster; and crop plants are receiving genes that make them more resistant to disease.

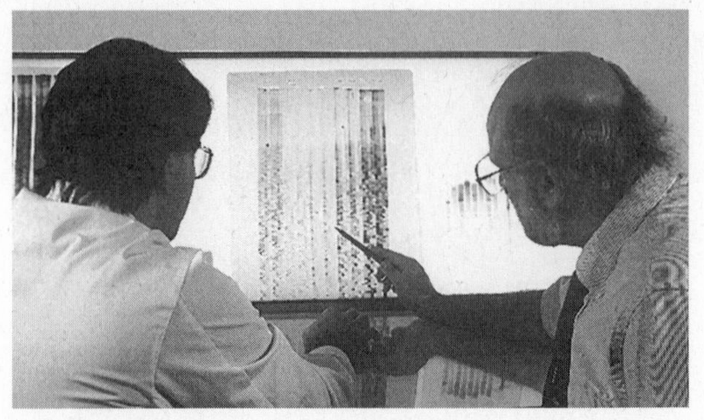

**Figure 20-5** Results from the analysis of DNA fingerprints can be used to argue the guilt or innocence of individuals in criminal trials.

## Chapter 20 Review

### Part A—Multiple Choice

**1.** Recombinant DNA technology involves
1 creating new DNA from their molecular subunits
2 inbreeding plants or animals with similar DNA
3 interbreeding plants or animals with different DNA
4 splicing pieces of DNA into other sections of DNA

**2.** Terms that describe the methods by which scientists change the genetics of organisms include all of the following *except*
1 biotechnology
2 genetic engineering
3 agricultural engineering
4 recombinant DNA technology

**3.** Humans produce more than 25,000 different proteins in their bodies. To make these proteins, every person must have approximately
1 5000 genes
2 15,000 genes
3 25,000 genes
4 50,000 genes

**4.** Scientists use restriction enzymes to
1 limit the length of DNA molecules
2 stop parts of DNA from replicating
3 prevent certain genes from being expressed
4 cut specific base-pair sequences out of DNA

**5.** The diagram below represents some steps in a procedure used in biotechnology. Letters X and Y represent the

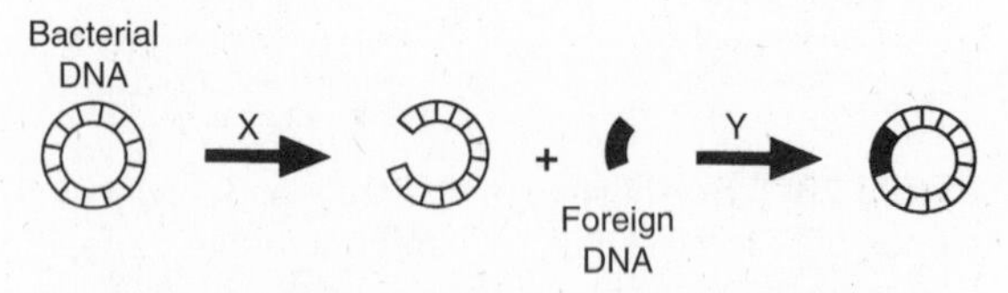


1 hormones that stimulate the replication of bacterial DNA
2 hormones that trigger rapid mutation of genetic information
3 enzymes involved in the insertion of genes into a new organism
4 enzymes required for genetic transcription and translation

**6.** The molecules that can move cut pieces of DNA from one organism to another are called
1 vectors
2 splicers
3 transformers
4 combiners

**7.** Genetic engineering has been used to improve crop varieties by
1 reproducing old genes for wild characteristics
2 removing genes that cause them to get diseases
3 inserting genes that make them disease resistant
4 adding animal genes that make them grow faster

8. Why do scientists insert human genes into bacteria?
   1 to give bacteria some human traits
   2 to make large amounts of human proteins
   3 to dispose of our defective genes
   4 to find out what the bacteria will do

9. When a human gene is inserted into a bacterial cell to become part of its DNA, the process is an example of
   1 DNA fingerprinting
   2 biotechnology
   3 karyotyping
   4 reproduction

10. Why are bacterial cells useful in recombinant DNA technology?
   1 They reproduce very quickly.
   2 They reproduce very slowly.
   3 They are almost identical to human cells.
   4 They can be placed within the human body.

11. The production of certain human hormones by genetically engineered bacteria results from
   1 inserting a specific group of amino acids into the bacteria
   2 interbreeding two different species of harmless bacteria
   3 splicing a piece of human DNA into a vector and then inserting it into bacteria
   4 deleting a specific amino acid from human DNA and inserting it into bacterial DNA

### Part B—Analysis and Open Ended

12. How is recombinant DNA technology different from the traditional practice of selective breeding of plants and animals?

13. Describe three examples of how recombinant DNA technology is being used today. Your answer must include at least:
   - *one* example for plants
   - *one* example for animals
   - *one* example for humans

*Base your answers to questions 14 and 15 on the following passage and on your knowledge of biology.*

For a number of years, scientists at Cold Spring Harbor Laboratory in New York have been working on the Human Genome Project to map every known human gene. By mapping, researchers mean that they are trying to find out on which of the 46 chromosomes each gene is located and exactly where on the chromosome the gene is located. The scientists want to be able to improve the health of people. By locating the exact positions of defective genes, scientists hope to cure diseases by replacing defective genes with normal ones, a technique known as gene therapy. Scientists can use specific enzymes to cut out the defective genes and insert the normal genes. They must be careful to use the enzyme that will splice out only the target gene, since different enzymes will cut DNA at different locations.

14. Using one specific example, state why the Human Genome Project is considered important.

15. Explain why scientists must use only certain enzymes when inserting genes or removing genes from a cell.

16. What discovery was made in the 1970s that advanced genetic engineering?

17. Explain why this discovery was so important for recombinant DNA technology.

18. What are vectors and how are they used in genetic engineering?

19. The following diagram represents a procedure used in biotechnology. Name a specific substance that can be produced by this technique and state how humans have benefited from the production of this substance.

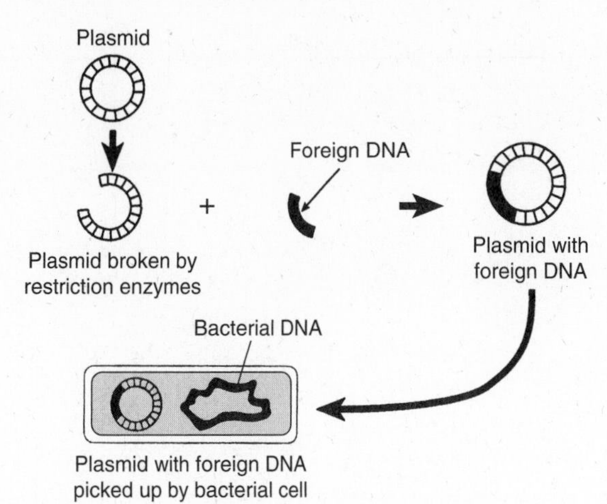

**20.** The diagrams below illustrate a process used in biotechnology. For each step (*a* through *d*) shown, use one of the following phrases to label what the structure represents: *DNA fragment with useful genes is chosen; DNA fragment has been inserted into bacterial DNA; Free ends of bacterial DNA are exposed; Restriction enzyme makes cuts in bacterial DNA.*

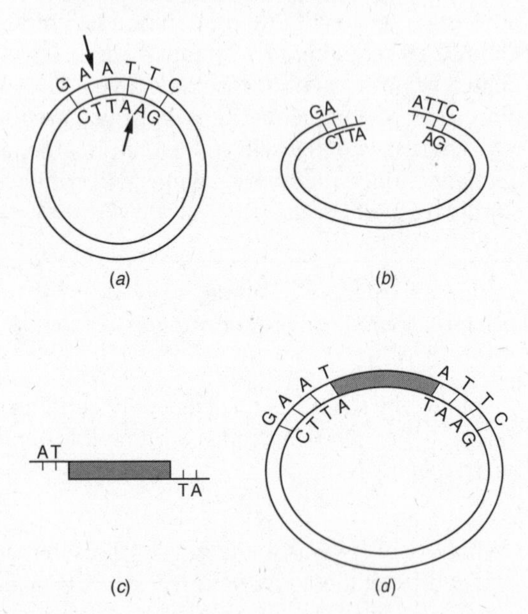


**21.** Biological research has generated knowledge used to diagnose genetic disorders in humans. Explain how a specific genetic disorder can be diagnosed. Your answer must include at least:

- a name of *one* genetic disorder that can be diagnosed
- a name or description of *one* technique used to diagnose the disorder
- a description of *one* characteristic of the disorder

**22.** How is biotechnology used to detect a disease even before its symptoms appear?

*Refer to the set of diagrams below to answer questions 23 and 24.*

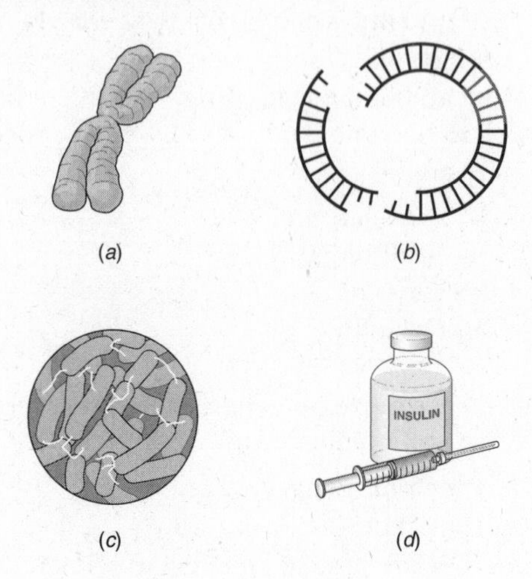


**23.** For each step (*a* through *d*) illustrated, write one sentence to explain which part of the genetic engineering process it represents.

**24.** Which one of the following titles would best describe the set of diagrams?

1 Restriction Enzymes and Genetic Advances
2 Agricultural Uses for Genetic Engineering
3 Genetic Engineering and Medicine Production
4 Genetic Engineering and DNA Fingerprinting

**25.** The bacteria that are used in recombinant DNA experiments have been changed so that they can survive only under special conditions in the laboratory. Why do you think scientists have included this safety precaution in their work?

### *Part C—Reading Comprehension*

*Base your answers to questions 26 through 28 on the information below and on your knowledge of biology.* Source: *Science News*, Vol. 163, p. 141.

> Silk cocoons could become puffs of valuable human proteins if a new bioengineering method developed by Japanese scientists pans out.
>
> In the past few decades, various biotechnology research teams have devised ways to mass-produce medically or industrially useful proteins by modifying the DNA of organisms. The animals create the proteins in their cells, milk, urine, or eggs.
>
> Now, Katsutoshi Yoshizato of Hiroshima University and his colleagues have genetically altered silkworms to produce a partial form of human collagen in their silk. Collagen is the structural protein in skin, cartilage, tendons, ligaments, and bones.
>
> Given that silkworms worldwide annually spin about 60,000 tons of silk, the technique could lead to inexpensive, high-volume manufacture of collagen for artificial skin grafts. The method might also produce the blood-serum component albumin and other proteins, the scientists say.
>
> In the January *Nature Biotechnology*, Yoshizato and his team report attaining concentrations of 0.8 percent collagen in the altered silkworms' cocoons. "If we raised the yield to 10 percent per total protein weight, we could produce it cheaply enough," Yoshizato predicts.

**26.** State the basic method by which biotechnology researchers have learned to make large quantities of useful proteins.

**27.** Explain why it would be valuable to produce large amounts of the protein collagen.

**28.** How might silkworms become involved in the inexpensive production of collagen?

# THEME VI

## Evolution: Change Over Time

# 21 The Process of Evolution

### Vocabulary

| | | |
|---|---|---|
| adaptations | genetic variations | radiation |
| artificial selection | hereditary | species |
| competition | natural selection | theory of evolution |
| extinct | pesticides | |

## THE THEORY OF EVOLUTION

Discoveries in modern science have shown that, over many thousands of years, populations of living things change. In 1859, the English naturalist Charles Darwin proposed a theory to explain *how* organisms change over time. It is called the *theory of evolution* (by means of natural selection). (See Figure 21-1.)

The **theory of evolution** explains how the immense variety of living things on Earth has developed from ancestral forms during the past three billion years. This theory is considered to be the most important unifying idea in biology. It offers an explanation, based on the fossil record and other scientific evidence, of how Earth has come to be populated by the millions of different species now alive and of how these species share common ancestry. (A **species** is usually defined as a group of related organisms that can breed and pro-

**Figure 21-1** Charles Darwin proposed the theory of evolution to explain how organisms change over time.

duce fertile offspring.) As a result of the changes that occur in living things and in the environment over time, nearly all species that have ever existed on Earth are no longer living today; that is, they have gone **extinct**.

## A STRUGGLE FOR EXISTENCE

During its lifetime, a female elephant may produce six offspring. Darwin calculated that, over hundreds of years, millions of elephants could descend from one original pair of elephants. Similarly, if a plant produced only two seeds a year, in 20 years there could be a million new plants descended from the original parent plants. However, this does not happen. In fact, only a relatively small number of offspring of any species survive to produce their own offspring. Darwin realized that Earth cannot support huge increases in populations; thus, they do not increase that dramatically. He concluded that there is a "struggle for existence" in which only a few offspring of any type survive to maturity and get to reproduce themselves.

In this struggle for existence, there is **competition** among organisms for various resources. Lack of food and space are two factors that can limit an organism's chances to survive and reproduce. Plants must have minerals, sunlight, water, and space in order to make their own food and grow. Animals also need resources such as food, water, space, and shelter in order to survive. Competition for available resources exists among individuals of the same species (for example, between two lions) and between members of different species living in the same area (for example, between a lion and a hyena). (See Figure 21-2.)

**Figure 21-2** This female lion—a predator—competes with other lions for food.

## GENETIC VARIATION

Through the process of reproduction, characteristics are passed on from parents to offspring. The resulting offspring resemble their parents. However, all the offspring produced by one pair of parents are not identical. The various characteristics that are passed on from one generation to the next are known as **hereditary** traits. The differences among the offspring that inherit these characteristics are called **genetic variations**.

It is very important to recognize the difference between hereditary, or inherited, traits and changes that occur to an individual during its lifetime. For example, eye color is inherited; it is a hereditary trait. By contrast, becoming extremely muscular due to weight lifting is an *acquired* trait; it is not inherited. A characteristic that is hereditary can be passed along to offspring. An acquired characteristic cannot be passed on to offspring. Thus, only changes that are in the sex cells, or *gametes*, of the parents can become the basis for evolutionary change.

As you have learned, there are two types of reproduction. In *asexually* reproducing organisms, one parent organism splits, or buds, in two to produce two new organisms. (Refer to Figures 12-4 and 12-7.) In *sexually* reproducing organisms, one male organism and one female organism mate to produce offspring. Sexual reproduction involves the combining of genetic material from two individuals. (Refer to Figure 13-2.) The traits in the offspring are the result of a new assortment, or *recombination*, of traits inherited from both parents. Thus, sexual reproduction produces greater genetic variability among offspring than does asexual reproduction.

In addition to the recombination of genetic information that occurs during sexual reproduction, genetic variation can arise from mutations. A *mutation* is a sudden change that occurs in the genetic material of an organism. You saw in Chapter 16 that this involves a change in the *linear sequence* of nucleotides in DNA. These changes occur randomly and spontaneously, and may be caused by **radiation** or chemicals. A mutation in the gametes may produce a small change in the resulting offspring, a major effect in the offspring, or no noticeable effect in the offspring at all.

Darwin recognized that there is variation among individuals produced by the same parents. He wondered how the differences within a group of organisms could lead to differences between groups of organisms. Darwin asked this question about a diverse group of small birds, called finches, that he had observed on the Galápagos Islands. He realized that there probably were small differences among the first finches that arrived on the islands from the South American mainland. (See Figure 21-3.) How did these minor variations within the ancestral finch species lead to the significant differences that now exist between the various types of finches? In other words, how did they develop into separate species? Darwin recognized two important facts that play a role in the development of new species:

- ♦ There is a struggle for existence, which limits the number of offspring that survive.
- ♦ There are differences among offspring due to individual, inherited variations.

Perhaps more important, Darwin posed the question: What determines which individuals survive to reproduce and thus become the parents of the next generation of offspring? His answer to this question formed the basis for his theory of evolution and revolutionized our understanding of how various forms of life have come to be.

## NATURAL SELECTION

The special characteristics that make an organism well suited to a particular environment are called **adaptations**. How do organisms evolve the adaptations that enable them to survive so well in a particular environment? (See Figure 21-4.)

Darwin attempted to answer this question. He developed an answer by combining what he knew about the inheritance of traits with what

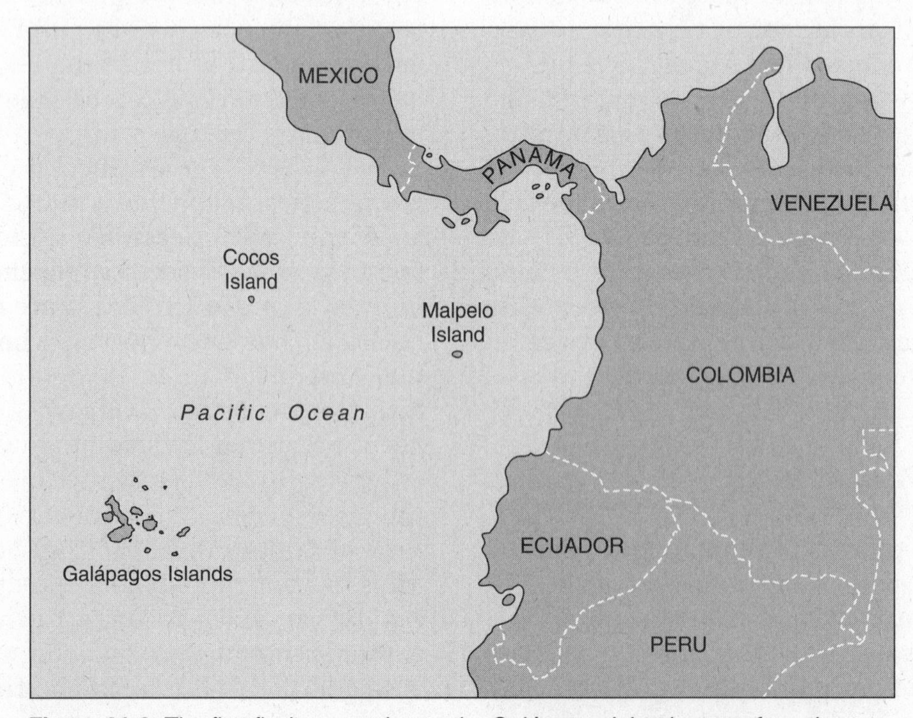


**Figure 21-3** The first finches to arrive on the Galápagos Islands came from the South American mainland, hundreds of kilometers away.

Ground finch eats seeds

Catches insects with beak

Tool-using finch digs insects from under bark with thorn

**Figure 21-4** The different beaks of these Galápagos finches show adaptions to different environments and different food sources.

he observed about an organism's struggle for existence. He concluded that whatever slight variations an organism had that gave it an advantage over other individuals in that environment would make it more likely to survive. That is what he meant by "survival of the fittest." An organism that was more fit and thus survived also would be more likely to reproduce and pass on its genetic variations to offspring. Those individuals that did not have such characteristics would be less likely to survive, reproduce, and pass on their characteristics. Darwin used the term **natural selection** to describe the way that environmental conditions determine—that is, *select*—which organisms survive to reproduce. Over time, the proportion of those more fit individuals would increase in a population. This is because of the increase in frequency of the genes responsible for those traits that give the surviving individuals the advantage.

## ANTIBIOTIC RESISTANCE IN BACTERIA—A CASE STUDY IN NATURAL SELECTION

Penicillin was the first antibiotic discovered. It is useful because it can kill some of the bacteria that cause disease in humans, leaving human cells relatively unaffected. Many more antibiotics with the ability to kill different kinds of harmful bacteria have also been discovered.

Over time, scientists noticed that some strains of bacteria that used to be killed by antibiotics were no longer affected. They had developed a *resistance* to penicillin and some of the other antibiotics. How did the bacteria develop this resistance? Were there genetic variations that made some bacteria naturally resistant to the antibiotic, without having prior exposure to it? (See Figure 21-5.)

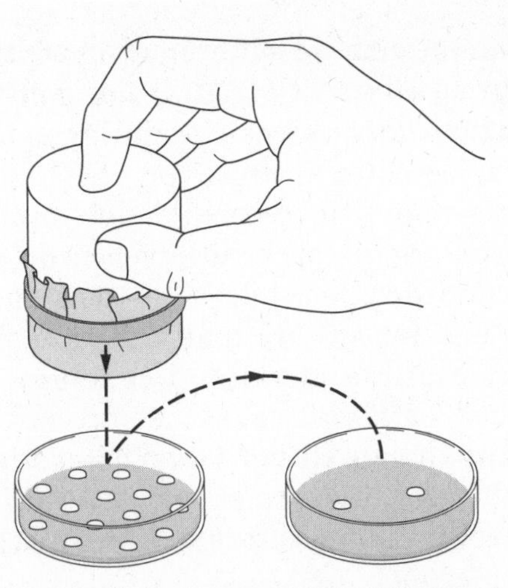

**Figure 21-5** In this experiment, colonies of bacterial cells were transferred to a dish that contained an antibiotic. The only bacteria that survived were those that already had a natural resistance to antibiotics.

If some bacteria were resistant to antibiotics from the start, they would have a survival advantage when such chemicals were added to their environment. In fact, this is what has happened. By killing off nonresistant bacteria, the antibiotics had decreased the competition for survival that existed in the original population. Without as much competition for food and space, many more resistant bacterial cells could grow and reproduce. The result would be a strain of bacteria that has resistance to antibiotics. Similar results have been noted in some insects exposed to **pesticides**; those that are resistant to the insect-killing chemicals can survive to reproduce. These are examples of natural selection at work—where external factors affect the survival of individuals within a population.

## ARTIFICAL SELECTION

People who raise dogs to perform certain tasks intentionally select, train, and breed those pups that have the characteristics best suited to their intended function. This intentional choice by people, to breed only organisms with specific characteristics, is known as *selective breeding*, or **artificial selection**. This process is similar to that of natural selection. However, in this case, humans—not the natural environment—select the organisms that have the desirable traits and decide which ones will survive, breed, and pass those traits on to their offspring. Plant and animal breeders have practiced artificial selection for centuries, resulting in a great variety of domestic livestock breeds and crops that are different from their wild ancestors. (See Figure 21-6.)

**Figure 21-6** The selective breeding, or artificial selection, by people of specific characteristics produces particular breeds of dogs, such as these.

## Chapter 21 Review

### *Part A—Multiple Choice*

**1.** How did Darwin explain the fact that only a small number of offspring of any species survive to reproduce?

1 Each species acts to limit the size of its own population.
2 Every species is limited to a certain number of offspring.
3 The parent organisms allow only specific offspring to reproduce.
4 Offspring must compete for the available resources to survive.

**2.** Which statement best describes competition? It exists

1 only among individuals within the same species
2 only between different species in the same area
3 only between different species living in different areas
4 among individuals of the same species and between different species living in the same area

**3.** Suppose two animals live in the same location and eat the same kind of food. What adaptation would decrease the competition between them?

1 Both animals eat at the same time.
2 Both animals breed at the same time.
3 One animal has hair and the other has feathers.
4 One eats during the day and the other eats at night.

**4.** Heredity is best described as

1 behavioral differences among offspring
2 the struggle for existence among living things
3 traits that are passed from one generation to the next
4 the slow and gradual change in organisms over millions of years

**5.** A couple has two children. One child has blue eyes and the other child has brown eyes. This difference is an example of

1 natural selection
2 artificial selection
3 genetic variation
4 acquired characteristics

**6.** Which description relates to an acquired characteristic?

1 Jamal is tall and thin.
2 Olivia has curly, blond hair.
3 Brittney has a widow's peak like her father.
4 Jose has large muscles from doing exercises.

**7.** What happens during asexual reproduction?

1 Two organisms join together to become one new organism.
2 A single parent organism splits to produce two organisms.
3 Two organisms mate to produce a new single offspring.
4 A single organism forms from the halves of two organisms.

**8.** In sexually reproducing organisms, the offspring inherit a combination of genetic traits from the

1 mother only
2 father only
3 mother and father
4 grandparents only

**9.** When compared to asexual reproduction, sexual reproduction produces

1 less genetic variation among offspring
2 greater genetic variation among offspring
3 offspring that are identical to the parents
4 offspring that are identical to one another

**10.** A mutation results from

1 artificial selection by humans
2 the fact that only the fittest organisms survive
3 a sudden change in the genetic material of an organism
4 competition for resources such as food and water

**11.** When mutations occur in body cells, they can be passed along to

1 sex cells only
2 other body cells only
3 offspring only
4 gametes only

**12.** Which statement best describes the current understanding of natural selection?

1 Natural selection influences the frequency of adaptive traits in a population.
2 Changes in gene frequencies due to natural selection have little effect on evolution.
3 Natural selection has been discarded as an important concept in evolution.
4 New mutations of genetic material are due to natural selection.

**13.** The Florida panther, a member of the cat family, has a population of fewer than 100 individuals and has limited genetic variation. Based on this information, which inference would be most valid?

1 The panthers will begin to evolve very rapidly.
2 The panthers can easily adapt to their environment.
3 Over time, the panthers will be less likely to survive in a changing environment.
4 Over time, the panthers will become more resistant to diseases.

**14.** Which statement represents the major concept of the biological theory of evolution?

1 A new species moves into a habitat when another species becomes extinct.
2 Present-day organisms on Earth developed from earlier, different organisms.
3 Every period of time in Earth's history has its own group of organisms.
4 Every location on Earth's surface has its own unique group of organisms.

**15.** Which concept is *not* a part of the theory of evolution?

1 Present-day species developed from earlier species.
2 Complex organisms develop from simple organisms over time.
3 Some species die out when environmental changes occur.
4 Change occurs according to the needs of an individual organism to survive.

**16.** Which statement best illustrates a rapid biological adaptation that has actually occurred?

1 Pesticide-resistant insects have developed in certain environments.
2 Paving large areas of land has decreased habitats for certain organisms.
3 Scientific evidence indicates that dinosaurs once lived on land.
4 The characteristics of sharks have remained unchanged over a long period of time.

**17.** When a breeder allows only the strongest and fastest horses to reproduce, she is practicing

1 artificial selection
2 natural selection
3 artificial mutation
4 asexual reproduction

**18.** Unlike in natural selection, in artificial selection

1 genetic information is passed from one generation to the next
2 humans, not the natural environment, decide which organisms will reproduce
3 the natural environment, not humans, decides which organisms will reproduce
4 mating is random and all organisms may pass their traits on to their offspring

**19.** People can develop new varieties of cultivated plants by carrying out

1 selective breeding for all traits
2 random breeding for all traits
3 selective breeding for specific traits
4 random breeding for specific traits

**20.** Selective breeding for particular traits can be used to

1 develop cultivated plants only
2 develop domesticated animals only
3 develop cultivated plants and domesticated animals
4 breed rare, wild animal species only

**21.** Which situation would most likely result in the highest rate of natural selection in a population?

1 reproduction of organisms by an asexual method in an unchanging environment
2 reproduction of organisms in an unchanging environment that has few predators
3 reproduction of organisms that have a very low mutation rate in a changing environment
4 reproduction of organisms that show genetic differences due to mutations in a changing environment

**22.** Some behaviors, such as mating and caring for the young, are genetically determined in most species of birds. The presence of these behaviors is most likely due to the fact that

1 birds do not have the ability to learn
2 these behaviors helped birds to survive in the past
3 individual birds need to learn to survive and reproduce
4 within their lifetimes, birds developed these behaviors

**23.** According to the theory of natural selection, why are some individuals more likely than others to survive and reproduce?

1 Some individuals pass on to their offspring new characteristics they have acquired during their lifetimes.
2 Some individuals do not pass on to their offspring new characteristics they have acquired during their lifetimes.
3 Some individuals are better adapted to exist in their environment than others are.
4 Some individuals tend to produce fewer offspring than others in the same environment.

**24.** According to modern evolutionary theory, genes responsible for new traits that help a species survive in a particular environment will usually

1 not change in frequency over time
2 decrease rapidly in frequency
3 decrease gradually in frequency
4 increase in frequency over time

**25.** Competition for food in a given area is most likely to exist between

1 lions and cheetahs
2 foxes and hares
3 mice and fleas
4 oak trees and squirrels

### *Part B—Analysis and Open Ended*

**26.** Suppose there are two types of fur color, brown and white, in a species of rabbit that lives in an area with very little snow all year. Most of the rabbits have brown fur. Then the environment changes so that there is snow much of the year. Based on your knowledge of natural selection, you might predict that the proportion of white fur to brown fur in the new climate would change so that

1 equal numbers of rabbits would have brown fur and white fur
2 more rabbits would have white fur than brown fur
3 more rabbits would have brown fur than white fur
4 more rabbits would have white fur with brown splotches

**27.** Briefly explain how the diversity of species alive today is related to the process of evolution.

**28.** How is the "struggle for existence" important to the study of evolution? Give an example.

**29.** In terms of evolution, why are the variations among individuals within a population more important than the similarities between them?

**30.** Explain the main difference between changes due to evolution and changes that are due to aging or that are acquired. Your answer must include the following:

- which type of change can be passed on to offspring
- why only that type of change can be passed on to offspring
- *one* example of a change due to evolution
- *one* example of a change due to aging or that is acquired

**31.** Which concept is best illustrated by the accompanying diagram, which shows changes in the body size and form of horses over time?

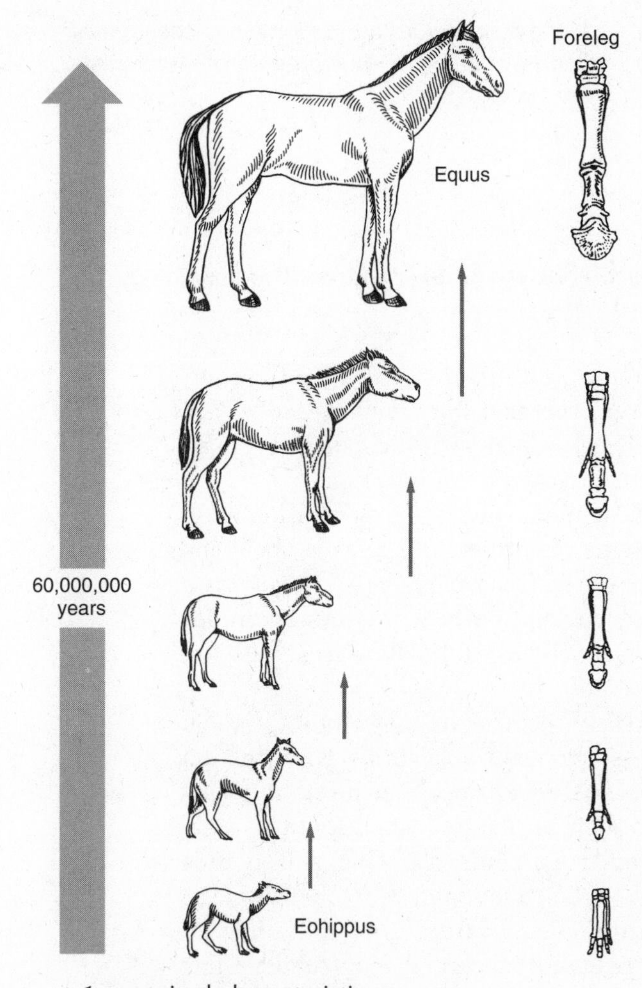


1 acquired characteristics
2 artificial selection
3 partial inheritance
4 evolution by natural selection

**32.** Why does sexual reproduction produce greater variation among offspring than asexual reproduction? Why is this important for the process of evolution?

**33.** Briefly explain why mutations are important to evolutionary change.

**34.** The best title for the chart below would be

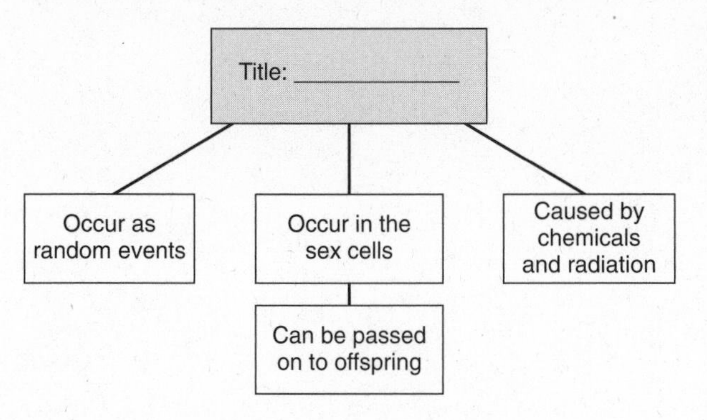


1 Types of Natural Selection
2 Characteristics of Mutations
3 Survival of the Fittest
4 Asexual Reproduction

**35.** Give two possible causes of genetic mutations. What cells would they have to occur in to be passed along to offspring? Explain.

**36.** Briefly describe four factors that lead to natural selection among organisms within a population.

**37.** The following terms relate to important factors in the evolutionary process: *struggle for existence*; *natural selection*; *environmental change*; *variation among offspring*. Which concept includes the other three? Explain.

**38.** Describe how natural selection can lead to the development of new species.

**39.** The best title for the chart below would be

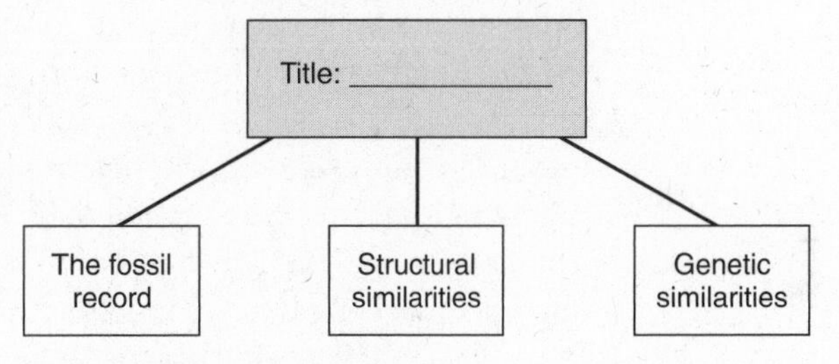


1 Evolutionary Pathways
2 Proof of Evolution
3 Natural Selection
4 Mutations in Evolution

**40.** A patient was given an antibiotic for an infection. The doctor told the patient to take the medicine for 10 days. But the patient took it for only two days, felt better, and then stopped taking it. After several days, the patient became sick with the same infection again. Why?

**41.** Write an essay in which you identify the adaptive value of a particular trait in a population. In your answer, be sure to:

- define, and give an example of, an adaptation
- explain how the adaptation helps an organism survive
- describe how the adaptation becomes more widespread

**42.** Many pesticides have been used to kill insects that destroy crops or that spread diseases such as the West Nile Virus. Unfortunately, some of these pesticides are no longer as effective as they once were for getting rid of insect pests. Explain why.

***Part C—Reading Comprehension***

*Base your answers to questions 43 to 45 on the information below and on your knowledge of biology.*

Antibiotics are used to treat infections in people and animals. Due to the enormous success of antibiotics, their use is very common worldwide. When we are ill, we have come to expect quick, effective treatment with antibiotics. Physicians often prescribe antibiotics at the earliest sign of an infection.

One result of the widespread use of these medicines is a growing number of antibiotic-resistant strains of bacteria. Some scientists have warned about the alarming possibility of infections that will not be treatable by the antibiotics we currently have. Already, one disease, tuberculosis—which was largely under control—has reappeared in a strain that is much more difficult to treat with antibiotics.

Recently, scientists became alarmed to find bacteria, in the food that is given to chickens, that are resistant to the most powerful antibiotics. Even though those particular bacteria were harmless, the finding raised the disturbing possibility that these bacteria could pass on their antibiotic resistance to disease-causing bacteria in chickens and, ultimately, in humans. One reason it is thought that such drug-resistant bacteria are being found more frequently is the heavy, routine use of antibiotics in farm animals.

This is an issue for everyone to be aware of and concerned about. Science has provided us with a group of wonder drugs to treat diseases that once killed many people. However, we must be thoughtful and wise in our use of antibiotics. The laws of nature—in this case, the process of natural selection that produces resistance to antibiotics—can never be ignored.

**43.** How is the use of antibiotics a matter of both good news and bad news?

**44.** Why should people be concerned about the use of antibiotics in farm animals?

**45.** How is knowledge of the process of natural selection necessary in order to understand the problem of overuse of antibiotics?

# 22
# Evidence for Evolution

## Vocabulary

biochemistry
fossils
homologous structures
vertebrates

No single idea explains the enormous diversity and complexity of life on Earth more powerfully than the theory of evolution by natural selection, as proposed by Darwin. Evidence that supports this theory includes fossils, the shapes and structures of living organisms, the chemicals that make up all living things, and the distribution of species on Earth today.

## EVIDENCE FROM FOSSILS

**Fossils** are traces or remains of dead organisms that have been preserved by natural processes. Usually only the hard parts of organisms—that is, the bones, shells, or teeth—become fossilized. A common way fossils are formed is through the gradual replacement of an organism's remains by other substances. In this process of fossil formation, the organism is usually buried in sediments, its hard tissues being slowly replaced by minerals dissolved in underground water. Over time, these minerals harden to form an exact copy of the original organism. In a series of undisturbed sedimentary layers, the oldest fossils are found at the bottom and the most recent are at the top. This fact helps scientists determine the approximate age of the fossils relative to one another. (See Figure 22-1.)

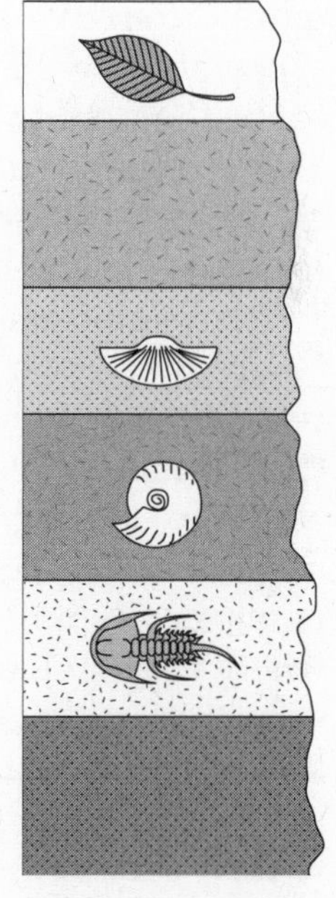

**Figure 22-1** Fossils in a series of layers of sedimentary rock.

Fossils can also form if the body of a plant or animal creates an impression in soft mud or clay. This process shows only the original external shape of the organism and not the internal structure, as do some other types of fossil formation. By studying fossils, scientists can see that species have changed over time and that most ancient life-forms no longer exist. (See Figure 22-2 on page 176.)

## EVIDENCE FROM COMPARATIVE ANATOMY

One way to determine the evolutionary relationships between different species is to find some similar structures, or characteristics,

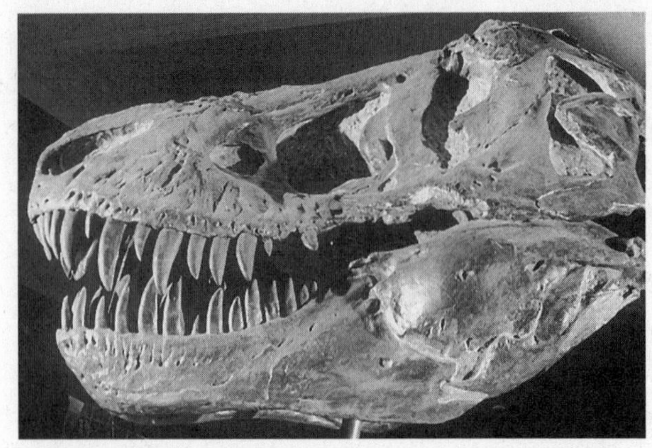

Figure 22-2 Dinosaur skull—a type of fossil that is formed when minerals slowly replace hard parts such as bones and teeth.

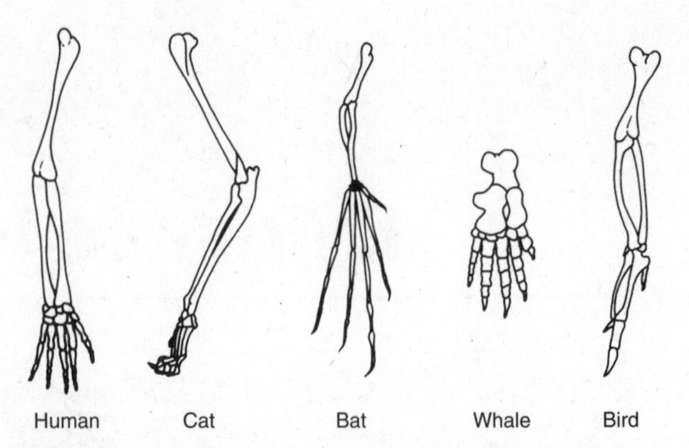


Figure 22-3 The similarities among the bones in each forelimb indicate that these animals shared a common ancestor.

that they both inherited from a common ancestor. Such features are known as **homologous structures.** For example, similarities exist in the forelimb bones of some very different animals. The wing of a bat, flipper of a whale, front leg of a cat, arm of a human, and wing of a bird—although they appear to be quite different—are all made up of the same types of bones. These bones are attached to each other and to other bones in similar ways. The forelimbs indicate that, long ago, these five animals all evolved from a common ancestor. (See Figure 22-3.)

Sometimes a homologous structure has little or no function in one species, but is clearly related to a more fully developed structure that does function in another species. Such a structure or organ is known as a *vestigial structure*. For example, the appendix in humans is a small sac attached to the place where the small and large intestines meet. In appearance, the appendix is a smaller copy of the cecum, which is a large pouch found in plant-eating mammals such as rabbits. In a rabbit, the cecum contains microorganisms that digest plant materials that are ingested. The fact that a similar organ is still useful in another species is evidence that humans evolved from an organism that also had this larger, functional structure.

## EVIDENCE FROM COMPARATIVE EMBRYOLOGY

Figure 22-4 depicts five animal embryos at very early stages in their development. Although all these embryos resemble one another, they are actually the embryos of a salamander, a chicken, a pig, a monkey, and a human. These similarities provide evidence that all **vertebrates** (that is, animals with backbones) follow a common plan in their early stages of development. For example, they all share a feature known as *pharyngeal slits*, or gills slits. In fish these structures become gills; in other vertebrates they become parts of throats, jaws, and ears. This shared feature is due to the fact that vertebrates have similar sets of genes, and this similarity comes from their having had common ancestors.

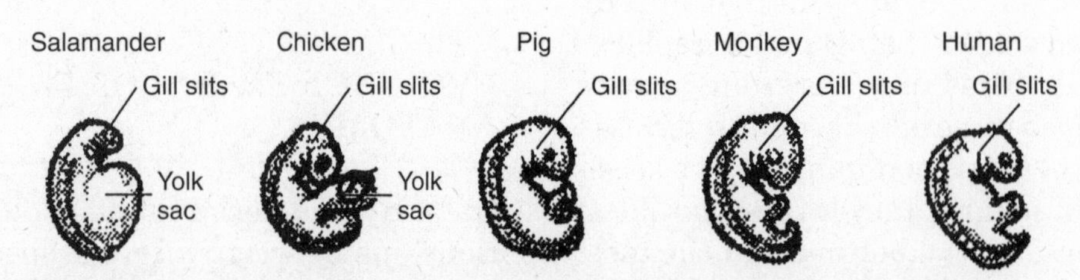


Figure 22-4 The very early embryos of these five animals show many similarities, such as gill slits and a tail—indications that they shared a common ancestor.

## EVIDENCE FROM COMPARATIVE BIOCHEMISTRY

The similar chemistry of living things, or **biochemistry**, provides some of the strongest evidence that organisms evolved from common ancestors long ago. All organisms store their genetic information—which is passed from one generation to the next in DNA molecules—in almost exactly the same manner. This genetic code shows that all organisms are related in fundamental ways.

As you have read, protein molecules are found in all living things, and they are made up of smaller units called *amino acids*. The same protein in two different species may be made up of similar but not identical amino acids. Biologists now know that a small number of amino acid differences in the same protein means that the two species are closely related in evolutionary terms. On the other hand, a large number of amino acid differences means that the two species are more distantly related. (Refer to Figure 16-4 on page 132.)

In addition, by comparing a DNA sequence from one organism with that of another organism, scientists can determine if the sequences belong to organisms of the same, closely related, or distantly related species. Again, the greater the similarity, the more closely related the species are; that is, the more recently the two species split from a common ancestor. (See Figure 22-5.)

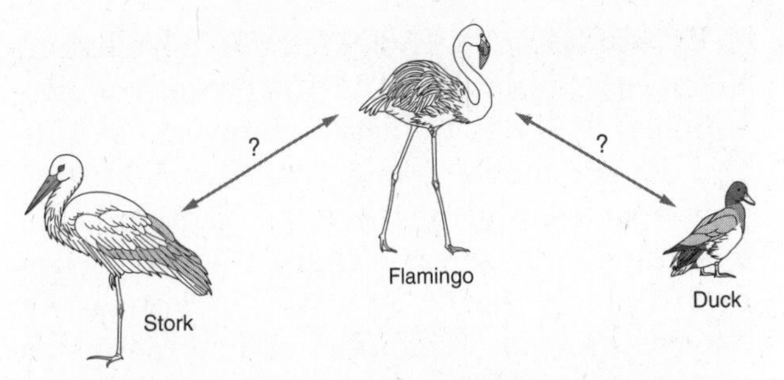


**Figure 22-5** By comparing the birds' DNA sequences, scientists have determined that flamingos are more closely related to storks than to ducks.

## EVOLUTION WITHIN POPULATIONS

Evolution occurs all the time within populations as frequencies of inheritable traits change. The process occurs due to natural selection acting on genetic variations from one generation to the next. This kind of small-scale change within populations is known as *microevolution*.

The peppered moth, carefully studied in England for more than a century, provides one of the best-known examples of microevolution. When the peppered moth was first studied, most of its population was light colored. The moths were well camouflaged when they rested on trees and rocks covered with light-colored lichens.

In 1845, a dark-colored peppered moth was observed for the first time. At that time, soot and smoke produced by coal-burning factories had begun to pollute the air. The trees and rocks became dark with soot and the lichens began to die. As a result, the light-colored moths were easily seen against the darker backgrounds and became the easy prey of insect-eating birds. By 1900, most of the peppered moth population was dark colored. Why did this happen? The darker moths were better camouflaged when resting on the tree trunks and rocks blackened by soot.

It is important to understand that the light-colored moths did not change color; they were replaced by increasing numbers of dark-colored moths through natural selection. The darker moths were less likely to be preyed on, and so were more likely to survive to pass on their genetic traits. Interestingly, as air pollution decreased, more light-colored moths were once again seen in the study area. (See Figure 22-6.)

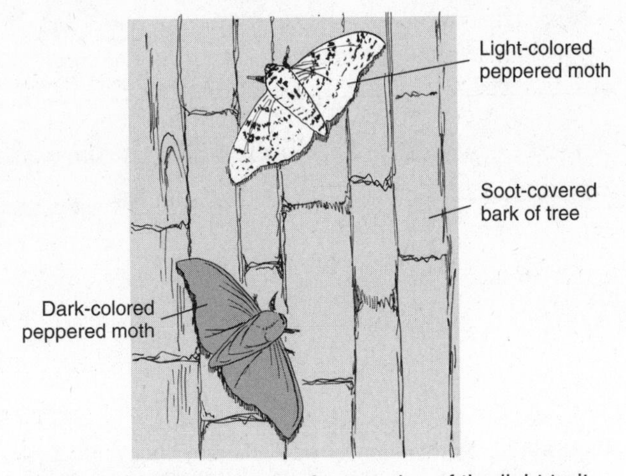


**Figure 22-6** The changing frequencies of the light trait and dark trait in the peppered moth population are due to natural selection.

By studying examples of microevolution, biologists can learn about the process of evolution that leads to bigger changes, such as the development of new adaptive features, new species, and new groups of organisms. This kind of large-scale change, which occurs over a long period of time, is known as *macroevolution*. Examples of new adaptive features include the legs of an amphibian (such as a salamander), the shell-encased eggs of a reptile (such as a turtle), and the large brain of a primate (such as an ape). Examples of entirely new groups of species that arose are the mammals and the flowering plants. Another example of large evolutionary developments is both the appearance and, millions of years later, the extinction of all dinosaur species.

## CONSTRUCTING A FAMILY TREE

Simple diagrams, such as the family tree in Figure 22-7, can be used to represent evolutionary relationships between different living and extinct species. Suppose that letters A, B, C, and D represent four living species. The letters E, F, and G represent ancestral forms of the species that are already extinct. In this case, organisms B and C are more closely related because they split off and evolved from their common ancestor E most recently. Organisms B and C are both equally related to A; they all share the more distant common ancestor F. Organism D is the least closely related to the others because it split off and evolved from their common ancestor G the longest time ago. A family tree, also called a *phylogenetic tree,* is based on evidence from the fossil record, morphology, embryological development, biochemistry, and genetic studies. (*Note:* The term *phylogeny* refers to the evolutionary history of a group of genetically related organisms.) An example of a phylogenetic tree that illustrates evolutionary relationships between bears and some other closely related carnivore species is shown in Figure 22-8.

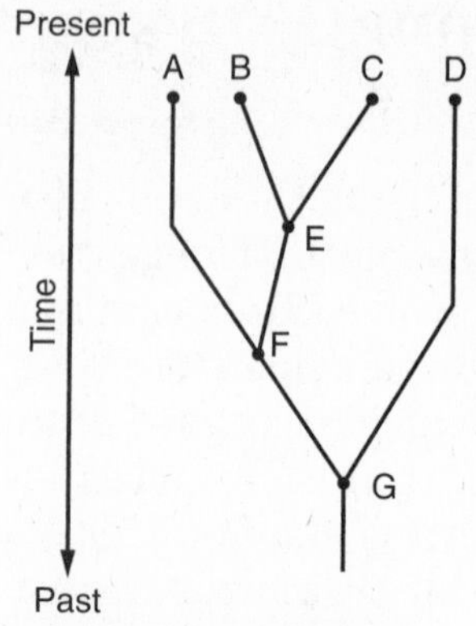


**Figure 22-7** A diagram can be used to represent evolutionary relationships among different living and extinct species.

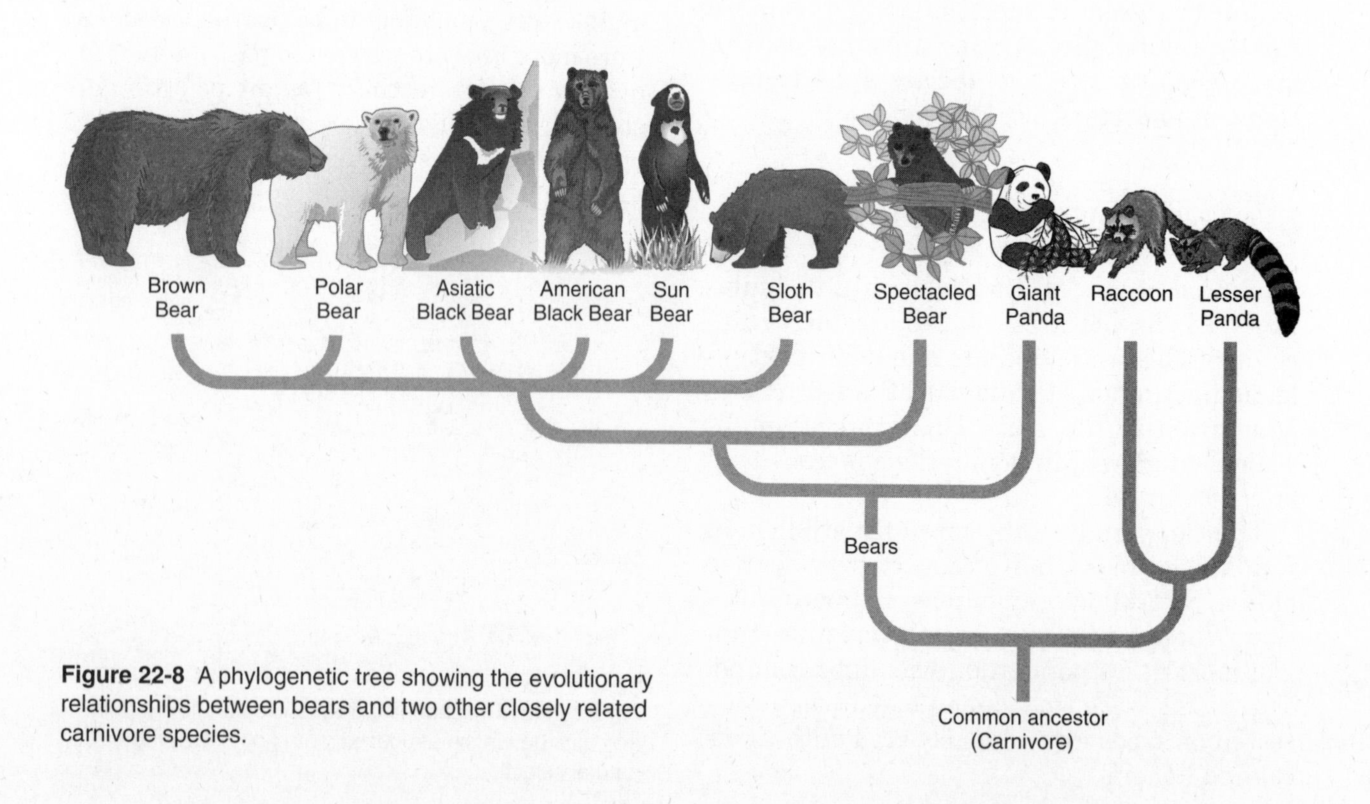


**Figure 22-8** A phylogenetic tree showing the evolutionary relationships between bears and two other closely related carnivore species.

# Chapter 22 Review

### *Part A—Multiple Choice*

**1.** All the following can be used as evidence to support Darwin's theory of evolution *except* the

1 similarity of chemicals in all living things
2 distribution of species on Earth today
3 shapes and structures of living organisms
4 distribution of mountain ranges on Earth's surface

**2.** Which statement is best supported by the fossil record?

1 Many organisms that lived in the past are now extinct.
2 The struggle for existence between organisms results in changes in populations.
3 Species occupying the same habitat have identical environmental needs.
4 Structures such as leg bones and wing bones can originate from the same type of tissue found in embryos.

**3.** Over time, fossils can be formed when an organism is

1 buried in sediment; then its hard tissues are replaced by dissolved minerals
2 buried in sediment; and then an impression of its internal parts is formed
3 entirely preserved in soft mud or clay, including its internal organs
4 buried in mud; then its bones dissolve and its internal organs remain intact

*Base your answer to question 4 on the diagrams below, which show the forelimb bones of three different mammals.*

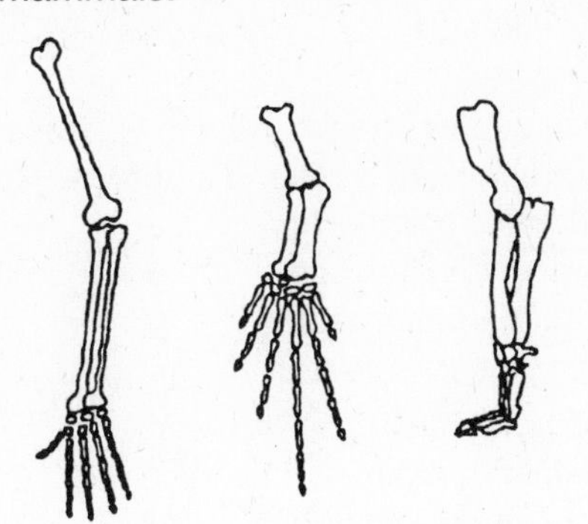

**4.** For these mammals, the number, position, and shape of the bones most likely indicates that they

1 developed in the same environment
2 have an identical genetic makeup
3 developed from a common earlier species
4 have identical methods of obtaining food

**5.** Suppose a scientist suggests that humans are related to rabbits because the human appendix resembles the cecum of a rabbit. The scientist is probably using evidence from

1 fossil remains
2 embryology
3 comparative anatomy
4 comparative biochemistry

**6.** The information below was printed on a calendar of important events in the field of biology.

| 1859<br>Darwin Publishes<br>*On the Origin of Species by Natural Selection* |
|---|

The title of the book that was published is most closely associated with an explanation of the

1 change in mineral types in an area due to ecological succession
2 structural similarities observed between diverse living organisms
3 reasons for loss of biodiversity in various habitats on Earth
4 effect of carrying capacity on the size of animal populations

**7.** Different organisms store their genetic information in the form of DNA in

1 their own unique way
2 a very similar manner
3 a way that differs between groups
4 one way in plants and another way in animals

**8.** Two species that have only a small number of amino acid differences in the same protein probably

1 are identical in their appearance
2 share the same parent organisms
3 are closely related in evolutionary terms
4 are distantly related in evolutionary terms

**9.** A scientist comparing organisms, in terms of biochemistry, might analyze

1 their DNA sequences
2 the stages of their embryos
3 the organs that have similar uses
4 the shapes of their fossilized footprints

**10.** After the Industrial Revolution in England, the number of light-colored moths decreased and the number of dark-colored moths increased. How can this be explained in terms of natural selection?

1 The dark-colored moths chased the light-colored moths away from the soot-covered trees.
2 The light-colored moths changed their colors in order to blend in with the darker trees.
3 Once the trees were dark, light-colored moths had a genetic variation that gave them an advantage over dark-colored moths.
4 Once the trees were dark, dark-colored moths had a genetic variation that gave them an advantage over light-colored moths.

**11.** Which is an example of an evolutionary change at the population level?

1 the development of legs on amphibians
2 the evolution of large brains in primates
3 the replacement of light-colored moths by dark-colored moths
4 the appearance of the flowering plants group

### *Part B—Analysis and Open Ended*

*Base your answer to question 12 on the diagram below, which shows the evolutionary relationships of several living and extinct mammals.*

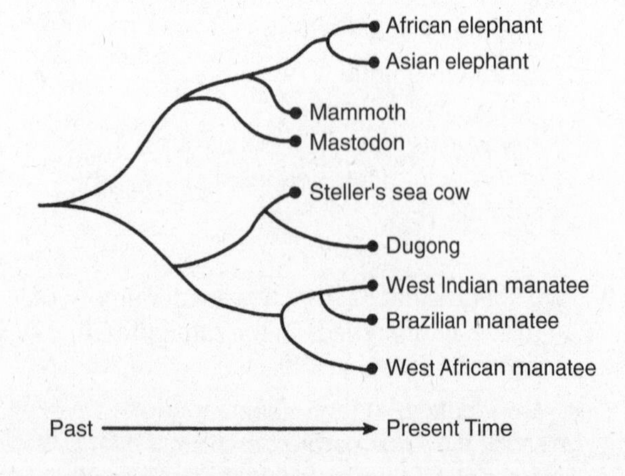


**12.** According to the diagram, which statement about the African elephant is correct?

1 It is more closely related to the mammoth than it is to the manatees.
2 It is not even remotely related to the Brazilian manatee or the mammoth.
3 It is more closely related to the West Indian manatee than it is to the mastodon.
4 It is the common ancestor of the Steller's sea cow and the dugong.

**13.** Explain how a family tree can be used to show evolutionary relationships between organisms.

*Use the diagrams below, which illustrate the forelimb bones of three different mammals, to answer question 14.*

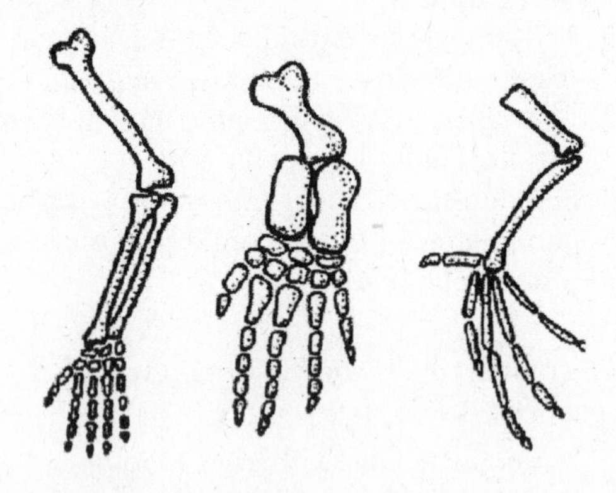

**14.** Differences in the bone arrangements support the hypothesis that these animals

1 are probably members of the same species
2 have adaptations for different environments
3 most likely have no ancestors in common
4 all contain the same genetic information

**15.** Why is an "evolutionary bush" considered a more accurate way to illustrate relationships between species than just a linear, or ladder-like, diagram?

**16.** The diagrams at right show the bones in the forelimbs of a cat and a human. The similarities between these appendages suggest that humans and cats

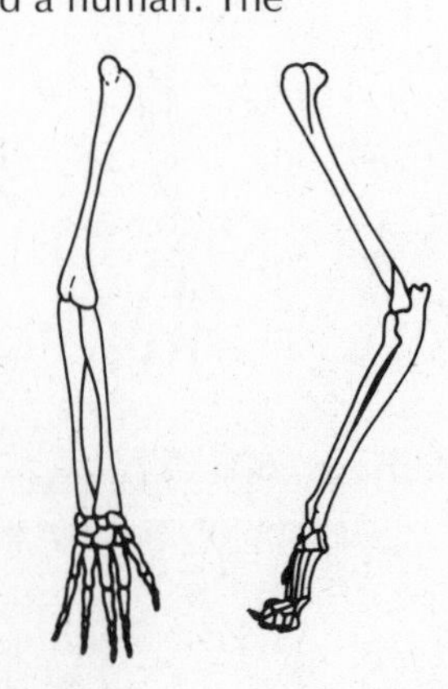

1 have identical genetic material
2 have the same direct ancestor
3 once shared a common ancestor
4 evolved in the same environment

**17.** Describe two ways fossils are formed, and tell which parts of an organism are usually fossilized. Your answer should explain the following:

- how fossils form through replacement by minerals
- how fossil imprints, molds, or casts are formed

**18.** The diagram at right represents a series of undisturbed sedimentary rock layers in a given area. Several layers show representative fossils of organisms. Relative to the other layers, the fossils of older, more primitive organisms would be found in the

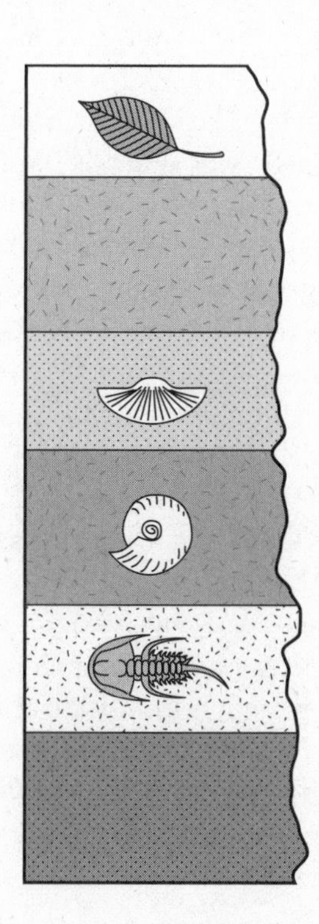

1 top layers only
2 bottom layers
3 middle layers
4 top and bottom layers

**19.** Explain why the presence of a body structure with no current function can provide evidence of an evolutionary relationship. Give an example.

**20.** The diagrams below show the embryos of three different vertebrate species. It is thought that they provide proof of evolution based on their similar

Turtle

Chicken

Pig

1 sizes
2 fossils
3 structures
4 molecules

**21.** Explain how studies of similarities in the biochemistry of proteins can be useful in determining evolutionary relationships between organisms.

**22.** The three species shown below have similar enzymes, hormones, and proteins. This supports the idea that they share a common ancestor, based on their similar

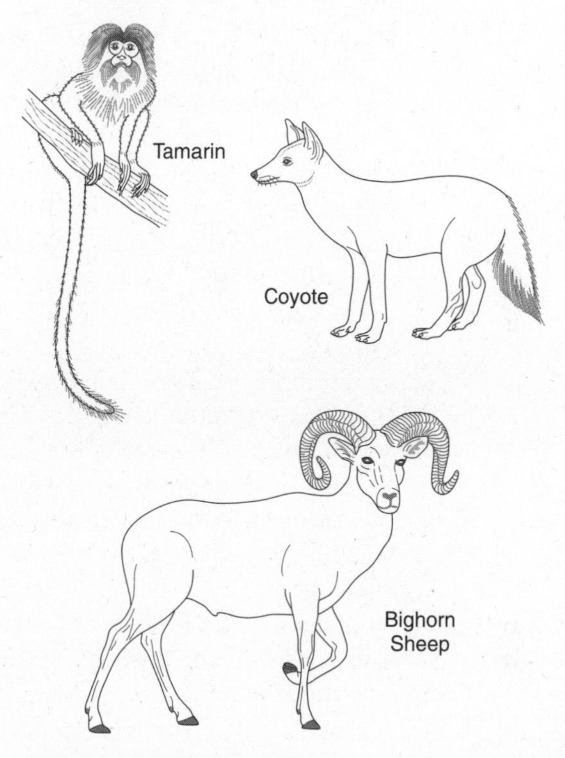


1 external structures
2 biochemistry
3 convergent evolution
4 behavioral patterns

**23.** Refer to Figure 22-6 on page 177, which shows the light-colored and dark-colored peppered moths. Over time, depending on changes in the environment, the percentage of each color type in their population has varied. Are these changes in frequency due to natural selection, artificial selection, or acquired characteristics? Explain.

**24.** As stated in the text, sometimes evolution leads to "the development of new adaptive features." Describe how such a change, over a period of time, may result in the development of a new species. Give either a real or an imagined example.

## *Part C—Reading Comprehension*

*Base your answers to questions 25 to 27 on the information below and on your knowledge of biology.*

In the 1920s, large amounts of limestone were being dug from the ground in Taung, an area of South Africa. Many human fossils were also being dug up in the limestone quarry. Raymond Dart, an Australian teaching at the medical school in Johannesburg, heard about the fossils. Dart was an expert on the anatomy of the human head and was anxious to examine the fossils—a natural curiosity. Dart contacted the owner of the quarry and, in time, two large boxes of fossils arrived at his home.

When he examined the material in the boxes, Dart found a dome-shaped piece of stone and immediately recognized that it was shaped like a brain. In this fossil, Dart saw the folds of tissue that make up the brain and even the blood vessels on the surface. Dart realized what had happened many years before. Long ago, someone had died in the vicinity of this present quarry. Sand and water that contained minerals had entered the skull; eventually these materials hardened into rock in the exact shape of the brain.

On close examination, Dart felt that the fossil brain looked like it had come from an ape, but he recognized that the fossil also had some similarities to a human brain. The skull might provide some clues to the brain's origin. Dart looked again in the box that contained the fossilized brain. Much to his amazement and delight, he found pieces of the lower jaw and the skull.

However, the front of the fossil skull—the face—was covered by layers of rock. In a procedure that took several months, Dart chipped away at the rock layers. What he eventually revealed was the face of a young, humanlike creature, later dubbed the "Taung Child," which Dart believed was an early ancestor of the human species. His find turned out to be one of the most important fossil discoveries ever made, adding crucial details to our understanding of human evolution.

**25.** Why might it be valuable for a scientist studying human evolution to have a background in medicine or human anatomy?

**26.** How did the bones of a humanlike skull come to be preserved as the fossils that were studied by Dart?

**27.** Imagine that you are Raymond Dart and you are writing a letter to a friend to describe your discovery. In the letter, explain to your friend why you think you have discovered evidence of an early ancestor of the human species.

# 23

# Origin and Extinction of Species

**Vocabulary**

adaptive radiation
extinction
hibernation
kingdom
niche
speciation
taxonomy

## ADAPTATIONS TO THE ENVIRONMENT

Every species lives in a particular place. And every place on Earth has specific conditions, such as average air temperature, monthly rainfall, kinds of minerals in the soil, and wind speeds. Darwin's theory of evolution states that those organisms that are best suited to tolerate the conditions of their environment—that is, the ones that have the most beneficial traits—will be most likely to survive and pass their traits on to their offspring. Since environments on Earth are constantly changing, however slowly, evolution of living things is an ongoing process.

Different adaptations in living things occur by chance as a result of the genetic variations within a population. Sometimes an adaptation works well for an organism—that is, it helps it survive—and sometimes it does not. The *adaptive value* of a trait is determined by the specific conditions of the environment. For example, at first, the dark coloration of some peppered moths did not aid survival, because it made those moths more visible on lichen-covered trees and rocks. The dark moths tended to be eliminated by natural selection. Yet that same dark coloration gave some moths a survival advantage when pollution caused the trees and rocks to darken. In that environment, such a trait was "chosen" by natural selection, and more dark moths lived to pass on their genes. If the main predator of moths did not hunt by sight, the dark coloration of the moths would not have provided an advantage or a disadvantage.

## TYPES OF ADAPTATIONS

Some of the most common types of adaptations are those that involve the shape and structure of organisms or the parts of organisms. Leaves, for example, are adapted in form and structure to the conditions of their environment, such as temperature, amount of sunlight, and water. These physical adaptations develop over time as plant populations adapt to changes in the environment. (See Figure 23-1 on page 184.)

An adaptation to conditions in the environment may also involve the functions and behaviors of an organism. For example, to survive the long, cold winters during which food is scarce, some animals—such as black

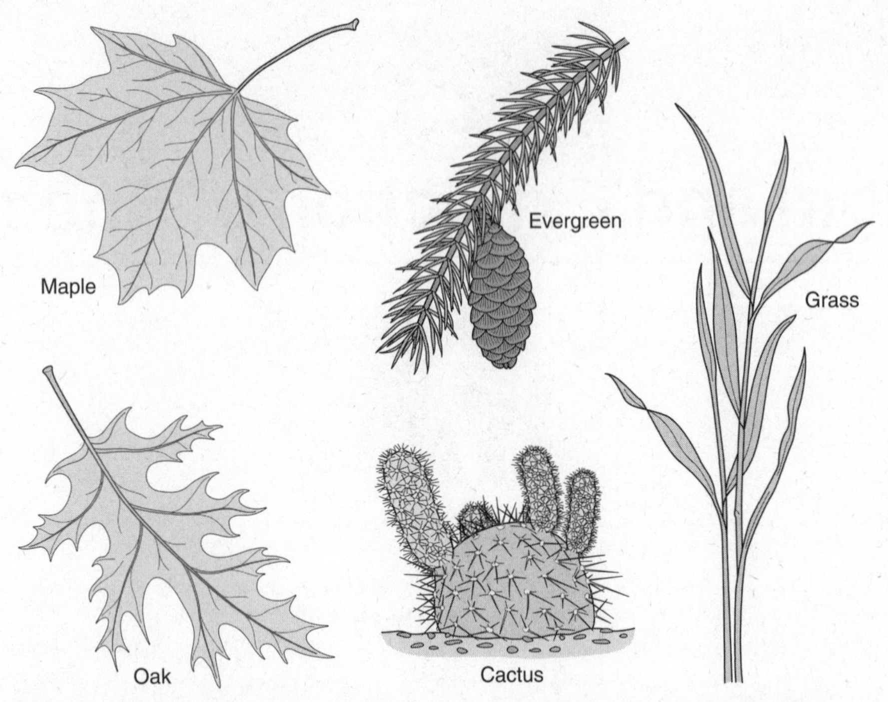


**Figure 23-1** The different shapes and structures of leaves are physical adaptations of plants to their environment.

bears, skunks, woodchucks, hedgehogs, bats, turtles, and frogs—slow their metabolic functions. This adaptation leads to **hibernation**, a type of behavior in which the animal retreats to a secluded place within its habitat, such as a den or a cave, to sleep during the winter months.

## SPECIATION

In general, for a population to change, it must be physically separated from other populations of its kind, and usually for a long time. As a result of evolution, such a population may continue to change until its members are no longer able to reproduce with members of any other population. This is known as *reproductive isolation*. The isolated population has undergone **speciation**; that is, it has become a new species.

It is also known that a new species may evolve to fill a niche in the environment that has become available. A **niche** includes all the things an organism does to survive, such as how it gets its food, reproduces, and avoids predators. A new species in a niche will have some physical and/or behavioral advantages that help it survive best in that particular niche.

## GEOGRAPHIC ISOLATION

The most common type of separation that leads to the formation of new species is *geographic isolation*. An actual physical barrier, such as a river or a mountain, can prevent organisms from moving between related populations. For example, the Galápagos Islands, which were colonized by finches from mainland South America, all have different environmental conditions. Due to natural selection, the finches changed in different ways on each of the islands and evolved into several species. Those individuals that had advantageous traits (for each particular environment) survived to produce offspring. This process, by which several populations evolve from an original parent population, each adapted to different niches, is known as **adaptive radiation**. (See Figure 23-2.)

In addition to islands, there are other types of geographically isolated areas in which speciation can occur. Examples include mountaintops, lakes, and forests that are separated by different types of terrain. (See Figure 23-3.)

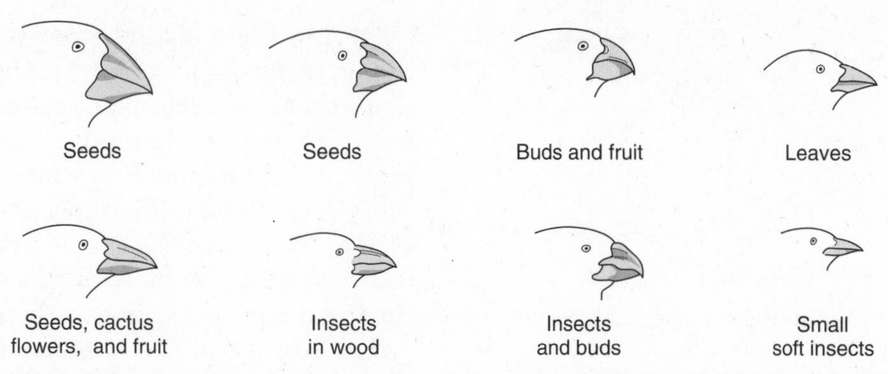


**Figure 23-2** Geographic isolation of the ancestral finches led to the formation of several species of finches, each with a different type of beak for a different type of food.

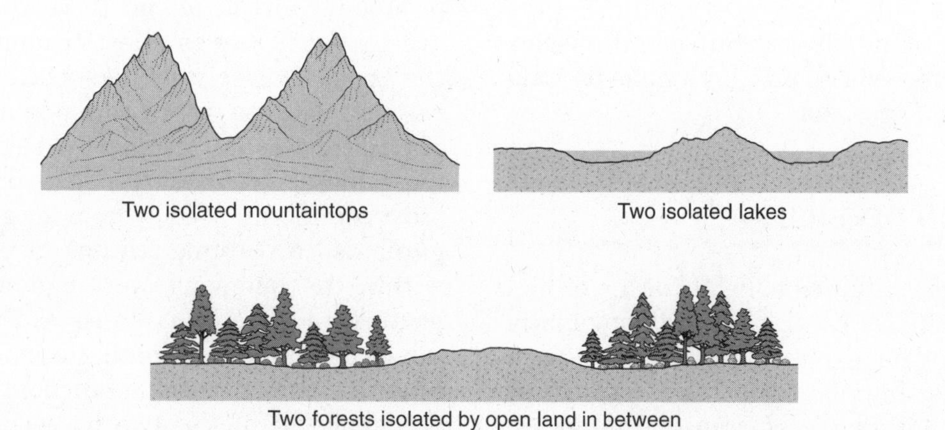


**Figure 23-3** Three examples of physical barriers that can cause geographic isolation.

## EXTINCTION

Another natural part of the process of evolution is **extinction**, the disappearance of species from Earth. Extinction occurs when a species no longer produces any more offspring. This inability to reproduce may occur when members of the population cannot adapt to changes in their environment. For example, a significant decrease in rainfall or a drop in average temperature are two environmental changes that may affect an organism's ability to survive and reproduce. Nowadays, global climate change is seen as a possible threat to species' survival.

Usually, before extinction occurs, a species' population reaches critically low levels. At that point, the species is said to be *endangered*. This can happen due to a change or loss of habitat, which is happening to many plant and animal species around the world today. The extinction of a species may also arise from problems that occur within a population. Harmful genetic traits that become widespread in a population may cause a species' extinction. (See Figure 23-4.)

At particular times in Earth's history, *mass extinctions* have occurred. These extinctions are usually due to natural events that drastically change the planet's climate, such as the impact of an asteroid or large comet. In a mass extinction, a large number of species disappears forever. Largely due to these mass

**Figure 23-4** The woolly mammoth went extinct about 10,000 years ago, most likely due to changes in its environment to which it could not adapt.

**Figure 23-5** Flying reptiles, such as the pterodactyl, died out more than 60 million years ago during the mass extinction that killed off their relatives, the dinosaurs. Such mass extinctions result from major changes in Earth's climate.

extinctions, about 99 percent of all species that have ever existed on Earth have become extinct. (See Figure 23-5.)

## UNITY AND DIVERSITY

The Earth is estimated to be about 4.5 billion years old. Life on Earth is thought by many scientists to have begun about 3.5 billion years ago as simple, single-celled organisms. About one billion years ago, through the process of evolution, increasingly complex multicellular organisms began to appear.

During the constant process of change in living things, some characteristics are lost while other characteristics are gained. As a result, Earth is populated by many different kinds of organisms. Scientists estimate that there are several million species alive today. To study so many species, biologists have organized them into numerous groups, classifying the organisms according to their similarities.

**Taxonomy** is the science of naming and classifying organisms according to their evolutionary relationships and shared characteristics. Similar organisms that are capable of producing fertile offspring with each other are placed in the smallest taxonomic group, the *species*. Closely related species that most recently evolved from a common ancestor are placed in the same *genus*, and so on up to the most inclusive level, the **kingdom**. In the most commonly used system for grouping organisms, all living things are classified within the following six kingdoms: Archaebacteria, Eubacteria, Protista, Fungi, Plantae, and Animalia. Note that previously, when scientists followed a five-kingdom system, organisms within the two bacteria kingdoms were placed together in the Kingdom Monera. (See Figure 23-6.)

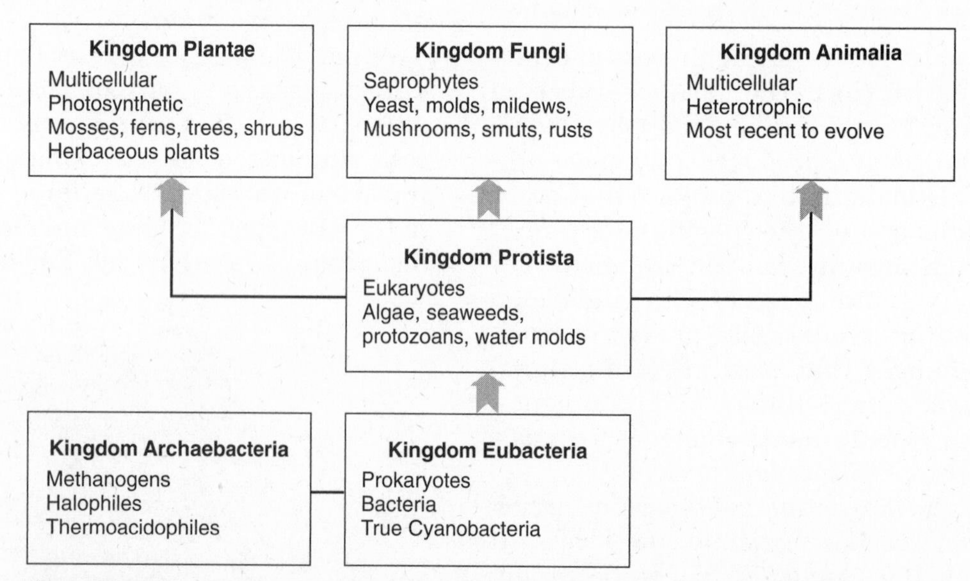


**Figure 23-6** The most commonly used classification system groups all living things within six main kingdoms.

# Chapter 23 Review

### Part A—Multiple Choice

**1.** Which example describes an adaptation that aids survival?

1 The height of a giraffe enables it to feed on the leaves of trees that other grazing animals cannot reach.
2 A person's poor eyesight makes it difficult for him to see without glasses.
3 A white peppered moth is clearly visible against the background of a dark-colored tree.
4 The broad leaves of a maple tree shrivel up when placed in the hot climate of a desert.

*Base your answer to question 2 on the following diagram, which illustrates the change that occurred in the physical appearance of a rabbit population over a 10-year period.*

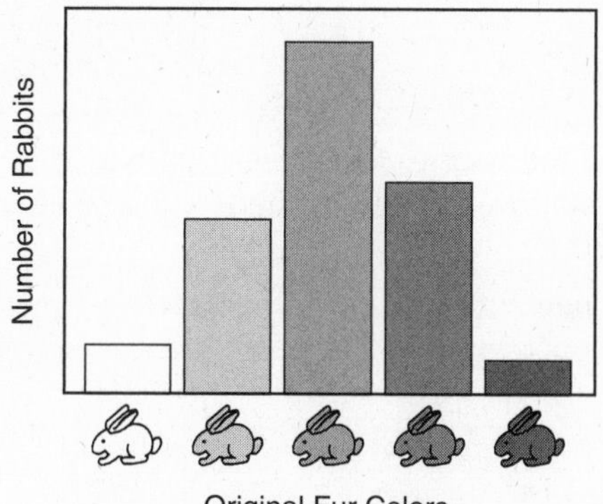


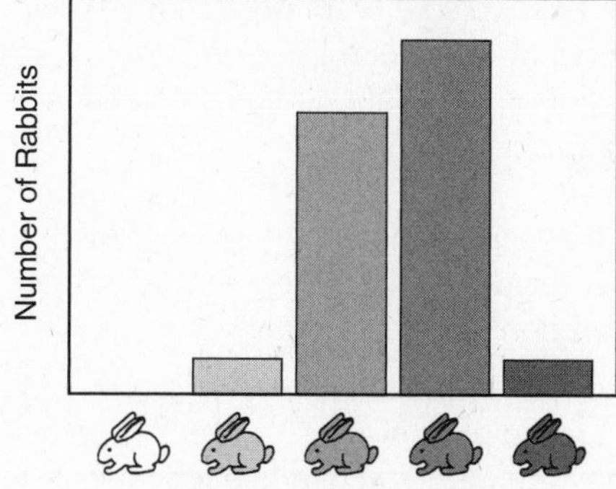


**2.** Which circumstance would explain this change over time?

1 a decrease in the mutation rate of the rabbits with black fur
2 an increase in the advantage of having white fur
3 a decrease in the advantage of having white fur
4 an increase in the chromosome number of the rabbits with black fur

**3.** The physical adaptations of an organism involve its

1 shape and behavior
2 shape and structure
3 structure and behavior
4 behavior and ecology

**4.** In an area in Africa, temporary pools form where rivers flow during the rainy months. Some fish have developed the ability to use their ventral fins as "feet" to travel on land from one of these temporary pools to another. Other fish in these pools die when the pools dry up. What can be expected to happen in this area after many years?

1 The fish using ventral fins as "feet" will be present in increasing numbers.
2 The fish using ventral fins as "feet" will develop real feet.
3 The "feet," in the form of ventral fins, will develop on all fish.
4 All of the varieties of fish will survive and produce many offspring.

**5.** One explanation for the variety of organisms present on Earth today is that over time

1 new species have adapted to fill available niches in the environment
2 each niche has changed to support a certain variety of organism
3 evolution has caused the appearance of organisms that are similar to each other
4 the environment has remained unchanged, causing rapid evolution

**6.** The Galápagos finches eventually formed 13 new species as a result of

1 the eruption of volcanoes on the islands
2 their geographic isolation on different islands
3 the extinction of other species on the islands
4 artificial selection by researchers who visited the islands

**7.** According to modern evolutionary theory, genes responsible for new traits that help a species survive in a particular environment will usually

1 not change in frequency
2 decrease rapidly in frequency
3 decrease gradually in frequency
4 increase in frequency

*Base your answer to question 8 on the following information.*

As the Colorado River formed the Grand Canyon, a population of squirrels gradually became separated. The conditions on the northern portion of the canyon were different from those on the southern portion. The squirrels on the northern portion evolved into a different species of squirrel.

**8.** This scenario is an example of

1 extinction, in which a species could no longer survive on Earth
2 migration, in which a species traveled to a new environment
3 distribution, in which members of the same species are distributed randomly
4 isolation, in which a new species evolved as a result of a physical separation

**9.** The complete disappearance of a species from Earth is known as

1 isolation
2 extinction
3 speciation
4 classification

**10.** A species will become extinct when its individuals

1 adapt to new environmental conditions
2 can no longer adapt and reproduce
3 move to a new environment and adapt
4 have survived beyond a specific period of geologic time

*Base your answer to question 11 on the diagrams below.*

**11.** According to some scientists, patterns of evolution can be illustrated by diagrams such as those shown below. Which statement best explains the patterns seen in these diagrams?

1 The organisms at the end of each branch can be found in the environment today.
2 Evolution involves changes that give rise to a variety of organisms, some of which continue to change through time while others die out.
3 The organisms that are living today have all evolved at the same rate and have all undergone the same kinds of changes.
4 These patterns cannot be used to illustrate the evolution of extinct organisms.

**12.** Organism X appeared on Earth much earlier than organism Y did. Many scientists think that organism X appeared between 3 and 4 billion years ago, and that organism Y appeared approximately 1 billion years ago. Which row in the chart below most likely describes both organisms X and Y?

| Row | Organism X | Organism Y |
|---|---|---|
| (1) | Simple multicellular | Unicellular |
| (2) | Complex multicellular | Simple multicellular |
| (3) | Unicellular | Simple multicellular |
| (4) | Complex multicellular | Unicellular |

1 (1)
2 (2)
3 (3)
4 (4)

**13.** Of all the species that have ever existed on Earth, approximately what percentage are now extinct?

1 5 percent
2 20 percent
3 50 percent
4 99 percent

**14.** The system of classification used today is based mainly on

1 what an organism typically eats
2 the relative size of an organism
3 where an organism lives on Earth
4 evolutionary relationships between organisms

**15.** Which statement about the rates of evolution for different species is in agreement with the theory of evolution?

1 They are identical, since all species live on the same planet.
2 They are identical, since each species is at risk of becoming extinct.

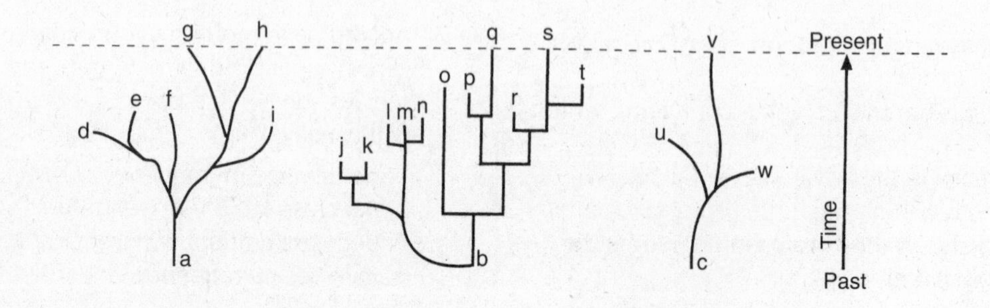

3 They are different, since each species has different adaptations for its environment.
4 They are different, since each species has access to unlimited resources.

**16.** The largest division of the classification system is a

1 species 3 kingdom
2 class 4 genus

**17.** Norway maples, sugar maples, and red maples are probably classified in

1 the same species and same genus
2 the same species but different genuses
3 different species but the same genus
4 different species and different genuses

**18.** Which process is correctly matched with its explanation?

| | Process | Explanation |
|---|---|---|
| (1) | Extinction | Adaptive characteristics of a species are not adequate |
| (2) | Natural selection | The most complex organisms survive |
| (3) | Gene recombination | Genes are copied as a part of mitosis |
| (4) | Mutation | Overproduction of offspring takes place within a certain population |

1 (1) 3 (3)
2 (2) 4 (4)

**19.** Organisms that are alike and capable of producing fertile offspring with each other are placed in the same

1 genus 3 species
2 family 4 kingdom

### *Part B—Analysis and Open Ended*

**20.** Why is the evolution of living things considered to be an ongoing process?

**21.** What determines the "adaptive value" of a trait within a population? Give an example, either real or imagined.

*Base your answer to question 22 on information in the following table and on your knowledge of the process of evolution.*

| Habitat | Number of Toes | Type of Horse Species |
|---|---|---|
| Plains | One toe (hoof) | Modern horse (*Equus*) |
| Forest | Four toes | Ancestral horse (*Eohippus*) |

**22.** You could hypothesize that modern horses have fewer toes than ancestral horse species had because the

1 changed habitat wore down their side toes as they ran fast over the plains
2 ancestral horse species mated with several mutant one-toed horses
3 people who first rode them preferred horses that had one large hoof per leg
4 changed habitat favored survival of faster horses, which had reduced toes

**23.** In what way is the behavior of an organism related to the evolutionary process? Your answer should describe or explain the following:

- *one* type of behavioral adaptation of an organism
- how natural selection affects the development of behaviors
- why a population's behavior may change over time

*Base your answer to question 24 on page 190 on the information in the paragraph and map below.*

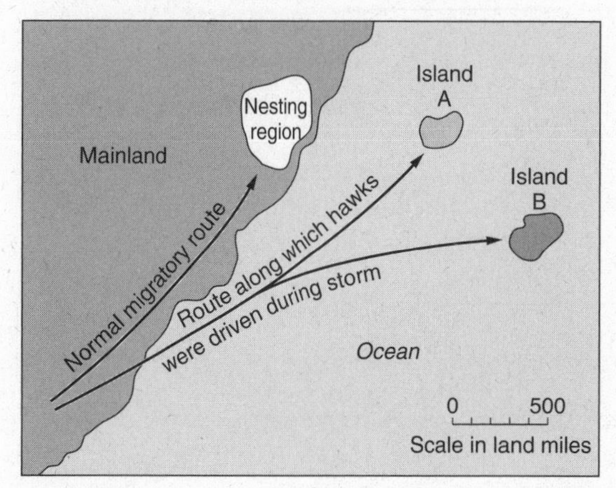


Thousands of years ago, a large flock of hawks was driven from its normal migratory route by a severe storm. The birds scattered and found shelter on two distant islands, shown on the map above. The environment of Island A is very similar to the hawk's original nesting region. The environment of Island B is very different from that of Island A. The hawks have survived on these islands to the present day with no migration between the populations.

**24.** Which statement most accurately predicts the present-day condition of these island hawk populations?

1 The hawks that landed on Island B have evolved more than those on Island A.
2 The populations on Islands A and B have undergone identical mutations.
3 The hawks that landed on Island A have evolved more than those on Island B.
4 The hawks on Island A have given rise to many new species of hawks.

**25.** The following diagrams indicate that the frequency of unspotted beetles is decreasing relative to the frequency of spotted beetles in this population of insects. Possible explanations for the changing frequencies of these traits include all of the following *except* that

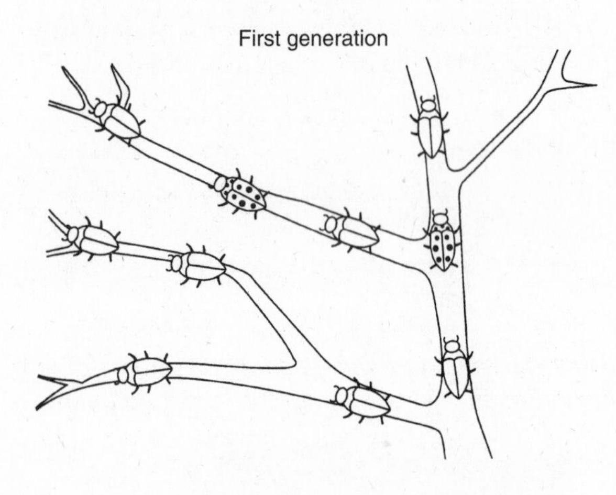


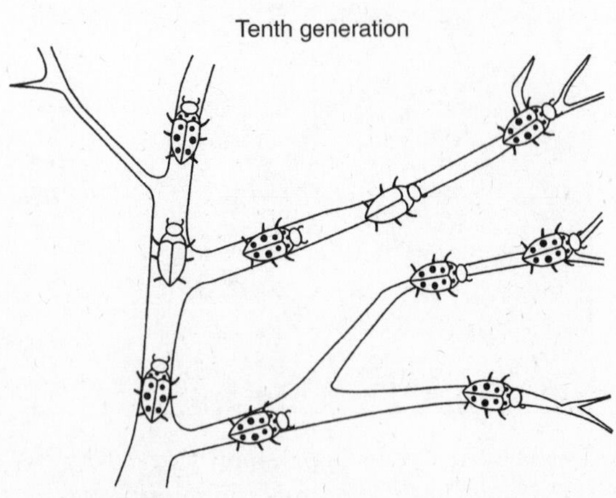


1 the beetles' environment has been changing over time
2 spotted beetles are better adapted to the changing habitat
3 unspotted beetles are better adapted to the changing habitat
4 natural selection is occurring, which affects survival rates

*Answer questions 26 and 27 based on the information in the paragraph below.*

The variation of organisms within a population increases the likelihood that at least some members of the species will survive changing environmental conditions. A large population of houseflies was sprayed with a newly developed, fast-acting insecticide. While most of the houseflies were killed off, some houseflies that were resistant to the new insecticide survived.

**26.** The changing environmental condition in this case was the

1 original population of houseflies
2 appearance of resistant houseflies
3 newly developed fast-acting insecticide
4 houseflies that were exposed to the spray

**27.** Which of the following items represents the variation that enabled some flies to survive the changing conditions?

1 The insecticide was new and fast acting.
2 Most of the flies were killed by the spray.
3 Some flies were resistant to the spray.
4 Only some of the flies were sprayed.

*Base your answers to questions 28 and 29 on the graph below, which illustrates changing percentages of two varieties within a species' population over time.*

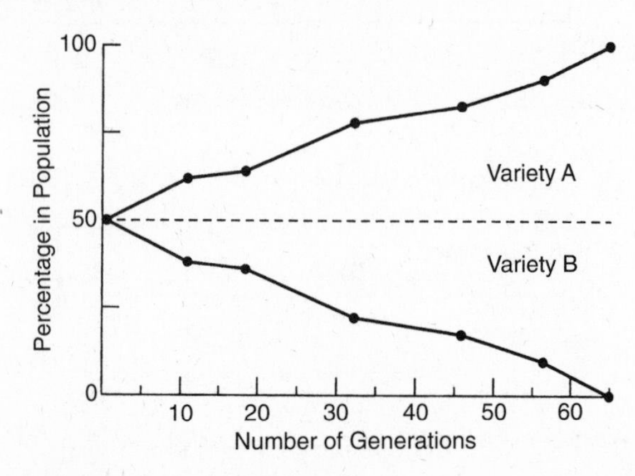


**28.** Which variety will most likely contribute to this population's traits in the future?

1 variety A only
2 variety B only
3 both variety A and variety B
4 neither variety A nor variety B

**29.** What is the probable reason that the percentage of variety A is increasing while the percentage of variety B is decreasing?

1 There is no opportunity for variety A to mate with variety B.
2 Variety A has some adaptive feature that variety B does not have.
3 Variety B has some adaptive feature that variety A does not have.
4 There is no genetic variation between variety A and variety B.

**30.** Distinguish between a situation in which variations within a population enable some organisms to survive changing conditions and a situation in which a mass extinction occurs. What is the main difference in each situation's cause and effect?

*Answer questions 31 and 32 based on the data in the chart below.*

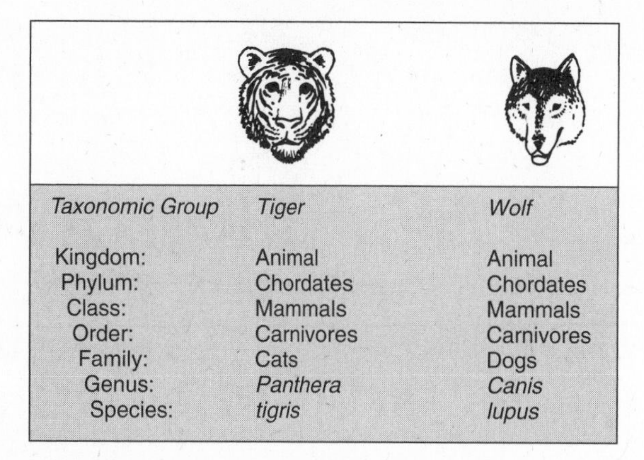

| *Taxonomic Group* | *Tiger* | *Wolf* |
|---|---|---|
| Kingdom: | Animal | Animal |
| Phylum: | Chordates | Chordates |
| Class: | Mammals | Mammals |
| Order: | Carnivores | Carnivores |
| Family: | Cats | Dogs |
| Genus: | *Panthera* | *Canis* |
| Species: | *tigris* | *lupus* |

**31.** The tiger and the wolf are classified within the

1 same species but different genuses
2 same genus but different species
3 same order but different families
4 same family but different orders

**32.** According to the chart, the largest division used in this classification system is the

1 species
2 family
3 class
4 kingdom

### *Part C—Reading Comprehension*

*Base your answers to questions 33 to 35 on the information below and on your knowledge of biology. Use one or more complete sentences to answer each question.* Source: *Science News*, Vol. 163, p. 46.

A road that a person strolls across with barely a thought proves a deadly barrier for many other creatures and can disrupt the usual traffic of their genes throughout a population.

Most of the recent studies of this effect have focused on vertebrates, note Irene Keller and Carlo R. Largiadèr of the University of Bern in Switzerland. These researchers turned their attention to the ground beetle *Carabus violaceus*.

This insect doesn't fly but proves to be a plucky traveler on the ground. Other researchers had observed that another species of ground beetle is extremely wary of crossing roads.

Keller and Largiadèr collected their beetles from eight places in a Swiss forest sliced by three roads. The biggest genetic differences appeared between beetles living on opposite sides of the roads. Also, beetle populations confined to specific forest areas by the roads seemed to have lost some of their genetic diversity, the researchers report in an upcoming *Proceedings of the Royal Society of London B*. The scientists warn that expanding networks of roads could be reducing healthful genetic diversity in invertebrates.

**33.** State one reason why a road may have an effect on a population of animals. What is the specific term for this kind of physical separation?

**34.** Explain why the researchers from Switzerland collected beetles from a variety of places in the Swiss forest.

**35.** State one discovery that the researchers described about the beetles in their report.

# 24
# Evolution of Animal Behavior

**Vocabulary**

| | | |
|---|---|---|
| behavior | migration | societies |
| fixed action patterns | reflex | territory |

## THE PURPOSE OF ANIMAL BEHAVIOR

Everything an organism does is its **behavior**. This includes finding a place to live, avoiding predators, finding food, mating and reproducing, and even dying. (See Figure 24-1.)

The purpose of animal behavior is to allow the organism to maintain homeostasis, survive, and reproduce. For example, birds build nests, keep their eggs warm until they hatch, then feed and protect their hatchlings. In Africa, huge herds of wildebeests travel northward and then southward again each year, following the seasonal rains, in a behavior pattern known as **migration**. When they are very young, discus fish swim around their parents' bodies and feed off a nutritious slime produced on their body surfaces. All these behaviors aid survival; and all of them have evolved over millions of years. (See Figure 24-2.)

**Figure 24-2** Animal behaviors, such as nest building, allow an organism to maintain homeostasis, survive, and successfully reproduce and raise offspring.

**Figure 24-1** Everything an animal does is its behavior. This frog, shown leaping with its tongue extended to catch its insect prey, is carrying out feeding behavior.

A variety of animal species—from insects to mammals—live in organized social groups, or **societies**, in which different members of the group have different tasks or levels of importance. Many animals also have a social behavior that involves defending the area in which they live, known as a **territory**. For example, to define their territory, many birds sing out a warning song and even attack other birds that cross the borders into their area. In general, these kinds of behaviors have evolved to aid survival, either of the individual or of the group as a whole. Territorial behavior, in particular, helps animals defend their access to valuable resources such as food, nesting sites, and potential mates.

## BEHAVIOR AND HOMEOSTASIS

Maintaining a constant temperature, water balance, and level of nutrition are all behaviors that control aspects of homeostasis. Homeostasis requires that the conditions within an organism remain relatively constant. Animals have developed a variety of behaviors that enable them to maintain these life-sustaining conditions.

Honeybees, which live in a social group called a *colony*, lay eggs in individual wax cells that make up the structure of the beehive. The eggs hatch and develop through larval and pupal stages in the cells before they develop into adults. Sometimes the developing bees die in their individual cells. Some types of honeybees remove young pupae that die in their cells. Other types of honeybees do not. How has this particular behavior developed? It has evolved over time, like any other adaptation that has survival value. Although requiring more effort, removing dead pupae may have the survival value of keeping the beehive free from disease—a condition that helps to maintain homeostasis. (See Figure 24-3.)

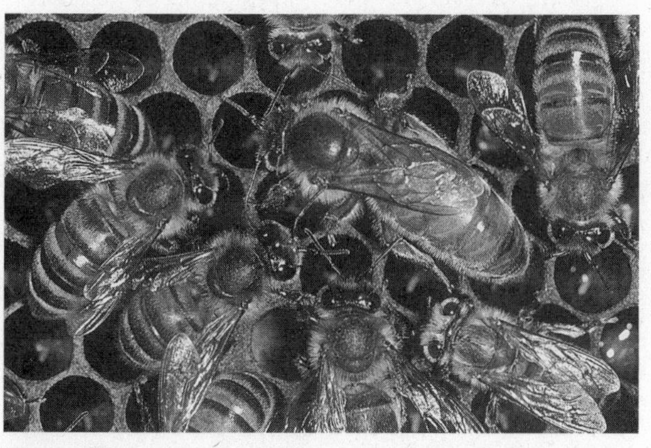

**Figure 24-3** The social behavior of the members of a honeybee colony serves to maintain stable conditions within the hive—conditions that enhance the health and survival of the bees.

## EVOLUTION OF ANIMAL BEHAVIOR

How have animals evolved all the behaviors they now have? Within a population of organisms, there are variations among the individuals—in behavior as well as in appearance. The behaviors may actually be determined by genetic traits already present at birth. Those individuals whose behavior makes them more likely to survive will pass those genetic traits on to their offspring. Other individuals, whose behavior is not as adaptive —particularly in a changing environment— will be less likely to survive and reproduce. In this way, the environment naturally selects certain behavioral traits (just as it selects physical traits). Thus, the evolution of behavior occurs over time. More and more individuals with adaptive behaviors live to reproduce; and these behaviors then appear with increasing frequency in the population. (See Figure 24-4.)

A behavior that occurs automatically, such as pulling your hand away from a hot stove, is called a **reflex**. In terms of evolution, reflex actions have been selected for since they help

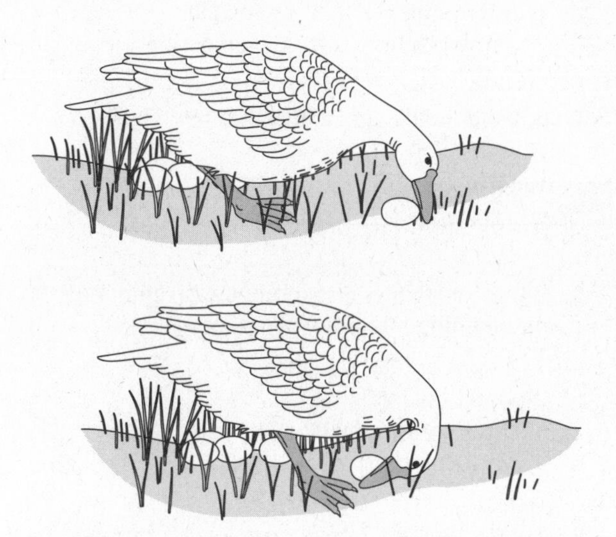

**Figure 24-4** Animal behaviors evolve over time because they have survival value. Here, a goose sees that an egg has rolled from the nest, so she rolls it back—an inborn behavior that increases her reproductive success, and which will be passed along to her offspring.

an organism survive. Scientists have observed a wide variety of more complex behaviors in animals that are also automatic. Animals are born with these behaviors, called *instincts*, which are a series of reflexes. Reflex behaviors do not have to be learned. All animals' instincts are inherited from their parents, just as their physical characteristics are inherited. These behaviors do not change, even if repeated many times; thus they are called **fixed action patterns**.

There are other types of behavior that an animal acquires by learning. An example of a learned behavior is *imprinting*, which occurs when an animal, such as a baby goose, learns to follow the first moving object it sees after birth, usually its mother; this is a behavior that aids survival.

## Chapter 24 Review

### Part A—Multiple Choice

**1.** An organism's behavior can best be described as

1 how the organism acts when it is angry
2 its response to a stimulus only
3 everything the organism does to survive
4 actions that are not related to its survival

**2.** Wildebeests of Africa migrate northward and southward each year because they

1 are forced to relocate by humans
2 need to follow the seasonal rains
3 try to add members to their herds
4 enjoy changes in body temperature

**3.** An example of a behavior that an animal learns soon after birth is

1 imprinting
2 a reflex action
3 an instinct
4 a fixed action

**4.** Some animals defend the area they live in, which is known as a

1 hometown
2 core area
3 territory
4 colony

**5.** Behaviors that control homeostasis include maintaining all of the following *except*

1 a constant temperature
2 water balance
3 level of nutrition
4 vocalization patterns

**6.** Removing dead pupae from a beehive may have the survival value of

1 making the beehive look cleaner
2 keeping the beehive free from disease
3 regulating the beehive's temperature
4 preventing wax buildup in the beehive

**7.** A behavior probably will be passed on to offspring if it

1 separates the organism from others in the population
2 makes the organism more likely to survive and reproduce
3 causes the organism to change its external environment
4 is a characteristic of the organism for a long period of time

**8.** Which of the following statements is true?

1 Physical traits only—*not* behaviors—can be naturally selected.
2 Behaviors only—*not* physical traits—can be naturally selected.
3 Behaviors as well as physical traits can be naturally selected.
4 Behaviors are *never* determined by genetic traits present at birth.

**9.** Some behaviors, such as mating and caring for the young, are genetically determined in birds. The presence of these behaviors is most likely due to the fact that

1 birds do not have the ability to learn new behaviors
2 these behaviors helped birds to survive in the past
3 individual birds need to learn to survive and reproduce
4 within their lifetimes, birds developed these behaviors

### *Part B—Analysis and Open Ended*

**10.** Give three examples of the general types of behavior in animals.

**11.** How does animal behavior help to maintain homeostasis?

*Refer to the following map to answer questions 12 and 13.*

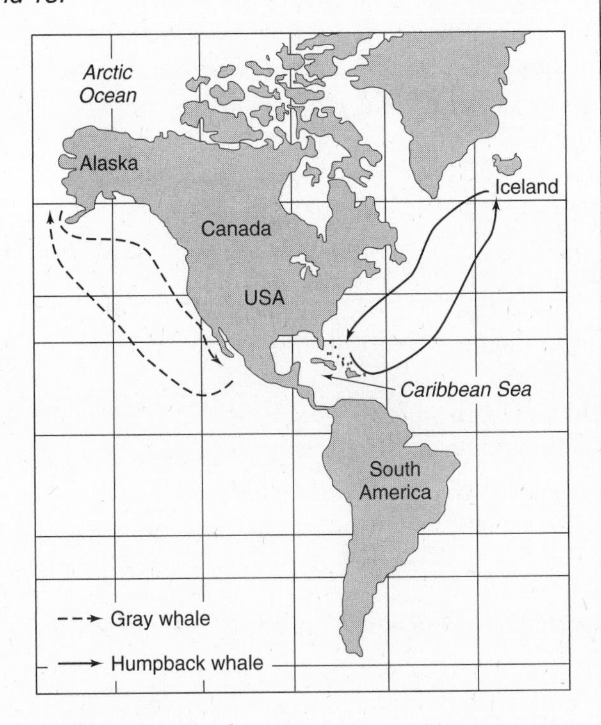


**12.** The arrows on the map represent the movements of groups of whales over the course of one year. This kind of behavior is known as

1 societal
2 territorial
3 migration
4 defensive

**13.** In what way might such long-range movements be adaptive for the whales?

**14.** Explain why defending the area in which they live might aid the survival of a group of animals.

*Refer to Figure 24-3 on page 193 to answer questions 15 and 16.*

**15.** Honeybees live together in a beehive, within a social group known as a

1 tribe
2 colony
3 nest
4 culture

**16.** How is this way of living similar to the way in which humans live?

**17.** What are some conditions in an animal's body that may affect its behavior (in terms of its need to maintain homeostasis)? Explain how.

**18.** Use the following terms to fill in the boxes of the concept map below: *low survival rate; high reproductive rate; natural selection; a changing environment.*

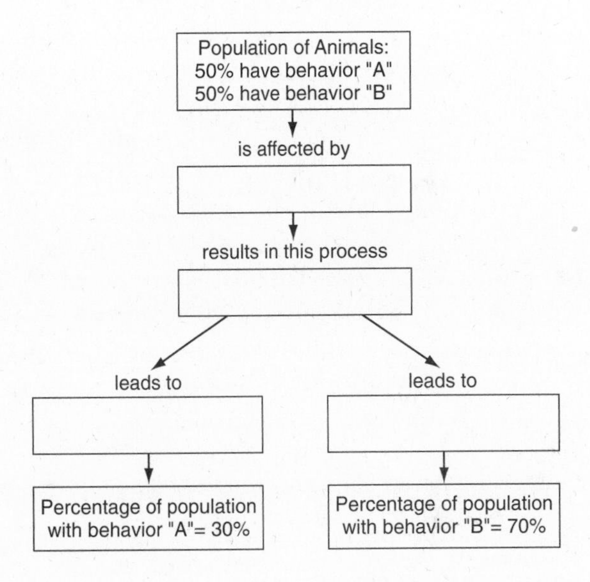


**19.** An appropriate title for this concept map probably would be:

1 Competition for Resources
2 The Defense of a Territory
3 Evolution of Animal Behavior
4 Behavior in Different Species

**20.** Describe how an animal's behavior may develop as a result of natural selection.

**21.** Suppose that a monkey starts to beg for food scraps from tourists in India. Later, after observing this, her offspring start to beg for food from tourists, too. The offspring grow well nourished and survive to reproduce. Their offspring, in turn, learn to survive by begging for food. Is this an example of natural selection acting on inherited behavioral traits? Explain why or why not.

### *Part C—Reading Comprehension*

*Base your answers to questions 22 through 24 on the information below and on your knowledge of biology.* Source: *Science News*, Vol. 164, p. 29.

The first measurements of energy use in migrating songbirds have confirmed a paradox predicated by some computer models of bird migration: Birds burn more energy during stopovers along the way than during their total flying time.

Martin Wikelski of Princeton University and his colleagues monitored 38 Swainson's and hermit thrushes during the nights of their spring migration through the northern United States. The researchers injected the radio-tagged birds with chemical-isotope tracers that enabled the scientists to measure the birds' metabolism. The team members spent their nights driving a car, trying to keep up with a tagged bird. "We got stopped by a cop just about every night, not because we were speeding, but because they wanted to know what somebody was doing in a little town in Wisconsin at 4 A.M. with a giant antenna on the roof of a car," says Wikelski.

A dozen birds took night flights covering up to 600 kilometers. The rest stayed put. The scientists determined that the birds that flew burned 71 kilojoules (kJ) of energy on an average night's flight of 4.6 hours. The birds that didn't fly burned energy at 88 kJ per day.

Since the birds spent about 24 days and nights on stopovers during a typical 42-day journey from Panama to Canada, actual flying consumed only 29 percent of the total energy budget for the migration, Wikelski and his coworkers report in the June 12 *Nature*.

**22.** How did researchers from Princeton University measure the metabolism of thrushes while they were migrating?

**23.** Explain what comparison the researchers were interested in making.

**24.** State the conclusion that was made by the scientists after they finished their investigation of the migratory birds.

# 25
# Human Evolution

## Vocabulary

bipedalism | geologic time | hominids

## THE SEARCH FOR HUMAN ORIGINS

The theory of evolution is a wonderful example of what science does best: tests ideas against evidence and observations in the real world to determine if the ideas are supported. The study of how the human species has evolved provides a good example of how science has tested the theory of evolution.

For about 130 million years, reptiles such as dinosaurs were the dominant large animals on Earth. Then suddenly, in terms of **geologic time** (which spans millions of years), the dinosaurs became extinct, about 65 million years ago. Although early mammals first appeared over 200 million years ago, it was not until the extinction of the dinosaurs that an enormous variety of mammals began to evolve. By that time, the earliest ancestors of today's major mammal groups had appeared. (See Figure 25-1.) One group of mammals that evolved—called *primates*—was adapted to a life in the trees. The primate group includes prosimians, monkeys, apes, and humans. The first monkeys evolved from their prosimian ancestors about 50 million years ago. Fossil evidence indicates that the apes later evolved from monkeys in Africa and Asia. The higher primates include all species of apes and humans. (See Figure 25-2.)

**Figure 25-1** The early mammals lived at the same time as the dinosaurs.

**Figure 25-2** Primates, such as this New World monkey, show adaptations to a life in the trees.

## HOMINIDS: THE EARLIEST HUMANS

Approximately 20 million years ago, large changes in climate and landforms caused forested areas to diminish in size. At that time, Asian and African apes became separated from one another and diversified. Between 8 and 14 million years ago, an African ape evolved that became the common ancestor of both chimpanzees and humans. This does *not* mean that our ancestors are chimpanzees. It *does* mean that we are most closely related to chimpanzees because we share a more recent common ancestor with them than with any other living animal. (See Figure 25-3.)

**Figure 25-3** The chimpanzee, our closest living relative, can walk upright on two feet for limited periods of time.

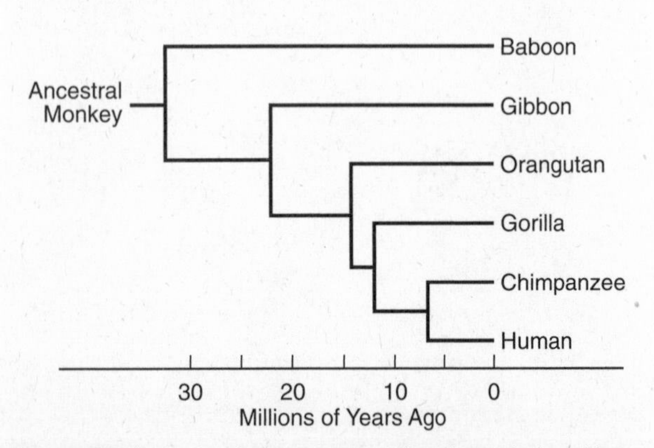


**Figure 25-4** An evolutionary tree showing relationships between humans and other closely related living primate species.

Scientists classify all human and ape species (living and extinct) in various closely related groups, generally going from the *hominoids* (humans and all apes) to the **hominids** (humans and African great apes) to the *hominins* (modern humans and all previous human ancestors). (See Figure 25-4).

## WHY DID EARLY HOMINIDS WALK ON TWO FEET?

Our earliest hominid ancestors lived in the trees. One of the first big steps on the path to becoming human occurred when early hominids started to walk—on two feet— upright on the ground. The shift to a life primarily on the ground may have occurred in response to changes in the environment, as forests gave way to savannas, or as a result of changes in hominid behavior.

The oldest hominid fossils that have been found so far, *Ardipithecus ramidus,* have been dated at 5.8 million years old. The fossils of these individuals, who lived in eastern Africa, show that the skull was balanced on top of the skeleton for upright walking. The ability to walk on two feet, called **bipedalism**, may have helped these hominids survive by freeing their hands to gather food and to carry their young more efficiently. Males with this advantage could bring food back to the females with whom they had mated and to their offspring, who would be well fed and most

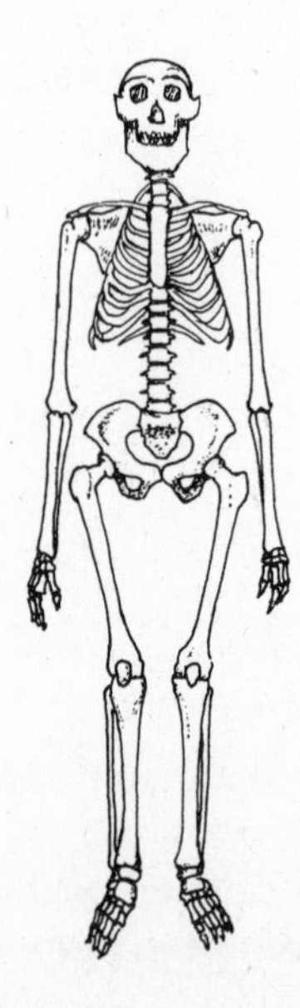

**Figure 25-5** This reconstruction of a four-million-year-old skeleton shows that, early on in their evolution, hominids were already upright and bipedal—an adaptation to life on the ground, not in the trees.

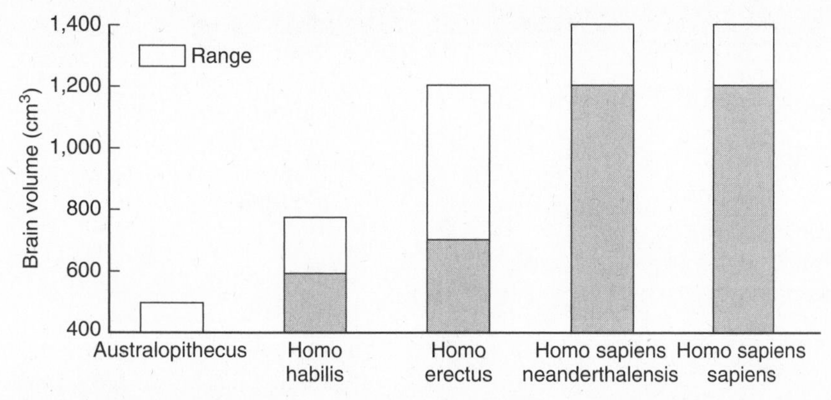


**Figure 25-6** A comparison of the brain volume of four ancestral hominids and of modern humans.

likely to survive. This advantage would have supported evolution of bipedal walking. Fossils of another early hominid species, *Australopithecus afarensis*, which date from 4 million years ago, show that these hominids were, in fact, adapted for a bipedal lifestyle. (See Figure 25-5.)

## OUR OWN GENUS

Hominids in the genus *Homo* are characterized by having a large brain, which sets them apart from earlier hominids as well as from all other primates. *Homo erectus* is considered to be the first ancestor within our genus. *H. erectus* existed for a long time, from about 1.8 million to 300,000 years ago. It had a body skeleton much like ours and walked erect just as modern humans do. The brain of *H. erectus*, at 700 to 1200 cc, was much larger than that of earlier hominid species and almost as large as that of modern humans. (See Figure 25-6.)

*H. erectus* was the first hominid species known to build fires, live in caves, wear clothes, and manufacture tools such as stone hand axes. With these skills, *H. erectus* was able to migrate north to colder regions beyond Africa and thus colonize the Eurasian continent. *H. erectus* was ancestral to our own species, *Homo sapiens*. (See Figure 25-7.)

**Figure 25-7** Our ancestor *Homo erectus* made and used stone hand axes such as the one shown in this photograph.

Today, there is much debate about human evolution. There are two main scientific views on the subject. One group of scientists sees it as a ladder, with one hominid species at a time leading to the next species. The other group sees it as a tree with several branches of hominids: one leads to modern humans; the others lead to extinct hominid species. (See Figure 25-8.)

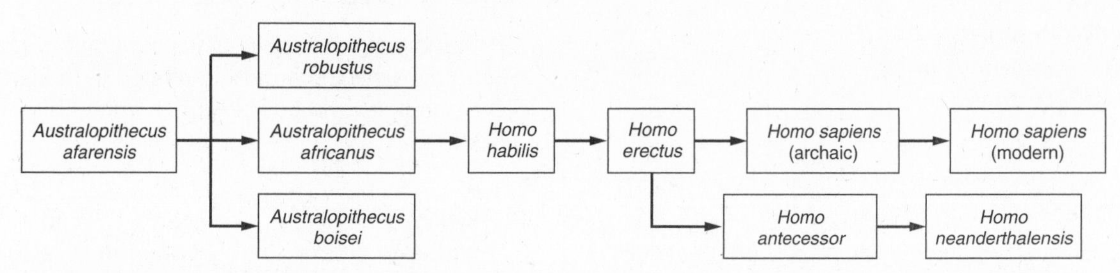


**Figure 25-8** An evolutionary tree showing several branches of hominid species: one leads to modern humans; the others lead to extinct hominid species.

# Chapter 25 Review

## Part A—Multiple Choice

**1.** For about 130 million years, the dominant large animals were

1 sea mammals such as whales
2 land mammals such as elephants
3 reptiles such as dinosaurs
4 primates such as apes

**2.** Based on the information about reptile and mammal evolution, it is correct to say that geologic time

1 is a very short-term measure of time
2 is a very long-term measure of time
3 is a very brief moment in time
4 really has nothing to do with time

**3.** Why did an enormous variety of mammals evolve after dinosaurs became extinct?

1 Mammals had been afraid of the dinosaurs.
2 The dinosaurs had been eating all the mammals.
3 Dinosaurs and mammals did not exist at the same time.
4 Without dinosaurs, there was less competition for resources.

**4.** Primates are mainly adapted for a life in the

1 water
2 trees
3 arctic
4 desert

**5.** Monkeys evolved from small primate ancestors about

1 1 million years ago
2 25 million years ago
3 50 million years ago
4 100 million years ago

**6.** Apes first evolved from monkeys in

1 Africa and Asia
2 South America
3 Europe and Asia
4 North America

**7.** What is the evolutionary connection between humans and chimpanzees?

1 Humans evolved from the first chimpanzees.
2 Humans have exactly the same DNA as chimpanzees.
3 Humans and chimpanzees both always walk on two feet.
4 Humans and chimpanzees share a common ancestor.

**8.** Walking on two feet helped hominids survive because it

1 made them look taller
2 made their brains grow larger
3 left their hands free to gather food and carry their young
4 let them swing from branch to branch more efficiently

**9.** The oldest hominid fossils indicate that our ancestors first evolved in

1 Asia
2 Africa
3 Eurasia
4 America

**10.** What characteristic separates hominids in the genus *Homo* from other primates?

1 large feet
2 an enlarged brain
3 large hands
4 a height of over 1.7 meters

**11.** *Homo erectus* were probably the first hominids to

1 build fires
2 walk on two feet
3 climb trees
4 eat meat

**12.** *Homo erectus* had skills that enabled them to

1 survive in a warm climate only
2 migrate to live in colder climates
3 keep a written history of their lives
4 grow grain and domesticate cattle

## Part B—Analysis and Open Ended

**13.** Based on your reading, state whether modern humans are more closely related to modern Asian apes or modern African apes. Give your reason.

**14.** Which statement is more correct: "humans and gorillas probably share a common ancestor" or "humans probably evolved from gorillas"? Explain your answer.

**15.** Study the following photograph, which shows a chimpanzee using two sticks. Explain the importance of this chimp's actions in terms of the following ideas:

- the significance of the chimp's behavior in relation to hominid evolution
- how the chimp's behavior is quite similar to that of humans
- how the chimp's behavior is different from that of humans

**16.** Describe how the differences between a grasslands environment and a forest might affect an animal's adaptations and survival.

**17.** Briefly explain the importance of bipedalism to early hominids. Your answer should describe the following:

- environmental changes that might have led to bipedalism
- possible effect of bipedalism on methods of food gathering
- possible effect of bipedalism on hominid social behavior

**18.** In what way do hominids in the genus *Homo* differ from earlier hominids and from other primates? Why do you think this difference is so important?

**19.** List some traits *Homo erectus* had that were more advanced than those of earlier hominids and similar to those of modern *Homo sapiens.* Your answer must include at least:

- *one* physical characteristic different from that of earlier hominids
- *two* physical characteristics similar to those of modern humans
- *three* behavioral characteristics similar to modern humans

**20.** The figure below illustrates what an early *Homo sapiens* (Neanderthal) might have looked like. List four features you can observe that set it apart from most other primates and make it closest to modern humans. Tell why these traits and abilities were important for the survival of early humans.

## Part C—Reading Comprehension

*Base your answers to questions 21 through 23 on the information below and on your knowledge of biology.* Source: *Science News*, Vol. 157, p. 302.

> Excavations in Wyoming have yielded the partial skeleton of a 55-million-year-old primate that probably was a close relative of the ancestor of modern monkeys, apes, and people. The creature was built for hanging tightly onto tree branches, not for leaping from tree to tree, as some scientists had speculated, based on earlier fragmentary finds. Also, despite expectations, the ancient primate didn't have eyes specialized for spotting insects and other prey.
>
> Jonathan I. Bloch and Doug M. Boyer, both of the University of Michigan in Ann Arbor, unearthed the new specimen. It belonged to a group of small, long-tailed primates that lived just before the evolution of creatures with traits characteristic of modern primates—relatively large brains, grasping hands and feet with nails instead of claws, forward-facing eyes to enhance vision, and limbs capable of prodigious leaping.
>
> The new find, in the genus *Carpolestes*, had long hands and feet with opposable digits, Bloch and Boyer report in the Nov. 22 *Science*. The animal grew nails on its opposable digits, and claws on its other fingers and toes. Unlike later primates, *Carpolestes* had side-facing eyes and lacked hind limbs designed for leaping.

**21.** Describe *two* characteristics of the 55-million-year-old Wyoming primate (fossil) that are different from what the scientists had expected.

**22.** Identify *two* or more characteristics the fossil lacked, which are considered to be those of modern primates.

**23.** What kind of environment did the ancient Wyoming primate inhabit, and what physical features did it have for locomotion in that environment?

# THEME VII

## Interaction and Interdependence

# 26

# Introduction to Ecology

### Vocabulary

| | | |
|---|---|---|
| abiotic | biotic | habitat |
| acidity | ecology | limiting factors |
| adaptations | ecosystem | niche |
| algae | | |

## ECOLOGY AND ECOSYSTEMS

An aquarium is a self-contained miniature world of life. Like the living things in an aquarium, every organism on Earth lives within its surroundings, or environment. All living things interact—they affect other living things and their environment; and all living things depend on each other and their environment—they are interdependent. These relationships of *interaction* and *interdependence* between organisms and their environment are studied in the branch of biology known as **ecology**. (See Figure 26-1.)

**Figure 26-1** An aquarium is like a miniature ecosystem—the living things interact with one another and with the nonliving parts of their environment.

Every organism has to live somewhere. The environment in which an organism lives is defined by many different *factors*, such as availability of food and water, amount of sunlight, temperature, and type of soil. Conditions that involve other living organisms are known as **biotic** factors. For a fish in the aquarium, the biotic factors could include other fish, snails, algae, and plants. Conditions that involve nonliving things are known as **abiotic** factors. For that same fish, the abiotic factors could include the water, air bubbles, gravel, temperature, light, and **acidity** (the pH level).

The interaction and interdependence of organisms and their environment can be understood by examining specific places. All the living and nonliving factors that interact in one specific place are the parts of an **ecosystem**. A pond is an ecosystem. A forest is an ecosystem. Even the little aquarium is an ecosystem that can be used to study ecology.

**Figure 26-3** Many rain-forest trees have wide, woody supports at the base of their trunks to keep them upright in the shallow tropical soil.

## ADAPTATIONS AND EVOLUTION

Camels have extremely wide, two-toed feet to avoid sinking into the desert sand. (See Figure 26-2.) Many rain-forest trees have wide supports at the base of their trunks to keep them upright in the shallow tropical soil. (See Figure 26-3.) Wherever we look on this planet, we see many examples of the features that enable organisms to survive Earth's great variety of living conditions. How did all this happen?

**Figure 26-2** Organisms have adaptations that help them survive in particular environments. The camel's wide, two-toed feet enable it to walk in the desert sand without sinking.

No individual organism intentionally changed to survive in a particular environment. What did happen is that there have always been some differences among individuals in populations. Two dogs from the same litter, for example, might be very different in size or color. Sometimes the differences are not obvious. Yet a slight difference in the biological makeup of an organism, such as the ability to make a particular enzyme, might provide it with a survival advantage over other organisms. These variations are due to inherited differences. Because of these differences, some individuals are better suited than others to certain environmental factors.

It is through the process of natural selection that species, not individual organisms, evolve. This is the basis of evolution. Over time, a species' traits make a remarkable fit with its environment. If they do not, the

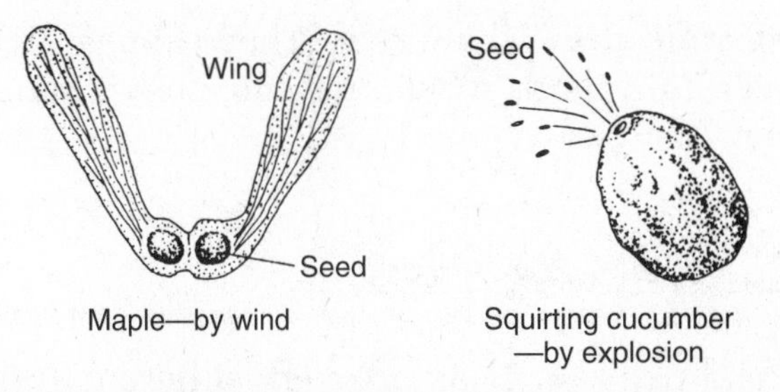


**Figure 26-4** Plants have many different adaptations for releasing and dispersing their seeds.

species will probably not survive in that particular environment. The characteristics of an organism that give it this fit are called **adaptations**. Adaptations can be in physical traits, such as the size, shape, or color of the organism. They also can be in the behavior of an organism, such as the building of nests by birds or the release of seeds once a year by plants. (See Figure 26-4.)

## AN ORGANISM'S HABITAT AND NICHE

Every organism is adapted through the process of evolution to live in a particular place. Each species of organism is adapted to a specific set of conditions. The place where an organism lives is its **habitat**. The habitat of a bullfrog is a pond. The habitat of a giant anteater is open grassland. An organism's habitat is its "address."

To understand an organism's relationship to its environment, we must know more than its address, or *where* it lives. We must also know its "occupation," or what it does and *how* it lives. The occupation of an organism is called its **niche**. The niche of an organism includes how it gets food, reproduces, avoids predation, and so on. The behavioral adaptations of an organism make up its niche. These adaptations are the result of evolution just as are its physical adaptations. The niche of an organism determines its habitat. In other words, the ways that an organism has evolved to survive will also determine where it can live. For example, a woodpecker cannot live in grasslands; its niche involves finding its insect food in the trunks of trees. So woodpeckers need a habitat that has trees, and the insects that live in them, to survive. (See Figure 26-5.)

**Figure 26-5** The niche of an organism determines where it can live. For example, the woodpecker needs to live in a forested habitat because its niche involves finding insects that live in trees.

## ENVIRONMENTAL FACTORS

Every type of organism is adapted to the specific conditions in its environment. Since organisms do have specific habitat requirements, different environmental conditions can limit where they live. These conditions are called **limiting factors**.

For example, the availability of sunlight is an important limiting factor for plants. The amount of sunlight in oceans and lakes varies but does not reach below a certain depth. Below that depth, it is too dark for aquatic plants to grow. Temperature, of course, is another major limiting factor. Each species of plant or animal has a fairly narrow temperature range that it prefers. In other words, the organism can tolerate temperatures only within this range. Other environmental factors, such as chemical nutrients in the habitat, may be less obvious, but are still very important for a species' survival. In general, organisms have a tolerance range for a variety of environmental factors, sometimes a very narrow one, usually neither too low nor too high. The tolerance range determines the best conditions for a specific type of organism in a specific location.

All life depends on water, and organisms have a variety of adaptations that enable them to survive in very specific ranges of available moisture. For example, water constantly moves out of openings on the surface of leaves. So some species of trees, such as pines, have evolved ways to save water—they have narrow leaves, or needles, from which little moisture is lost. In areas with abundant rainfall, water loss is not a problem, so the trees have large, flat leaves.

## AQUATIC ECOSYSTEMS

Because water is so essential to life, we will first look at aquatic ecosystems. A map of the world shows individual oceans; however, all the world's oceans are actually connected. Some ecologists consider this world ocean to be one tremendously large *saltwater ecosystem*. Three main limiting factors in the ocean are salinity, temperature, and sunlight. As the amount of salt in ocean water varies, the density of the water also changes accordingly; the more salt, the higher the density. The temperature of ocean water varies throughout the world, too, and affects water density. Cold water is denser than warm water, so the colder water sinks. The amount of sunlight also varies over different parts of the ocean, and it penetrates only to a certain depth. Because they need light to carry out photosynthesis, all plants and **algae** in the ocean live only in the top (photic) zone of the water. (See Figure 26-6.)

There are two main types of *freshwater ecosystems* on the surface of Earth: lakes and ponds, which are still bodies of water; and rivers and streams, which are running water. In the still-water ecosystems, temperature and light are the main limiting factors. In the running-water ecosystems, the temperature and light are fairly constant at any given point but vary along the length of the river or stream. For example, the water flows faster and colder at the start of a river than at its end. So, different kinds of plants and animals are adapted to survive in different parts of the river. In general, conditions in water are fairly constant over wide areas. They change little over time and, when they do, they change very slowly.

## LAND ECOSYSTEMS

Conditions on land are very different from those in the water. From season to season, and from one part of the day to another, temperatures on land can vary widely. Variations in temperature, moisture, soil type, length of days and nights, seasons, and altitude all work together to produce many different land ecosystems.

There are three main forest ecosystems on Earth, each with its characteristic types of trees: tropical rain forests, broad-leaved forests, and needle-leaved forests. Tropical rain forests exist in a wide band north and south of Earth's equator. They have large amounts of rainfall, warm temperatures, and a stable length of daylight throughout the year. Abundant life of all kinds exists in the tropical rain forests.

Farther north and south of the equator, climate patterns change. Definite seasons occur, with variations in temperature and rainfall. This creates the forest environment of broad-leaved trees, such as maple and oak. Here, fewer tree species are found, and the seasonal dropping of leaves is typical. Still farther north, where it is colder, there are the great forests of needle-leaved evergreen trees, such as spruce and pine. (See Figure 26-7.)

Farthest from the equator, the temperatures and amount of rainfall are too low for

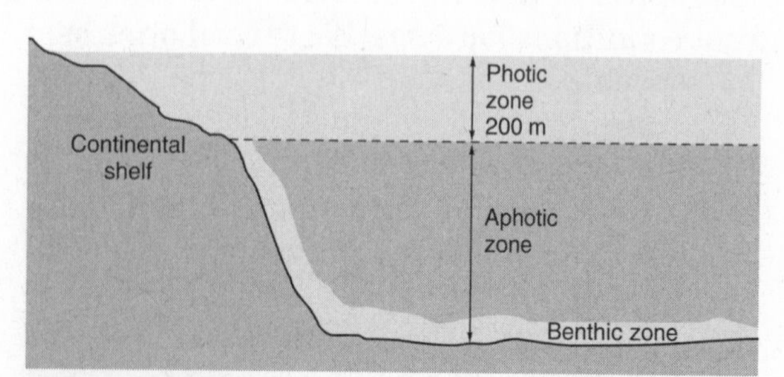


**Figure 26-6** Sunlight is a limiting factor for the growth of plants. Because sunlight penetrates only about 200 meters below the ocean's surface, aquatic plant life is restricted to that top (photic) zone.

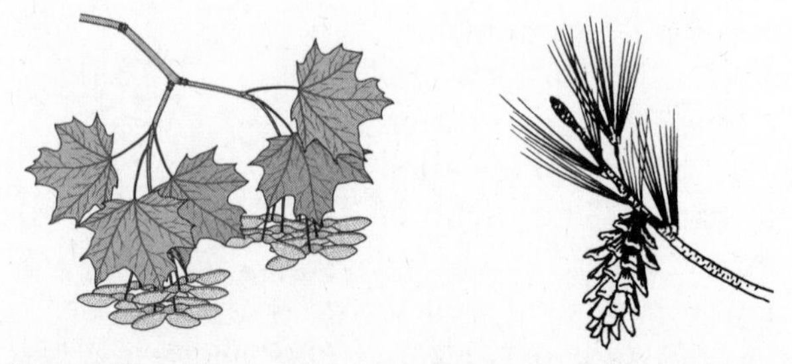

**Figure 26-7** Water is another important limiting factor for plants. In areas of abundant rainfall, plants such as the maple (left) have broad, flat leaves; in areas of limited rainfall, plants such as the white pine (right) have narrow, needle-shaped leaves to reduce water loss.

trees to grow. Only one or two months of the year are warm enough to support plant growth. Then, a thick layer of mosses, lichens, grasses, and low shrubs covers the surface.

In regions where precipitation, but not temperature, decreases, there are grasslands, because the limited moisture prevents the growth of trees. Finally, the driest places on Earth are the deserts. In a typical desert, plants include water-storing cactuses, shrubs with roots that grow deep to reach water, and wildflowers and grasses that flourish briefly after the infrequent rains. (See Figure 26-8.) Animals also show adaptations to the desert conditions. For example, the kangaroo rat lives underground for much of the day to avoid the heat, and desert predators such as coyotes and foxes are also more active at night.

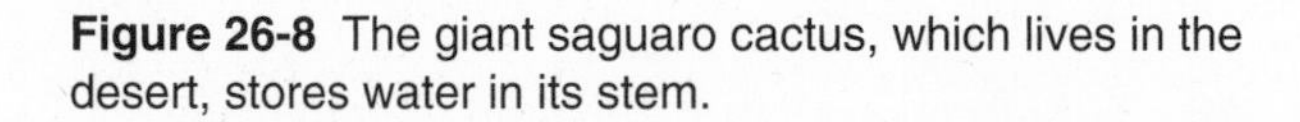

**Figure 26-8** The giant saguaro cactus, which lives in the desert, stores water in its stem.

## Chapter 26 Review

### Part A—Multiple Choice

**1.** Ecology can best be described as the study of

1 all the plants in a certain environment
2 living organisms and their environment
3 living factors that affect an organism
4 nonliving factors that affect an organism

**2.** A *biotic* factor in a snake's environment would be

1 sunlight
2 water
3 sand
4 a mouse

**3.** An *abiotic* factor in an eagle's environment would be

1 a tree
2 water
3 a snake
4 an insect

**4.** Which event illustrates the interaction of a biotic factor with an abiotic factor in the environment?

1 Water temperature affects water density in the ocean.
2 The lamprey eel survives as a parasite on other fish.
3 Shorter daylight hours cause maple trees to lose leaves.
4 A gypsy moth caterpillar eats the leaves of an oak tree.

**5.** An ecosystem is best described as the

1 type of food that an organism eats
2 type of home an organism builds
3 group of organisms in a particular place
4 living and nonliving factors in one place

**6.** Which of the following can be considered an ecosystem?

1 a large rock
2 a bird's nest
3 a rain forest
4 a rain cloud

**7.** Natural selection is important because it

1 allows an individual organism to evolve
2 enables a species to adapt and evolve
3 allows humans to control animal traits
4 changes an environment to fit the organisms

**8.** An adaptation is a body part or behavior that

1 is no longer used by an organism
2 prevents an organism from surviving
3 helps an organism survive and be fit
4 is useful to only one organism at a time

**9.** A habitat can be described as an organism's

1 average size
2 main function
3 wild behavior
4 natural address

**10.** A frog's habitat would be the

1 pond it lives in
2 sounds it makes
3 insects it eats
4 color of its skin

**11.** An organism's niche is most similar to a person's

1 character
2 occupation
3 address
4 personality

**12.** In a forest community, a shelf fungus and a slug live on the side of a decaying tree trunk. The fungus digests and absorbs materials from the tree, while the slug eats algae growing on the outside of the trunk. These organisms do not compete with one another because they occupy

1 the same habitat, but different niches
2 the same niche and the same habitat
3 the same niche, but different habitats
4 different habitats and different niches

**13.** In a pond, a carp eats decaying matter from the base of an underwater plant, while a snail scrapes algae from the leaves and stems of the same plant. These animals can both survive in the same pond because they occupy

1 the same habitat and the same niche
2 the same niche, but different habitats
3 the same habitat, but different niches
4 different habitats and different niches

**14.** Light, temperature, and water are examples of environmental

1 habitats
2 niches
3 limiting factors
4 adaptations

**15.** Why do plants live only in the top zone of the ocean?

1 There is too much salt in deeper water.
2 They automatically float to the top.
3 Plants cannot grow underwater.
4 Plants need sunlight to make food.

**16.** Which factor has the greatest influence on the variety of species that live in different regions of a marine habitat?

1 depth of light penetration
2 size of predators
3 daily fluctuations in temperature
4 average annual rainfall

**17.** The main limiting factors in freshwater ecosystems are

1 temperature and light
2 salt and temperature
3 density and light
4 altitude and depth

**18.** Ecosystems vary more on land than in water because

1 there is more land than water on Earth's surface
2 conditions on land vary more than they do in water
3 evolution of new species does not occur in water
4 organisms that live in water become extinct sooner

**19.** Earth's three main forest ecosystems vary because they

1 experience different temperatures and rainfall
2 are located near different major cities
3 differ greatly in size
4 are in different stages of development

### *Part B—Analysis and Open Ended*

**20.** Identify five conditions that may be included as part of an organism's environment.

**21.** What is the difference between biotic and abiotic factors in an ecosystem? Your answer should include the following:

- the definition of *biotic*
- the definition of *abiotic*
- *two* examples of biotic factors
- *two* examples of abiotic factors

**22.** Provide one example of an interaction between a biotic factor and an abiotic factor that you have observed in your local environment.

*Refer to Figure 26-1 on page 203 to answer questions 23 and 24.*

**23.** Why is the aquarium a good example to use when studying about ecosystems?

**24.** List at least two biotic and two abiotic factors that are present in this aquarium.

**25.** Identify one *abiotic* factor that would directly affect the survival of organism A, shown in the diagram below.

**26.** How does natural selection explain the close fit of organisms to their environment?

**27.** Why does the niche of an organism determine its habitat? Give one example.

**28.** Using an example such as sunlight or moisture, explain the idea of a limiting factor. How are such factors related to the diversity of environmental conditions and life-forms on Earth?

*Base your answers to questions 29 to 32 on the information below and on your knowledge of biology.*

Duckweed is one of the smallest flowering aquatic plants. It grows floating on still or slow-moving fresh water. Duckweed species are found throughout the world, except in very cold regions, and are the subject of much scientific research. The plants are used to study basic plant biochemistry, plant development, and photosynthesis.

Environmental scientists are using duckweed plants to remove hazardous substances from water. Fish farmers use them as an inexpensive food source for the fish they raise. As with other aquatic plants, duckweed grows best in water containing high levels of nitrates (nitrogen compounds) and phosphates. The level of iron-containing compounds is often a limiting factor. A cover of duckweed on a pond shades the water below and reduces the growth of algae. A key for identifying duckweed species is shown below.

**Duckweed Identification Key**

| Description of Duckweed | Plant Has No Roots | Plant Has One or More Roots |
|---|---|---|
| Plant body is flat | *Wolffiella* | |
| Plant body is oval (less than 1 mm) | *Wolffia* | |
| Plant has one mid-sized root | | *Lemma* |
| Plant has two or more large-sized roots | | *Spirodela* |

**29.** Describe the value that duckweed has for heterotrophic organisms living in ponds where it grows.

**30.** Explain what is meant by the statement, "The level of iron-containing compounds is often a limiting factor."

**31.** State *one* way in which shading the water below it causes duckweed to affect the growth of algae.

**32.** Explain why *Spirodela* would most likely absorb more hazardous substances from the water than would the other duckweed species identified in the key.

**33.** Describe the three main limiting factors in the ocean. Why do they have an effect on life in the sea?

**34.** Suppose a dam is constructed near the headwaters (that is, the point of origin) of a river that flows from a mountain to the sea. How might the river change both upstream and downstream from the dam? How would the changes affect the biotic and abiotic factors in the river?

**35.** Use the following terms to fill in the table at right, which outlines the three types of forest ecosystems: *broad leaves that drop seasonally; equatorial regions; colder temperatures and less rainfall; north and south of the equator; wide bases that support trunks in shallow soil; warmer temperatures and more rainfall; needle-shaped leaves that conserve moisture; variations in temperature and rainfall; northern regions.*

| Characteristics | Tropical Rain Forests | Maple/ Oak Forests | Spruce/ Pine Forests |
|---|---|---|---|
| Type of Trees | | | |
| Type of Climate | | | |
| Geog. Location | | | |

### *Part C—Reading Comprehension*

*Base your answers to questions 36 to 38 on the information below and on your knowledge of biology.*
Source: *Science News*, Vol. 162, p. 253.

The textbook case of how to survive in a desert may have important details wrong, according to new studies of kangaroo rats.

Species in the genus *Dipodomys*, nocturnal rodents that scurry through North America's deserts, have epitomized toughness in punishing climates, says Randall Tracy of the University of Connecticut in Storrs. Earlier researchers, he says, marveled at how the creatures apparently got water by metabolizing seeds and avoided overheating by staying in cool burrows until late at night.

In an (upcoming) issue of *Oecologia*, Tracy and Glenn E. Walsberg from Arizona State University in Tempe challenge those views. They made their observations near Yuma, Arizona, in the Sonoran desert.

The animals' burrows get hotter than expected, the researchers found. For more than 100 days of the year, soil temperatures rose to over 30°C at depths of 2 meters. Yet during most of the summer, the kangaroo rats remained less than a meter deep, where it's about 35°C. Nor did the animals emerge only in the cool part of night; they ventured above ground right after sundown. Also, forget the seeds-only menu. The rodents ate a considerable amount of green plant tissue, presumably a substantial water source during tough times.

**36.** Explain the *two* methods that scientists had first thought kangaroo rats used to survive in the desert.

**37.** State the research findings that challenge the traditional explanation of how kangaroo rats stay cool.

**38.** Explain why scientists now doubt that kangaroo rats get their water from the metabolism of seeds.

# 27
# Populations and Communities

## Vocabulary

| | | |
|---|---|---|
| carrying capacity | parasite | prey |
| community | population | succession |
| competition | predator | symbiosis |
| host | | |

## THE STUDY OF POPULATIONS

Ecologists are more interested in groups of organisms than in individuals. All the organisms of one species that live in one place at a particular time make up a **population**. No population ever lives alone. Other organisms—plants, animals, fungi, and microorganisms—are also present. For example, a field has a population of mice, but it also has populations of wildflowers, insects, mushrooms, birds, and snakes. All the populations that interact with each other in a particular place make up a **community**. In large part, the study of ecology is about populations and communities.

## FACTORS THAT AFFECT POPULATION GROWTH

Most organisms are able to reproduce rapidly. Even if a pair of individuals produced only two offspring each year, the growth rate would be enormous if all offspring survived. Suppose a population was founded by two individuals. By doubling each year, after only 10 years there would be a population of more than 1000 individuals. Many organisms produce even greater numbers of offspring, such as fish that lay thousands of eggs per year. If these eggs all hatched and survived to reproduce, the number of resulting fish would be enormous. Trees also produce many thousands of seeds per year. With these growth rates, Earth could quickly be overcrowded with organisms. Of course, this does not occur. Thus, an important area in ecology is the study of what controls population growth.

A key factor in population growth is *density*, which is the number of individuals in a population in a given area. A population with a small number of individuals in a particular area has a low density. When a population's density is low, there is usually sufficient food and space for existing organisms. The birth rate increases, while the death rate drops. As a result, the density of the population begins to increase at a faster rate. But this rate of increase cannot last forever. At some point, the population gets too crowded. (See Figure 27-1 on page 212.)

The most basic needs of organisms from their habitats are food and space. However, every habitat has limits. When the population

**Figure 27-1** You can visualize the population density of New York City from this photograph, taken at night by the crew of a space shuttle. The brighter areas are those where population density is higher.

density increases too much, available food and space decrease. At that point, the population density has reached a maximum for the particular habitat. The death rate increases, while the birth rate drops. The size of a population that can be supported by any ecosystem is called the **carrying capacity**. Population growth slows and may reach zero growth as the population size approaches an area's carrying capacity. Zero growth means the population size is no longer increasing—the birth rate and death rate are about equal. The rate at which a population grows is shown in Figure 27-2.

Factors that limit the size of a population include **competition** for food and space. This competition increases as population density increases. Another problem with high population density is that the more crowded a population is, the easier it is for predators to find prey. In a crowded population, there is also a greater chance for disease to spread among individuals.

## COMMUNITY INTERACTIONS

Organisms that live in the same environment are always interacting with each other. The survival of many plants and animals often depends on the relationships they have with other organisms. (See Figure 27-3.) Some important types of relationships are discussed below. Competition is one of the main interactions between organisms. For example, if two different species of birds ate the same species of butterflies from treetops in the same forest, there would be competition between them. This competition between two different species would be intense, in this case, because of the overlap of the birds' niches. However, if the two bird species fed on different insects in the same treetops, there would be less overlap in their niches and less competition between them.

The greatest competition usually occurs between members of the same species, because such individuals most likely share identical

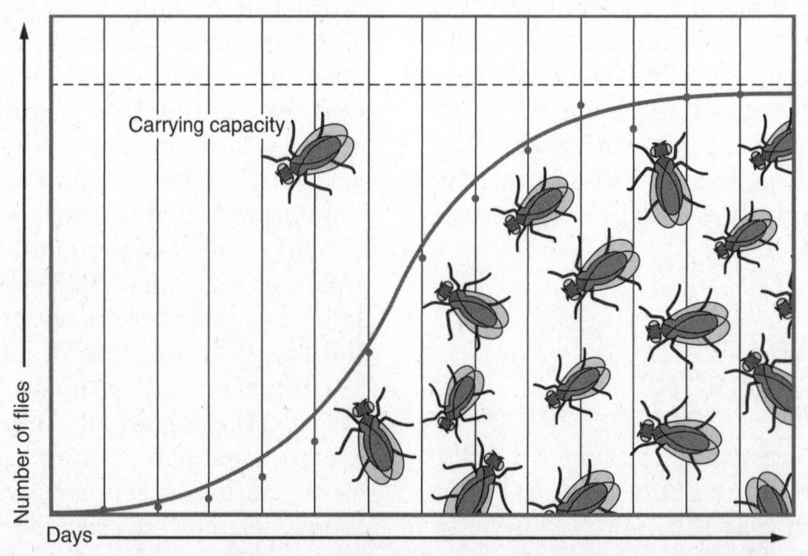


**Figure 27-2** A typical population growth curve. A new population increases slowly at first, because there are few individuals. Then, as the population increases, the growth rate increases rapidly as more individuals reproduce. After some time, it levels off to zero growth as the population approaches the area's carrying capacity.

**Figure 27-3** A community, such as that illustrated living in and around this pond, is made up of many different organisms. The survival of most plants and animals depends on the relationships they have with the other organisms in their environment.

niches. In other words, they live in exactly the same way and compete most intensely for the same limited resources in their area. Competition results in natural selection. Recall the thousands of eggs produced by the fish—most will not survive to maturity. Only the most fit individuals live to pass on their genes to offspring. Therefore, competition is an important force in the process of evolution.

Predation is one of the most basic relationships that occurs between organisms. In nature, living things either "eat or are eaten." That is, they are either the **predator** or the **prey**. In predation, members of one population are the food source for members of another population. Even grass seeds can be considered prey when a mouse—the predator, in this case—eats them. Of course, the mouse becomes the prey when a snake, another predator, eats it. (See Figures 27-4 *a* and *b*.)

## TYPES OF SYMBIOTIC RELATIONSHIPS

As stated above, almost no organism lives entirely alone. In fact, most organisms have close relationships with at least one other type of organism. This close relationship is called **symbiosis**. In this relationship, one type of organism can live near, on, or even in another organism. Each partner in the relationship can either help, harm, or have no effect on the other partner.

A **parasite** lives on or in another organism that it uses for food and, sometimes, for shelter. The organism the parasite uses is called the **host**. In this type of interaction, called *parasitism*, the parasite is helped, while the host is harmed. This relationship is different from the *predator-prey relationship*. In that relationship, the prey is usually killed right

**Figure 27-4*a*** The mouse, which is a rodent, is considered the predator when it preys on plants.

**Figure 27-4*b*** The snake is the predator when it eats a rat or mouse, which is then the prey.

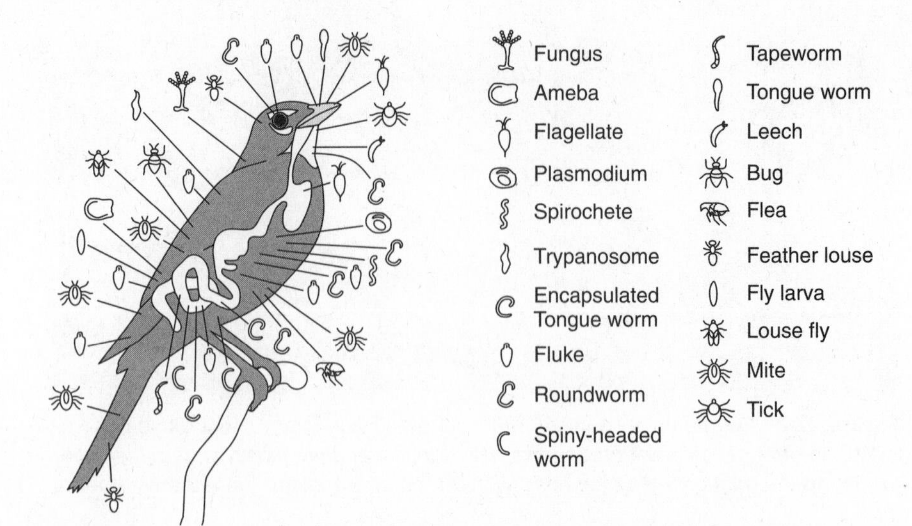


**Figure 27-5** One bird may be host to many different types of parasites (although not all of these at once). Many parasites have evolved to be specifically adapted to their host, which is typically harmed but not killed by the parasites.

away. In parasitism, the host organism usually continues to live, but it is harmed. (See Figure 27-5.) Parasites usually evolve together with their host and have characteristics that make them specifically adapted to it. For example, the human tapeworm is a parasite that is adapted to live inside the human intestines.

In many *symbiotic relationships*, one organism benefits, while the other organism remains unaffected. For example, when the Cape buffalo walks through the African plains to browse, it disturbs insects in the grass. Birds called cattle egrets gather around the buffalo and feed on the insects. The birds are helped, and the Cape buffalo is unaffected; this is called *commensalism*.

Finally, there are symbiotic relationships in which both parties benefit, called *mutualism*. An example of this is the relationship between a type of acacia tree and a species of stinging ants. The trees produce hollow thorns. The ants make their nests in these thorns and feed on sugars produced by the plant. If any other insect lands on the acacia, the ants quickly surround and kill it. The ants have shelter and a source of food, and the tree is protected from other plant-eating insects.

## CHANGING COMMUNITIES

A complex variety of interactions exists within an ecosystem. Many different populations live together and affect each other. In a forest, for example, there are populations of grasses, shrubs, trees, fungi, insects, worms, bacteria, birds, reptiles, and mammals that make up the community. Yet this community will probably not remain exactly the same forever.

**Figure 27-6** This photograph shows that regrowth and succession have taken place in a formerly fire-ravaged forest. Such rapid regrowth after a disaster provides scientists with the opportunity to understand the normally slower process of succession.

Some populations of organisms may disappear entirely, while other populations may move in from somewhere else. Existing populations may increase or decrease in number. The community may even change with the seasons. Despite these kinds of changes, however, the forest—especially a very old forest—may remain essentially the same over many years. If so, it has become a *stable* community.

But then a sudden, profound change may occur. For example, a fire may destroy much of the forest. Some animals may die, while others are able to escape. Did the fire permanently destroy the forest? No. Instead, the ecosystem undergoes what scientists call the process of ecological **succession**. Usually, succession is a series of slow changes that occur in an area until a stable community is formed. Most naturally occurring successions in an area take longer than a person's lifetime. However, after a sudden disturbance, a formerly stable community quickly begins to go through a series of changes. These changes often follow similar patterns wherever the same kind of disturbance has occurred. (See Figure 27-6.)

Succession also occurs when a new environment appears for the first time, such as when a volcano produces new rock. The gradual succession of communities that appear in these places can tell us a great deal about how communities on Earth evolved during ancient times.

## Chapter 27 Review

### Part A—Multiple Choice

**1.** A population can best be described as all the

1 plants in a particular place
2 animals in a particular place
3 different organisms in one place at a particular time
4 organisms of one species in one place at a particular time

**2.** An example of a population in a lake is all the

1 lake trout
2 lake trout and brown trout
3 plants and trout
4 soil, plants, and fish

**3.** A community can best be described as all the

1 plant species in one particular place
2 organisms of one species in a particular place
3 populations that interact in a particular place
4 animals that interact in a particular place

**4.** Which ecological term includes everything represented in the illustration at the top of the next column?

1 ecosystem
2 population
3 community
4 species

**5.** When population density is low, the

1 birth rate increases and the death rate drops
2 death rate increases and the birth rate drops
3 birth rate and the death rate both increase
4 birth rate and the death rate both decrease

**6.** An environment can support only as many organisms as the available energy, minerals, and oxygen will allow. Which term is best defined by this statement?

1 biological feedback
3 carrying capacity
2 homeostatic control
4 biological diversity

**7.** The carrying capacity of a given environment is *least* dependent upon

1 recycling of materials
2 the availability of food and water
3 the available energy
4 daily temperature fluctuations

**8.** When a population experiences zero growth, the

1 death rate increases faster than the birth rate does
2 birth rate increases and the death rate drops
3 death rate increases and the birth rate drops
4 birth rate and the death rate are about equal

**9.** Competition between organisms can best be described as an interaction in which the organisms

1 rely on the same resources
2 work together to find food
3 live in the same place but eat different food
4 eat the same food but live in different places

**10.** Competition can occur between members of

1 the same species only
2 different species only
3 both the same and different species
4 two different communities

**11.** Why is competition important in terms of evolution?

1 Organisms evolve to be in greater competition with each other.
2 Organisms evolve to work together without any competition.
3 Among competing organisms, only the fittest survive to reproduce.
4 Competition causes the environment to increase its resources.

**12.** Although three different bird species inhabit the same type of tree in the same forest, there is very little competition among them. The most likely reason for this is that the birds

1 are unable to interbreed
2 have different ecological niches
3 have a limited supply of food
4 share food with each other

**13.** A parasitic relationship differs from a predator-prey relationship in that a

1 host organism is killed right away, whereas prey is not
2 prey organism is killed right away, whereas a host is not
3 parasite helps its host, whereas a predator kills its prey
4 prey organism benefits, whereas a parasite's host does not

**14.** In the symbiotic relationship between cattle egrets and Cape buffalo,

1 both species benefit
2 both species are harmed
3 one species benefits, while the other is harmed
4 one species benefits, while the other is unaffected

**15.** Some crocodiles let small birds enter their mouths to pick bits of food from between their teeth. The crocodiles get clean teeth, while the birds get an easy meal. In this type of relationship,

1 both animals benefit
2 both animals are harmed
3 only the crocodiles benefit
4 only the birds benefit

**16.** The relationship between the crocodiles and birds could be described as a

1 predator-prey relationship
2 parasite-host relationship
3 symbiotic relationship
4 competitive relationship

**17.** In a certain ecosystem, rattlesnakes are predators of prairie dogs. If the prairie dog population started to increase, how would the ecosystem most likely regain stability?

1 The rattlesnake population would start to decrease.
2 The prairie dog population would increase rapidly.
3 The rattlesnake population would start to increase.
4 The prairie dog population would begin to prey on the rattlesnakes.

**18.** In the relationship between the hollow-thorned acacia trees and the stinging ants,

1 both the trees and the ants benefit
2 both the trees and the ants are harmed

3 the tree benefits, while the ants are harmed
4 the ants benefit, while the tree is unaffected

**19.** Succession in an ecosystem is *usually* a

1 sudden event that changes the ecosystem
2 series of very rapid changes in the area
3 series of slow changes that occur in the area
4 short period of rapid change followed by a stable period

**20.** A new island formed by volcanic action may eventually become populated by living communities as a result of

1 a decrease in the amount of organic material
2 the lack of abiotic factors in the area
3 decreased levels of carbon dioxide in the area
4 the process of ecological succession

**21.** Which statement concerning ecosystems is correct?

1 Stable ecosystems that are changed by a natural disaster will slowly recover and may become stable again if left alone for a long time.
2 Climatic change is the principal cause of habitat destruction in ecosystems in the last 50 years.
3 Competition does not influence the number of organisms that live in an ecosystem.
4 Stable ecosystems, once changed by a natural disaster, will never recover and become stable again.

**22.** Events that occurred in four different ecosystems are listed in the table below.

| Ecosystem | Ecological Events |
|---|---|
| A | A severe ice storm occurs during the winter, damaging trees and shrubs. No ice storms occur during the next 20 years. |
| B | A severe drought causes most of the leaves to fall from the trees during a single summer. There are no serious droughts during the next 20 years. |
| C | An island with a dense shrub population becomes submerged for three years. After the river lowers, the island does not become submerged again for the next 20 years. |
| D | A fire burns through a large grassy area. Fires do not occur in the area again for the next 20 years. |

Which ecosystem probably would require the most time for succession to restore it to its previous condition?

1 ecosystem A
2 ecosystem B
3 ecosystem C
4 ecosystem D

### *Part B—Analysis and Open Ended*

**23.** How would Earth be different if certain factors did not limit population growth?

**24.** Why is population growth more rapid when population density is low?

*Refer to the graph below, which shows the growth curve for a population of* Paramecium caudatum, *to answer questions 25 to 27.*

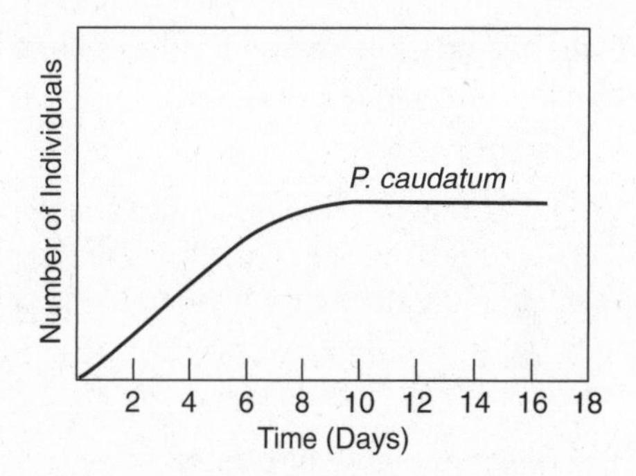


**25.** Why does the slope of the graph increase from the beginning to the middle?

1 The death rate begins to increase.
2 The growth rate slows after four days.
3 The population grows while it is below carrying capacity.
4 There is intense competition for resources.

**26.** The level (flat) portion at the top of the graph indicates that the population

1 is growing
2 is shrinking
3 is neither growing nor shrinking
4 no longer exists in that location

**27.** How does the size of the paramecium population change as it approaches carrying capacity?

*Refer to the following graph, which shows the growth curves for two different populations of paramecia species, to answer questions 28 and 29.*

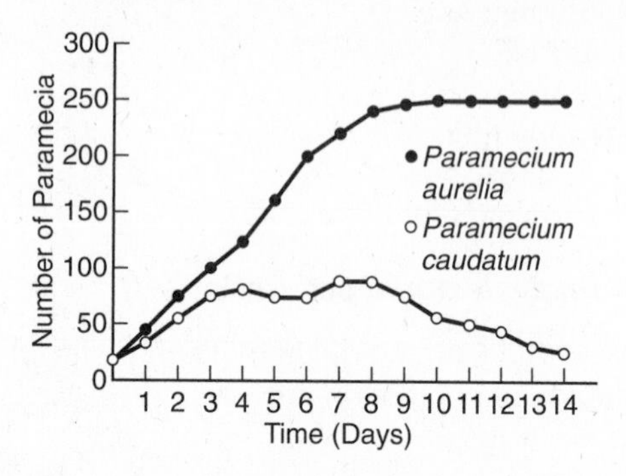


**28.** The two populations of paramecium species (*P. aurelia* and *P. caudatum*) were grown in the same culture dish for 14 days. Which ecological concept is best represented by this graph?

1 recycling
2 competition
3 equilibrium
4 decomposition

**29.** State *one* possible reason for the decline in the number of *P. caudatum*.

**30.** Why is competition for food and space usually greatest among members of the same species? How does this relate to the process of evolution?

**31.** The diagram below illustrates

1 competition between different types of plant life
2 rapid ecological succession after a forest fire
3 gradual succession from bare rock to stable forest
4 evolution of plant life on Earth over 2 billion years

**32.** Compare predation and parasitism. Your answer should include:

- the definitions of predation, predator, prey
- the definitions of parasitism, parasite, host
- *one* way predation and parasitism are similar
- *one* way predation and parasitism are different

*Refer to the diagram below to answer questions 33 and 34.*

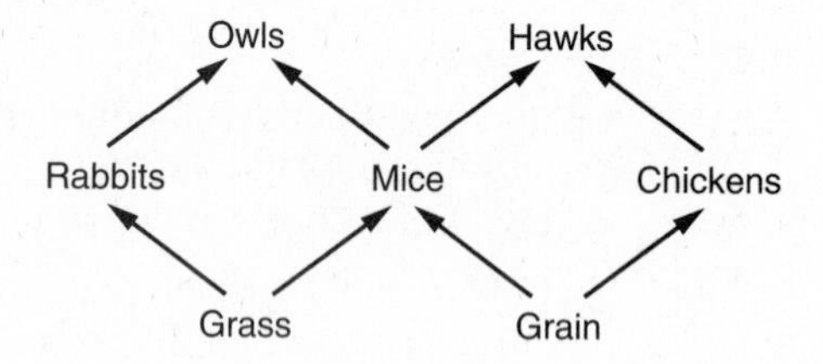


**33.** Based on the diagram, which of the following statements is true?

1 Rabbits and owls compete for grass.
2 Mice and chickens compete for grain.
3 Rabbits and chickens compete for grass.
4 Chickens and rabbits compete for grain.

**34.** Owls hunt at night, whereas hawks hunt during the day. This has the effect of

1 reducing competition for mice because the birds occupy separate forests
2 reducing competition for mice because the birds occupy separate niches
3 increasing competition for mice because the birds occupy the same niche
4 reducing competition for rabbits and chickens because the birds eat more mice

*Base your answers to questions 35 to 37 on the information and graph below and on your knowledge of biology.*

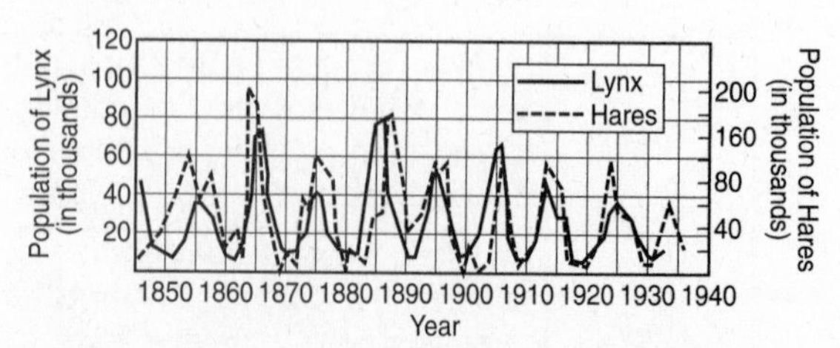


Scientists have hypothesized that the populations of both lynx and snowshoe hares should show cyclical changes—with increases in the predator population size lagging behind increases in the prey population size—if the assumption is made that snowshoe hares are eaten only by lynx.

Does this out-of-phase population cycle of predators and prey actually occur in nature? A classic example of such a cycle was observed by counting all the fur pelts (skins) from northern Canada lynx and snowshoe hares purchased by the Hudson Bay Company between 1845 and 1935. Population cycles of snowshoe hares and their lynx predators, based on the numbers of pelts, are shown in the graph on page 218.

As with any field investigation, many variables could influence the relationship between hare and lynx. One problem is that hare populations have been shown to fluctuate even without lynx being present, possibly because the carrying capacity of their environment had been exceeded. To test this hypothesis about population cycles more scientifically, investigators turned to controlled laboratory studies on populations of small predators and their prey.

**35.** Identify *two* variables other than the size of the lynx population that can affect the size of the hare population.

**36.** The phrase "carrying capacity" refers to the

1 animals storing extra food for the winter
2 number of organisms a habitat can support
3 transporting of food to organisms in an area
4 maximum possible weight of an individual organism

**37.** Why would researchers want to conduct a laboratory study on populations of different predators and their prey?

**38.** Why is evolution important to the relationship between parasites and their hosts?

**39.** As shown in the following figure, the remora has a suckerlike disk on its head by which it attaches to the underside of a shark. The remora feeds on leftovers from the shark's meals, without taking anything from the shark's body. This is an example of a symbiotic relationship in which

1 both parties benefit by being able to catch more food
2 one party benefits and the other is directly harmed
3 one party benefits and the other is apparently unaffected
4 both parties are harmed by not being able to swim as fast

**40.** The diagram below shows changes that might occur over time in an area after a forest fire. Which statement is most closely related to the events shown in the diagram?

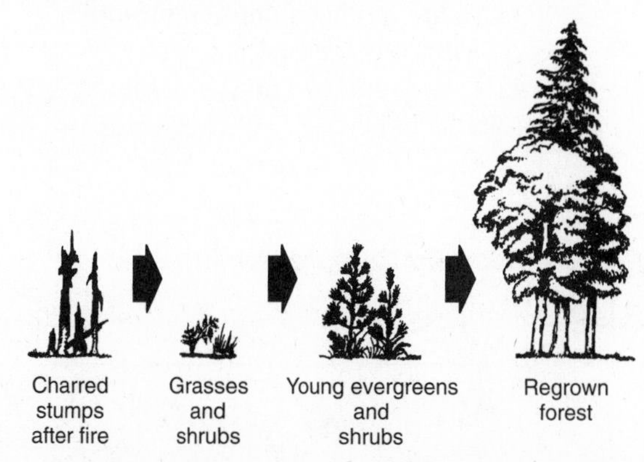


1 The lack of animals in an altered ecosystem speeds natural succession.
2 Abrupt changes in an ecosystem result from only human activities.
3 Stable ecosystems never become established after a natural disaster.
4 An abrupt environmental change can cause long-term changes in an ecosystem.

**41.** Explain how an event such as a forest fire provides scientists with the opportunity to observe ecological succession.

**42.** Explain why an ecosystem with a variety of predator species might be more stable over a long period of time than an ecosystem with only one predator species.

*Base your answers to questions 43 to 46 on the stages of succession shown below and on your knowledge of biology.*

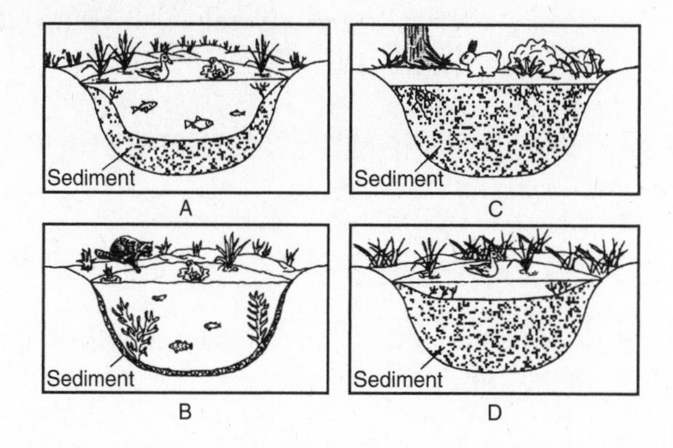

**43.** What is the correct sequence of these stages?

1 B → A → D → C
2 C → B → A → D
3 A → D → C → B
4 D → A → C → B

**44.** Which statement helps to explain this type of succession?

1 Species are replaced until an unstable ecosystem is established.
2 Species are replaced until a stable ecosystem is established.
3 Humans replace all species over time and fill all niches.
4 Animals control all changes in the plant species of an area.

**45.** Which population of organisms would be most harmed by the ecological changes occurring in this succession?

1 trees
2 fish
3 birds
4 rabbits

**46.** Identify *one* factor that could disrupt the final stage of this ecosystem.

### *Part C—Reading Comprehension*

*Base your answers to questions 47 to 49 on the information below and on your knowledge of biology.*

In 1869, the gypsy moth was imported from Europe into Massachusetts. Each gypsy moth caterpillar can eat more than 1 square meter of leaf tissue in its 8-week life, so by 1889 the residents of Boston began to notice many leafless trees.

Every few years, the population of gypsy moths rapidly increases in a season. In the course of two growing seasons, the number of eggs can range from 100 per acre to as many as 1 million per acre. In 1981, about 13 million acres of trees were defoliated (lost their leaves) in the American northeast, and many valuable oak trees died. Between 1979 and 1983, the cost of trying to control these pests totaled 24.2 million dollars. These attempts at control failed.

Rapid growth of a population occurs when there is an abundance of food or when an important environmental factor has been removed. Gypsy moth populations are normally kept in check by phenol chemicals that trees make and release into their leaves. These defensive chemicals stunt caterpillar growth and reduce the number of eggs a female can lay. After several years without caterpillars, the trees stop making these phenols. When this happens, the females eating the phenol-free leaves grow bigger and lay more eggs. Suddenly, a gypsy moth outbreak occurs again, and the cycle is repeated.

When a gypsy moth outbreak occurs, the surrounding ecosystem begins to change as well. Cuckoos, starlings, grackles, mice, and skunks feast on the extra caterpillars, and their numbers increase. Yet all these natural enemies cannot stop the gypsy moth. Trees are stripped of their leaves, weaker trees die at once, and others grow a second set of leaves. If the trees that survive are attacked repeatedly, they also may be weakened beyond recovery.

**47.** Describe *one* condition that might cause the gypsy moth population to increase rapidly.

**48.** State *one* reason that a rapid increase in a gypsy moth population may cause some species of herbivores to vanish or be reduced in number.

**49.** Identify *one* way the surrounding ecosystem changes due to an increase in the gypsy moth population.

# 28 Ecosystems

## Vocabulary

| | | |
|---|---|---|
| biodiversity | food chain | producers |
| carbon | food web | residue |
| carnivores | herbivore | scavenge |
| consumers | hydrogen | stability |
| decomposers | nitrogen | trophic levels |
| energy pyramid | oxygen | |

## THE BASIC CHARACTERISTICS OF ECOSYSTEMS

An ecosystem is made up of living and nonliving factors. In other words, *biotic factors*, such as plants and animals, and *abiotic factors*, such as water, air, and soil, function together in an ecosystem. For organisms to survive there must be a source of energy. The flow of energy between organisms and their environment is a basic characteristic of an ecosystem. Organisms are made up of matter. The flow of matter between organisms and their environment is another basic characteristic of an ecosystem.

What is the source of energy for almost all ecosystems on Earth? It is the sun. While energy is constantly reaching Earth from the sun, matter is not. The amount of matter on Earth remains constant. However, matter moves back and forth between organisms and the environment in all ecosystems.

## ENERGY FLOW THROUGH ECOSYSTEMS

In most ecosystems, energy arrives as sunlight. Some organisms are able to use this energy directly. Other organisms use it indirectly; that is, they get their energy by eating other organisms. Scientists describe and group all organisms in a system of **trophic levels**; the term *trophic* refers to "feeding." (See Figure 28-1 on page 222.)

On the first level are organisms that use energy, such as sunlight, directly from the environment. These first-level organisms are called **producers**. Green plants and algae are producers, or *autotrophs*, because they use the process of photosynthesis to make their own food with water, carbon dioxide, and sunlight. Organisms that feed on producers are in the next trophic level; they are called **consumers**, or *heterotrophs*. A caterpillar that eats oak leaves is a consumer. It is

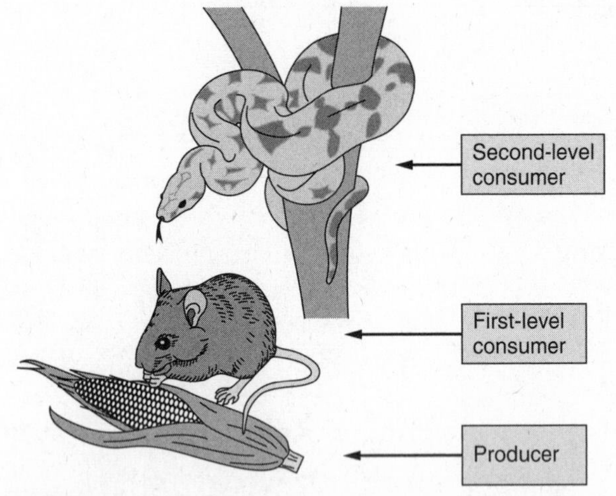


**Figure 28-1** The trophic levels describe the flow of energy in an ecosystem, from the producer to the different levels of consumers.

also a type of **herbivore**, because it feeds on plants. Additional levels exist in which consumers feed on consumers. For example, a small bird may eat a caterpillar. A large hawk may then eat the bird. These animals are called **carnivores**, because they eat other animals. Each of these steps is called a trophic level because it describes the source of the organisms' food. We can describe the flow of energy in an ecosystem by using these trophic levels.

## FOOD CHAINS AND FOOD WEBS

Energy enters an ecosystem at the producer level and is passed along from an organism in one trophic level to an organism in a higher trophic level. This transfer of food energy from one organism to the next is called a **food chain**. From oak leaf to caterpillar to small bird to hawk is a food chain. But in a real ecosystem, a simple food chain like this is never found. Caterpillars are not the only animals that eat oak leaves; small birds are not the only animals that eat caterpillars; and so on. Food chains are actually interconnected in a complex pattern called a **food web**. In a

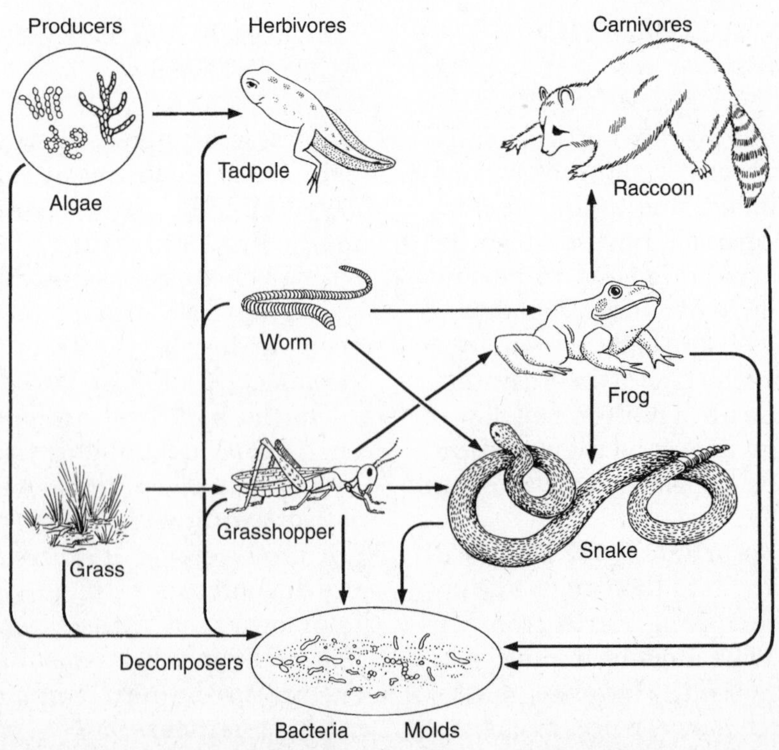


**Figure 28-2** Food chains actually interconnect in complex patterns to form a food web, in which the energy passes between many different organisms.

food web, the energy is passed between many different organisms. (See Figure 28-2.)

However, no matter how complex a food web is, energy always moves in one direction—from a lower to a higher trophic level. It does not get recycled. As energy moves through each trophic level, some of it is used and some of it is lost as heat. The most energy is present at the lowest trophic level (producers); the least is present at the highest trophic level (upper-level consumers). For this reason, additional energy must constantly enter an ecosystem, mainly in the form of sunlight.

Unfortunately, there is a hidden danger in many food chains. If a long-lasting chemical such as DDT enters the environment, it may get passed on from one trophic level to the next. The level of the chemical in each organism increases as it is moved along the food chain, a process called *biological magnification*. For example, little fish may contain some DDT, the larger fish even more, and finally the fish-eating birds the most. While harmless at very low levels, the chemical may have serious effects at the highest levels. That is why the use of DDT (a type of pesticide), which almost destroyed populations of fish-eating eagles and ospreys, was banned years ago in the United States. (See Figure 28-3.)

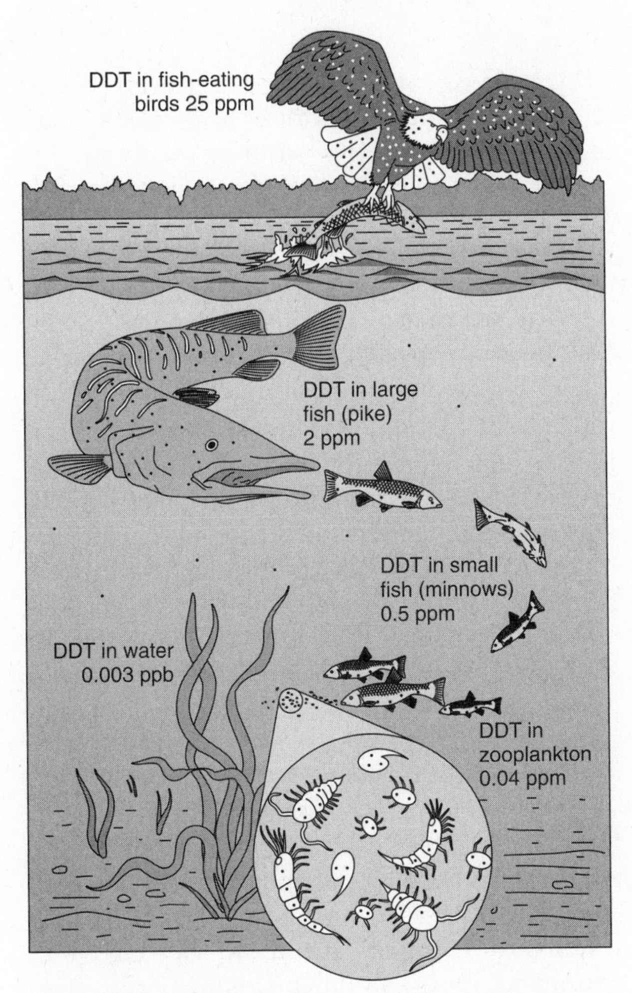


**Figure 28-3** A hidden danger may develop in food chains when certain chemicals enter the environment. For example, the level of DDT in each organism increases as the chemical moves up along the food chain. Harmful effects can occur at the highest trophic levels.

## THE ENERGY PYRAMID

Ecologists use an **energy pyramid** to describe the flow of energy through an ecosystem. The wide base of the pyramid represents the amount of usable energy in the producers, that is, the energy from the sun that is stored in all of the plants. The next step up in the pyramid shows the energy that the first-level consumers get from the producers. This layer is smaller than the energy layer for the producers. Why? Because only about 10 percent of all energy gets passed from one level up to the next. This is true as we move up the pyramid from each level of consumers to the next and then to the top level. (See Figure 28-4.)

An energy pyramid can provide an important lesson in how to feed the ever-increasing human population. Throughout the world, much more food energy is present at the producer level (crops) than at the consumer level (livestock). People may have to make choices based on such questions as: Which type of food is more abundant and available for

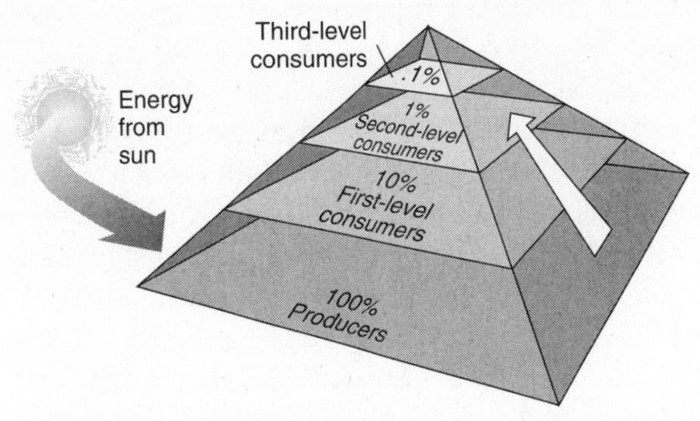


**Figure 28-4** This energy pyramid shows that the amount of available energy decreases at each higher trophic level because only 10 percent of all energy gets passed from one level to the next.

everyone? Which type of food makes a more efficient use of energy sources?

## THE RECYCLING OF MATERIALS IN ECOSYSTEMS

In many parts of the United States, people are now required to recycle consumer wastes such as glass and plastic bottles, newspapers, and metal. *Recycling*, although a new idea for people, is not a new idea in nature. Natural ecosystems have recycled materials since life began on Earth. In fact, life would not continue without this recycling of matter.

All substances are made up of chemical elements. Of the dozens that occur naturally, only a few elements are found in significant amounts in organisms. These include **carbon, hydrogen, oxygen, nitrogen**, phosphorus, and sulfur. The amount of these elements on Earth today is about the same as when the planet formed. Because they are needed by living things and their supply does not increase, these elements have to be recycled again and again.

How do these elements get recycled in nature? Let's first look at carbon, since all organisms are made of molecules that contain this element. The carbon in organic molecules is obtained from $CO_2$ in the air. Producers such as grasses and trees take in $CO_2$ from the air during photosynthesis. They use the carbon from the $CO_2$ gas to build their carbohydrates (sugars and starches). Consumers obtain carbon from producers and sometimes from other consumers that serve as food. To complete carbon's recycling, plants and animals return carbon to the atmosphere through respiration when they release $CO_2$. (See Figure 28-5.)

Recycling of carbon also occurs after a plant or an animal dies. This important part of the recycling process occurs through the actions of **decomposers**, mainly bacteria and fungi. Decomposers are heterotrophs, organisms that are unable to make their own food. They get their food by feeding on the **residue**, or remains, of dead organisms. As they carry out their life processes, decomposers also release $CO_2$ into the atmosphere. Animals that **scavenge** the remains help recycle the carbon, too.

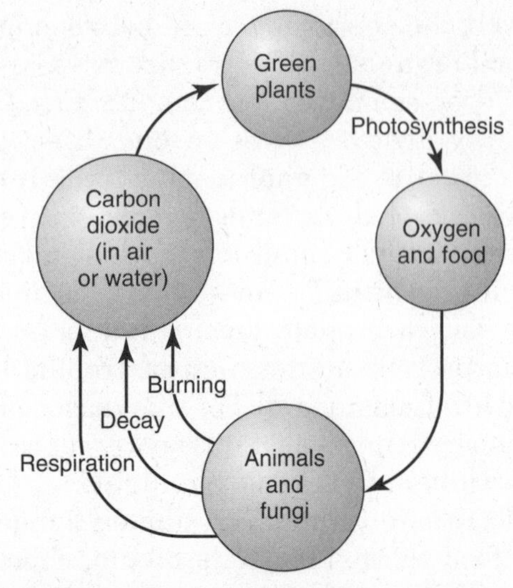


**Figure 28-5** Carbon and oxygen are recycled between living things (and the environment) through such processes as photosynthesis, respiration, and decay. Producers, consumers, and decomposers are all part of this natural recycling process.

Oxygen is also recycled between living things. Animals and other organisms need oxygen for respiration—the process that releases the chemical energy stored in food. Land animals obtain oxygen for respiration from the air they breathe. Aquatic animals like fish use the oxygen that is dissolved in the water they live in.

Almost all the oxygen in Earth's atmosphere originally came from the metabolic activities of plants. During photosynthesis, plants and algae give off oxygen as a waste product. Animals breathe in the oxygen given off by plants, just as plants take in the $CO_2$ released by animals and decomposer organisms. This is natural recycling.

Nitrogen, a gas that is common in the atmosphere, is another element essential for all living things. It is combined and recombined through another complex cycle between biotic and abiotic parts of the environment.

Phosphorus is another important mineral for the growth of plants and animals. Plant roots easily absorb phosphate minerals dissolved in water. In animals, phosphates are important in bone formation. Decomposers cannot easily break down this material; so it may take millions of years for phosphates to become available again. Therefore, phosphorus is often a limiting factor for plant growth in an ecosystem.

## CHANGE AND STABILITY: THE IMPORTANCE OF BIODIVERSITY

The tendency of an ecosystem to resist change and remain the same is known as **stability**. Ecologists have important questions to ask about the stability of ecosystems. Do entire communities in an ecosystem stay the same? What causes a particular community to change? Is the number of species that make up a community critical to its stability? The amount of variety in a community is called *species diversity*, or **biodiversity**. A community with only a few species of plants and animals has low biodiversity. A community with many species has great biodiversity. A tropical rain-forest community may have the greatest biodiversity of any community on Earth. (See Figure 28-6.)

Biodiversity is one major concern of ecologists today. Why? Many known species present on Earth just a few decades ago are already extinct. For the most part, human actions are the cause of these extinctions. As species disappear, biodiversity decreases. Scientists are concerned about the effects of decreased biodiversity on the functioning of ecosystems. It is generally thought that the greater the biodiversity, the greater the stability of an ecosystem.

Does an ecosystem need a certain number of species interacting with each other to remain stable? How many species can a community lose without being harmed? For example, if several species of insects in a forest died off, would the forest survive? Would the plants that the insects formerly ate grow too quickly? Would the bird populations suffer with fewer insects to eat? Finally, how much loss of biodiversity can occur before Earth's ecosystems stop functioning properly? This is a serious concern that affects the protection of species. Studies that investigate biodiversity and stability in specific communities are being conducted to try to answer important questions such as these.

**Figure 28-6** Tropical rain forests, such as this one in Central America, may have the greatest biodiversity of any ecosystem on Earth, containing millions of species of plants and animals.

## HABITAT DESTRUCTION

There is one main reason why biodiversity is decreasing. Many species are disappearing because of habitat loss. People are now using habitats in which wildlife formerly lived. For example, in the Midwest, many fields contained low-lying areas that remained filled with water all year. Migrating birds, such as ducks and geese, were able to rest and find food in these ponds during their journey each spring and fall. However, the farmers could not grow wheat and corn in these wet places. As a result, most of the ponds were filled in. This was a critical loss of habitat for the migrating water birds. Populations of ducks and geese decreased; and biodiversity was reduced.

This is only one example of the loss of a habitat affecting biodiversity. Habitat and species loss have occurred in many other places, such as on a river when a dam is built. Fish that survive only in moving water die in the still water of a lake that is formed behind a dam. Likewise, loss of habitat and species occurs in the ocean when coral reefs and seafloor communities are overfished and

**Figure 28-7** Habitat destruction threatens the biodiversity and stability of ecosystems. Part of this rain-forest habitat has been destroyed to make room for a banana plantation.

physically damaged in the process. Today, the greatest habitat destruction is occurring in the world's tropical rain forests. It is estimated that most of Earth's biodiversity will be lost if the rain forests are destroyed. Sadly, this is happening while scientists are trying to identify and classify the many organisms still being discovered in these forests. In addition, researchers fear that many tropical species, which may contain substances that could prove to be valuable medicines, are being lost forever before even being discovered. (See Figure 28-7.)

## Chapter 28 Review

### *Part A—Multiple Choice*

**1.** One basic trait of all ecosystems is the

1 flow of energy between organisms and the environment
2 lack of energy within organisms and the environment
3 flow of water between organisms and the environment
4 lack of water within organisms and the environment

**2.** The source of energy for most ecosystems is

1 rain 3 flowing water
2 wind 4 the sun

**3.** In an ecosystem, which component is *not* recycled?

1 nitrogen 3 energy
2 oxygen 4 carbon

**4.** The first trophic level consists of organisms that

1 use energy to make their own food
2 eat first-level producers only
3 eat producers and consumers
4 add matter to an ecosystem

**5.** Organisms that eat plants are called both consumers and

1 producers 3 carnivores
2 herbivores 4 scavengers

**6.** What is always transferred in a food chain?

1 toxins
2 energy
3 water
4 oxygen

**7.** Which list indicates a correct flow of energy?

1 herbivore→sun→carnivore
2 sun→producer→herbivore
3 producer→sun→carnivore
4 carnivore→herbivore→sun

**8.** One Arctic food chain consists of polar bears, fish, seaweed, and seals. Which sequence demonstrates the correct flow of energy between these organisms?

1 seals→seaweed→fish→polar bears
2 seaweed→fish→seals→polar bears
3 fish→seaweed→polar bears→seals
4 polar bears→fish→seals→seaweed

**9.** Which energy transfer is *least* likely to be found in nature?

1 consumer to consumer
2 host to parasite
3 producer to consumer
4 predator to prey

**10.** A spider stalks, kills, and then eats an insect. Based on this behavior, which ecological terms describe the spider's roles in a food chain?

1 producer, carnivore, consumer
2 carnivore, predator, consumer
3 predator, herbivore, consumer
4 scavenger, carnivore, consumer

**11.** In most habitats, the removal of predators will have the most immediate effect on a population of

1 producers
2 herbivores
3 decomposers
4 microbes

**12.** Which group contains terms that are *all* directly associated with the larger fish in the diagram below?

1 herbivore, prey, autotroph, host
2 carnivore, predator, heterotroph, multicellular
3 predator, scavenger, decomposer, consumer
4 producer, parasite, fungus, fish

**13.** In a food web, energy always moves

1 in a continuous cycle of trophic levels
2 back and forth between various trophic levels
3 from lower to higher trophic levels only
4 from higher to lower trophic levels only

**14.** A student could best demonstrate knowledge of how energy flows throughout an ecosystem by

1 labeling a diagram that illustrates ecological succession
2 drawing a food web using specific organisms living in a pond
3 conducting an experiment that demonstrates photosynthesis
4 making a chart to show the role of bacteria in the environment

*Base your answers to questions 15 and 16 on the diagram below and on your knowledge of biology.*

**15.** Which organism carries out autotrophic nutrition?

1 frog
2 snake
3 grass plant
4 grasshopper

**16.** The base of an energy pyramid for this ecosystem would include the

1 frog
2 snake
3 grass plants
4 grasshopper

**17.** A food web is more stable than a food chain because a food web

1 transfers all the producer energy to herbivores
2 includes alternative pathways for energy flow
3 reduces the number of niches in the ecosystem
4 includes more consumers than producers

**18.** Which trophic level contains the most available food energy?

1 producers
2 first-level consumers
3 second-level consumers
4 third-level consumers

**19.** The hidden danger in many food chains is that

1 some prey items taste better and thus are eaten too often
2 harmful chemicals can be passed from one level to another
3 some foods become poisonous after being eaten too often
4 the food chains interconnect to form enormous food webs

**20.** The diagram below represents a pyramid of energy in an ecosystem. Which level in the pyramid would most likely contain members of the plant kingdom?

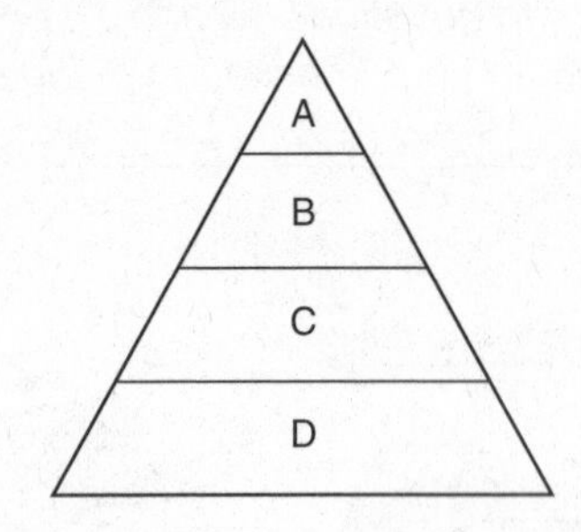


1 level A
2 level B
3 level C
4 level D

**21.** In an ecosystem, nutrients can be recycled if they are transferred directly from herbivores to carnivores to

1 hosts
2 decomposers
3 prey
4 autotrophs

**22.** Carbon is recycled in nature when

1 consumers take in carbon dioxide and then release oxygen
2 producers take in carbon dioxide and consumers release it
3 decomposers take in carbon dioxide and release oxygen
4 scavengers and decomposers take in carbon dioxide

**23.** Which set of statements best illustrates a material cycle in a self-sustaining ecosystem?

1 In summer, growing plants remove magnesium ions from the soil to make chlorophyll. In autumn, these plants release magnesium when they die and decompose. In spring, new plants will grow in this same area.
2 DDT is sprayed on a forest ecosystem to control the mosquito population. After a year, the level of DDT is found to be much higher in the tissues taken from a hawk than in the tissues taken from a mouse in this ecosystem.
3 Trees do not live in a desert ecosystem where there is not enough water present in the sandy soil to support their growth. Trees can live in a desert oasis.
4 Plants trap the sun's energy in the chemical bonds of organic molecules. This energy is then used for plant metabolic activities.

**24.** Which statement best describes what happens to energy and molecules in a stable ecosystem?

1 Both energy and molecules are recycled in the ecosystem.
2 Energy is recycled and molecules are continuously added to the ecosystem.
3 Neither energy nor molecules are recycled in the ecosystem.
4 Molecules are recycled and energy is continuously added to the ecosystem.

**25.** Decomposers release carbon dioxide as they

1 feed on bacteria and algae
2 carry out photosynthesis
3 build starches and sugars
4 carry out their life processes

**26.** The organisms that help recycle elements by breaking down organic matter include

1 grass and algae
2 bacteria and algae
3 bacteria and fungi
4 plants and fungi

**27.** Vultures, which are classified as scavengers, are an important part of an ecosystem because they

1 hunt herbivores, thus limiting their population size in an ecosystem
2 cause the decay of dead organisms, which releases usable energy to living organisms
3 feed on dead animals, which aids in the recycling of environmental materials
4 are the first level in food webs, making energy available to all the other organisms

**28.** Oxygen is needed for respiration, the process that

1 releases the chemical energy stored in food
2 uses carbon dioxide to produce sugars
3 breaks down the bodies of dead organisms
4 releases oxygen as a waste into the air

**29.** The oxygen that humans breathe is actually

1 a waste product of respiration
2 a waste product of photosynthesis
3 given off by decomposers
4 produced within the sun

**30.** The tendency of an ecosystem to stay the same is called

1 diversity
2 resistance
3 stability
4 sterility

**31.** Which condition would cause an ecosystem to become unstable?

1 Only heterotrophic organisms remain after a change in the region.
2 A variety of nonliving factors are used by the living factors.
3 A slight increase occurs in the number of heterotrophs and autotrophs.
4 The biotic factors and abiotic resources interact more often.

**32.** Which ecosystem has a better chance of surviving when environmental conditions change over a long period of time?

1 one with a great deal of genetic diversity
2 one with animals and bacteria but no plants
3 one with plants and animals but no bacteria
4 one with little or no genetic diversity

**33.** Unlike a desert, a tropical rain forest typically has

1 low biodiversity
2 great biodiversity
3 a small variety of organisms
4 a small number of organisms

**34.** The loss of biodiversity is often related to

1 the search for medical cures
2 too much rain in rain forests
3 a loss of natural habitat
4 evolution not occurring

**35.** Increased efforts to conserve areas such as rain forests are necessary in order to

1 protect biodiversity
2 exploit finite resources
3 promote extinction of species
4 increase industrialization

### *Part B—Analysis and Open Ended*

**36.** Briefly describe the *two* main components of any ecosystem.

**37.** How do producers differ from consumers?

**38.** Why are the levels of a food chain called "trophic" levels?

**39.** Why is a food web a more accurate description than a food chain of interactions in a community?

**40.** A food web is represented in the diagram below. Which organisms are correctly paired with their roles in this food web?

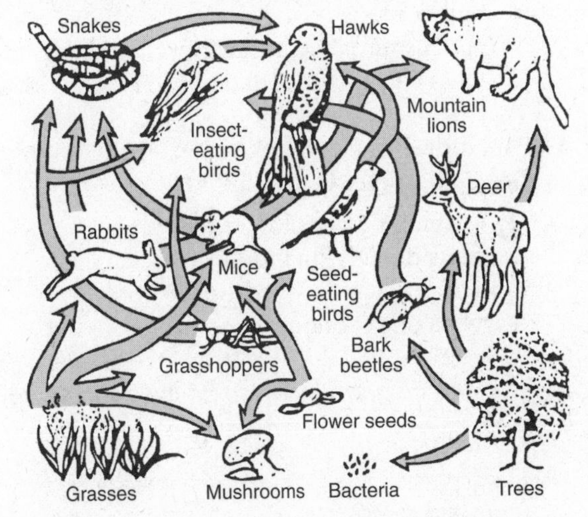


1 producers: mountain lions, snakes; heterotrophs: hawks, mice
2 consumers: all birds, deer; producers: grasses, trees
3 consumers: snakes, grasshoppers; autotrophs: mushrooms, rabbits
4 decomposers: seeds, bacteria; heterotrophs: mice, grasses

**41.** Explain why the amount of energy in trophic levels can be shown as a pyramid.

**42.** The diagram at the top of the next page represents a model of a food pyramid. Which statement best describes what happens in this food pyramid?

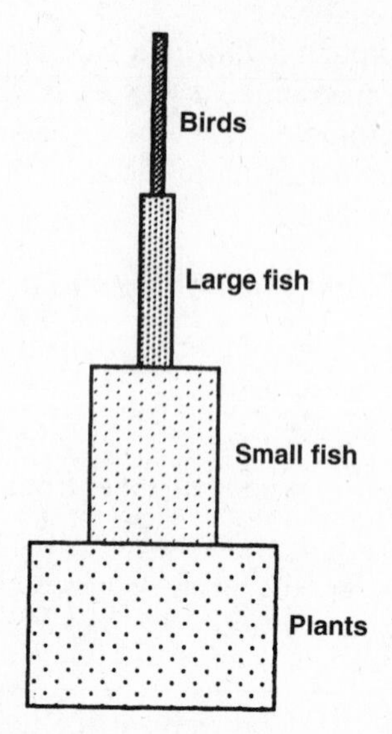


1 More organisms die at higher levels than at lower levels, decreasing the mass at the top.
2 When organisms die at higher levels, their remains sink to lower levels, increasing the mass at the bottom.
3 Energy is lost to the environment at each level, so less mass can be supported at each higher level.
4 Organisms decay at each level, and thus less mass can be supported at each higher level.

**43.** According to Figure 28-1 on page 222, a third-level consumer is one that

1 is capable of photosynthesis
2 feeds directly on producers
3 feeds on first-level consumers only
4 feeds on second-level consumers

*Refer to the diagram below to answer questions 44 to 47.*

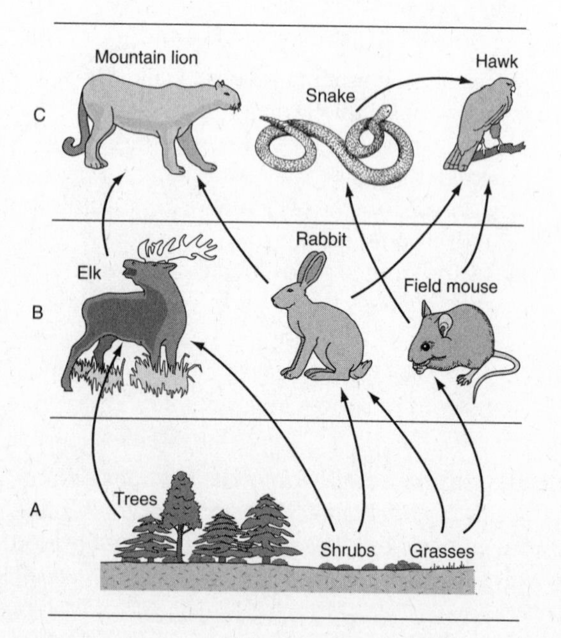


**44.** What does the diagram illustrate?

1 a food chain
2 a food web
3 a food cube
4 a food pyramid

**45.** The organisms shown in level B are classified as both

1 producers and prey
2 consumers and prey
3 scavengers and predators
4 decomposers and prey

**46.** Which populations would contain the greatest amount of available energy?

1 rabbits and field mice
2 hawks and rabbits
3 trees and shrubs
4 hawks and snakes

**47.** All these organisms living together make up a natural

1 population
2 species
3 community
4 consumer

**48.** Which statement about the producers in the marine food web shown below is correct?

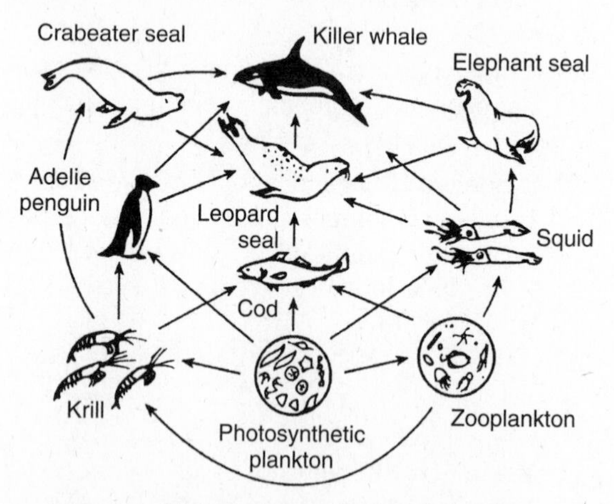


1 An increase in the producers would most likely decrease available energy for the squid.
2 If all the producers in this ecosystem were destroyed, the number of heterotrophs would increase and the ecosystem would reach a new equilibrium.
3 An important producer in this ecosystem is the zooplankton.
4 There is only one group of producers, so they must be numerous enough to supply the energy needed to support the food web.

**49.** What problem can occur in a food chain when a pollutant enters the ecosystem?

**50.** According to Figure 28-4 on page 223, as energy moves up each trophic level in an ecosystem, the amount of it that is available becomes

1 10 percent more than it was before
2 10 percent of what it was before
3 50 percent more than it was before
4 50 percent less than it was before

*Base your answers to questions 51 to 53 on the passage below, which was written in response to an article about eliminating predators.*

In nature, energy flows in only one direction. Transfer of energy must occur in an ecosystem because all life needs energy to live, and only certain organisms can change solar energy into chemical energy.

Producers are eaten by consumers, which are, in turn, eaten by other consumers. Stable ecosystems must contain predators to help control the populations of consumers. Since ecosystems contain many predators, exterminating predators would require a massive effort that would wipe out predatory species from barnacles to blue whales. Without the population control provided by predators, some organisms would soon overpopulate.

**51.** Draw an energy pyramid that illustrates the sentence, "Producers are eaten by consumers, which are, in turn, eaten by other consumers." Include *three* different, specific organisms in your energy pyramid.

**52.** Explain the phrase "only certain organisms can change solar energy into chemical energy," which appears in the first paragraph. In your answer be sure to identify:

- the type of organisms being described in the statement
- the type of nutrition carried out by these organisms
- the process being carried out in this type of nutrition
- the organelles (in the cells of these organisms) that are involved in this process

**53.** Explain why an ecosystem with many predator species still has a much smaller number of *individual* carnivores than of herbivores.

*Refer to the diagram of a food pyramid below to answer questions 54 and 55.*

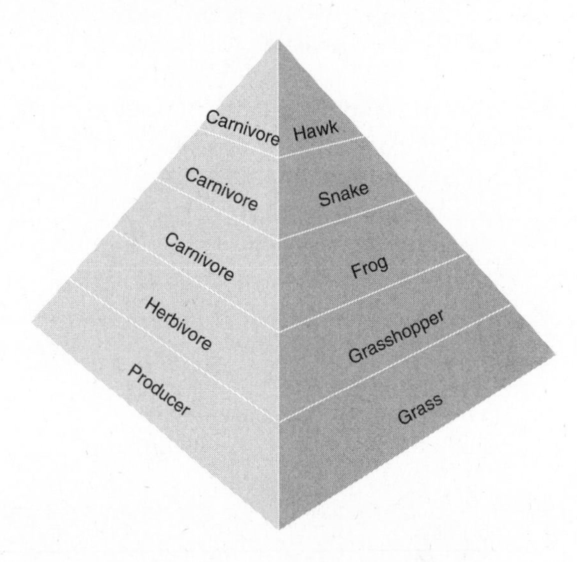


**54.** The consumer level that would have the largest amount of stored energy is that of the

1 hawk
2 snake
3 frog
4 grasshopper

**55.** The level that has the smallest amount of stored energy would be that of the

1 top carnivore
2 middle carnivore
3 herbivore
4 producer

*Base your answers to questions 56 to 59 on the diagram of a food web below and on your knowledge of biology.*

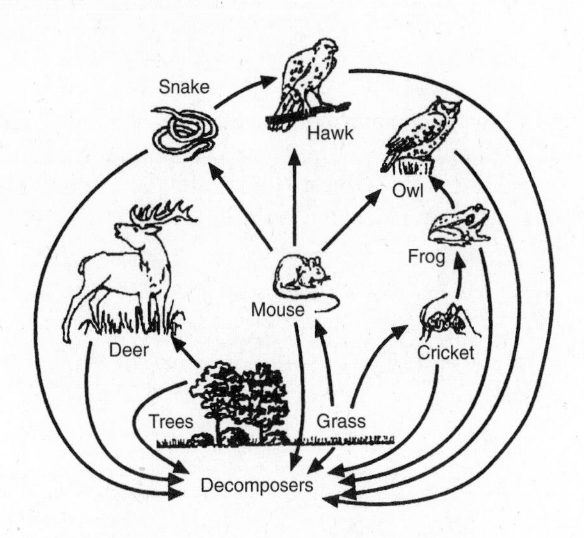


**56.** If the population of mice were reduced by disease, which change will most likely occur in the food web?

1 The cricket population will increase.
2 The snake population will decrease.
3 The grasses will decrease.
4 The deer population will decrease.

**57.** What is the original source of energy for this food web?

1 chemicals in sugar molecules
2 enzymatic reactions
3 energy from sunlight
4 chemical reactions of bacteria

**58.** Which microorganisms labeled in this diagram are essential to a balanced ecosystem?

1 heterotrophs
2 autotrophs
3 producers
4 decomposers

**59.** State *one* example of a predator-prey relationship shown in the food web. Indicate which organism is the predator and which is the prey.

**60.** Use the following terms to complete the categories in the flowchart below: *decomposers; producers; carnivores; herbivores.*

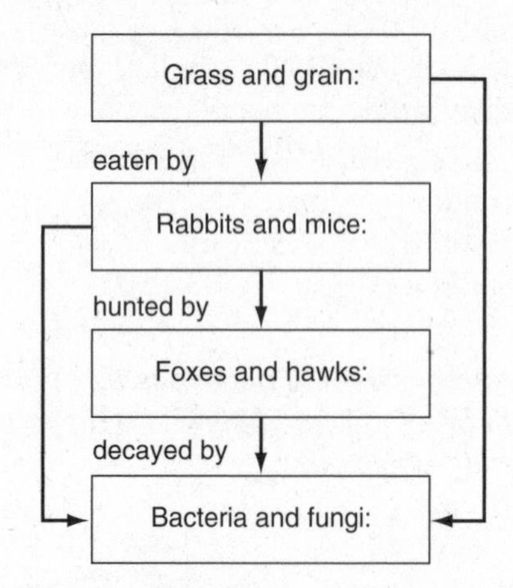


**61.** How is the amount of matter available on Earth very different from the amount of energy available? How is this related to the recycling of elements in nature?

*Refer to the following diagram to answer question 62.*

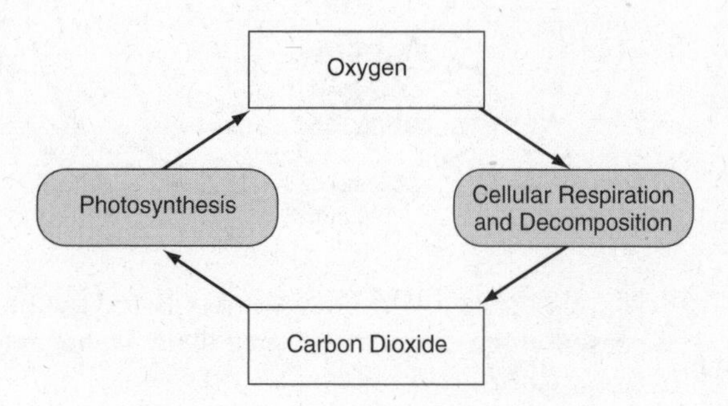


**62.** Explain how respiration and photosynthesis are involved in the cycling of oxygen and carbon dioxide between living things.

**63.** Why are ecologists concerned about the number of species in ecosystems?

**64.** Explain why most ecologists would agree with the following statement: "A forest ecosystem is more stable than a cornfield."

**65.** How might the biodiversity of an ecosystem be related to its stability?

*Base your answer to question 66 on the following passage and on your knowledge of biology.*

A tropical rain forest in the country of Belize contains over 100 kinds of trees, in addition to thousands of species of mammals, birds, and insects. Dozens of species living there have not yet been classified and studied. The rain forest could be a commercial source of food as well as a source of medicinal and household products. However, most of this forested area is not accessible because of a lack of roads and, therefore, little commercial use has been made of this region. The building of paved highways into and through this rain forest has been proposed.

**66.** Discuss some aspects of carrying out this proposal to build paved highways. In your answer be sure to:

- state *one* possible effect on biodiversity and *one* reason for this effect
- state *one* possible reason for an increase in the number of certain producers as a result of road building
- identify *one* type of consumer whose population would most likely increase as a result of an increase in a certain producer population
- state *one* possible action the road builders could take to minimize the human impact on the ecology of this region

**67.** In what ways are humans responsible for the current decrease in biodiversity?

### *Part C—Reading Comprehension*

*Base your answers to questions 68 to 70 on the information below and on your knowledge of biology.*
Source: *Science News*, Vol. 162, p. 269.

In an unusual test of a conservation strategy called wildlife corridors, strips of habitat boosted insect movement, plant pollination, and seed dispersal among patches of the same ecosystem.

Theory predicts that adding such corridors enhances the benefits of otherwise isolated preserves, says Joshua Tewksbury of the University of Washington in Seattle. He and his colleagues tested that strategy in South Carolina pine forests.

At eight locations, the researchers cleared mature vegetation and created open habitat on five 1-hectare plots—arranged as a central plot with four satellites. In each case, a 150-meter-long corridor connected the central plot to one outlier, while the [three] others remained isolated. The unlinked patches had dead-end corridors or additional area so they matched the habitat area of another patch and its connecting corridor. Thus, scientists could distinguish between effects of biological entities' ease of movement and of extra habitat.

Butterflies, pollen, and seeds all moved most often between the corridor-connected patches, the researchers report in an upcoming *Proceedings of the National Academy of Sciences*. Variegated fritillary and common buckeye butterflies that the researchers captured, marked, and released in the central patch proved two to four times as likely to show up in connected patches as in unconnected ones.

When researchers placed male holly plants in the center patches, females in connected patches showed an average increase in seed production of nearly 70 percent, compared with that of female hollies in unconnected patches. Also, bird droppings in connected patches harbored more berries from shrubs in the center patches than did droppings in patches not connected to the central patch.

This is the first test of a corridor's effect on plant-animal interactions, says Tewksbury.

**68.** Explain how, at all eight locations, *three* of the test plots were made different from the central and fourth plots.

**69.** What difference was observed in the number of butterflies that showed up in the connected plots as opposed to those that showed up in the isolated plots. How would you explain this difference?

**70.** Why did researchers study the droppings of birds in these test plots, and what was their conclusion from this study?

# 29
# People and the Environment

**Vocabulary**

atmosphere
depletion
global warming
ozone shield
thermal pollution

## PEOPLE CHANGING THE ENVIRONMENT

Up until about 10,000 years ago, all humans hunted and gathered their food. Then, people started planting crops and domesticating animals; this marked the beginnings of agriculture. As a result of agriculture, people use the land differently from before. When people cut down and burn trees to make room to plant crops and graze livestock, wild animals are often forced to leave the area—they lose their habitat.

Advances in science and technology have produced even greater changes in the environment. About 200 years ago, developments in science and technology led to the Industrial Revolution, which greatly increased the ways that humans affect the environment.

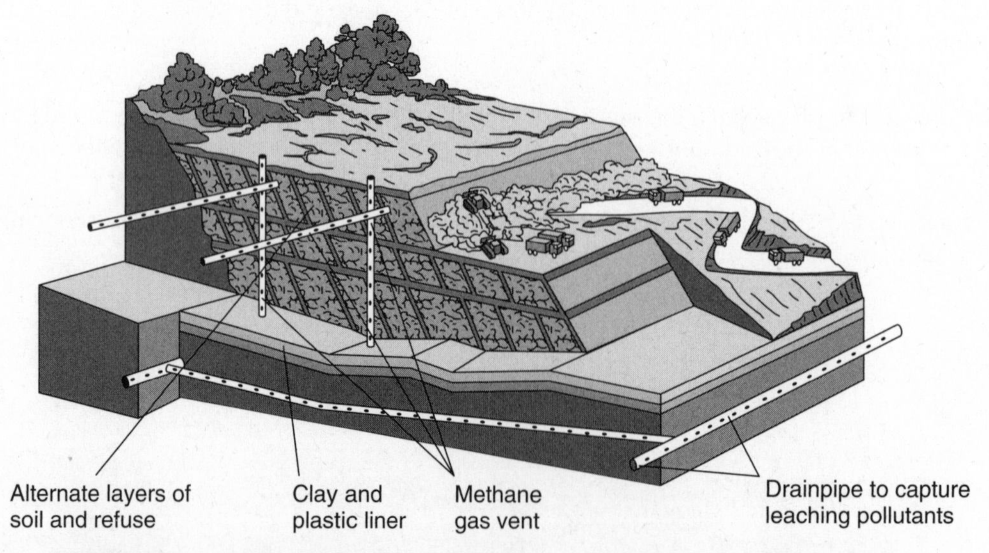


**Figure 29-1** A sanitary landfill is constructed in a way that limits the effects of the waste materials on the surrounding environment.

## CHANGES TO THE LAND—ADDING WASTES

All organisms produce wastes as a normal by-product of their life processes. However, since the time of the Industrial Revolution, the amount of wastes produced by humans has increased greatly. Also, since that time, the kinds of wastes have changed. Many of the wastes do not decompose and they often contain harmful chemicals. These waste materials, called *solid wastes*, are often deposited in landfills, areas in which garbage is buried. (See Figure 29-1.)

In a sanitary landfill, attempts are made to limit the effects of the wastes on the environment. Other kinds of landfills are much more harmful to the environment, such as a *toxic* waste dump. The most dangerous of all toxic wastes are radioactive substances. (See Figure 29-2.)

**Figure 29-2** Some landfills contain waste products that are particularly toxic; such sites may pose a health threat to nearby communities.

## CHANGES TO THE LAND—LOSING SOIL

Although soil is sometimes called dirt, it is actually a very valuable resource. In fact, this combination of organic and inorganic matter takes hundreds of years to form. Without good nutrient-rich soil, called *topsoil*, we could not grow food. Land ecosystems depend on this resource, too. (See Figure 29-3.)

Soil is now being lost because of human activities. For example, when toxic chemicals enter the ground, the soil becomes unusable. Poor farming practices and overgrazing by livestock can strip an area of all vegetation. The land becomes bare and, if weather patterns change and less rain falls, the land becomes a desert.

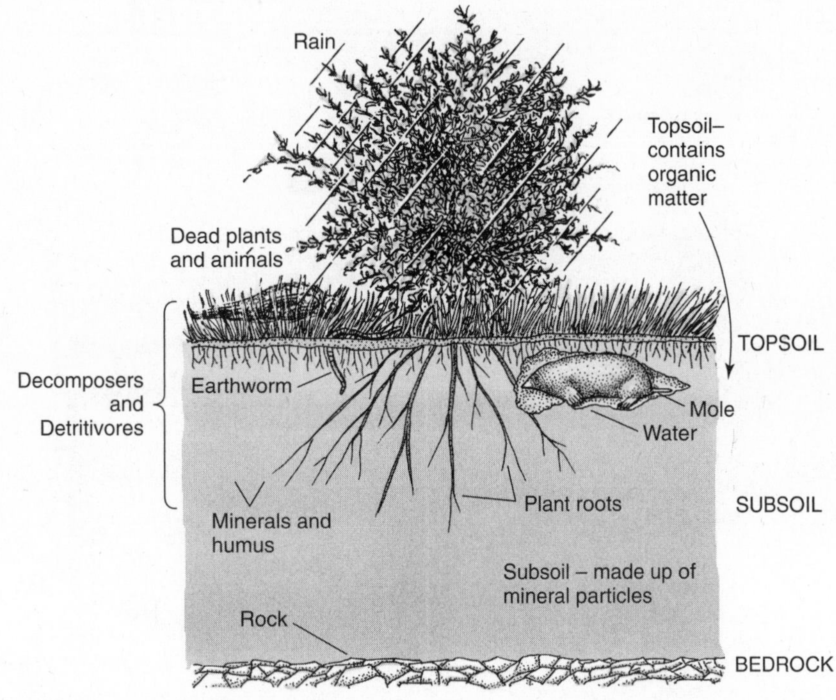


**Figure 29-3** Nutrient-rich topsoil takes a long time to form and is of great value to humans and wildlife. Crops cannot grow in soil that lacks adequate nutrients.

As people cut down forests or remove the plants that grow in an area, there is an increase in soil loss, or *erosion*. Both wind and water can cause erosion: rain washes away loose soil, and strong winds blow it away.

## CHANGES TO THE WATER

Because flowing water carries away wastes, a stream or river has always seemed the perfect place to dump garbage. Today, as the human population increases in size, many more wastes are placed into streams and rivers. There are simply more wastes introduced than the natural ecosystems can handle. The river or stream, once full of life, begins to lose its ability to support the same species of organisms as before. Wastes that are released into our waterways can harm marine ecosystems, too, since the water eventually flows from the rivers to the oceans. (See Figure 29-4.)

**Figure 29-4** When a community dumps sewage and industrial wastes into its streams and rivers, the local aquatic ecosystem—and those of communities living downstream—becomes polluted.

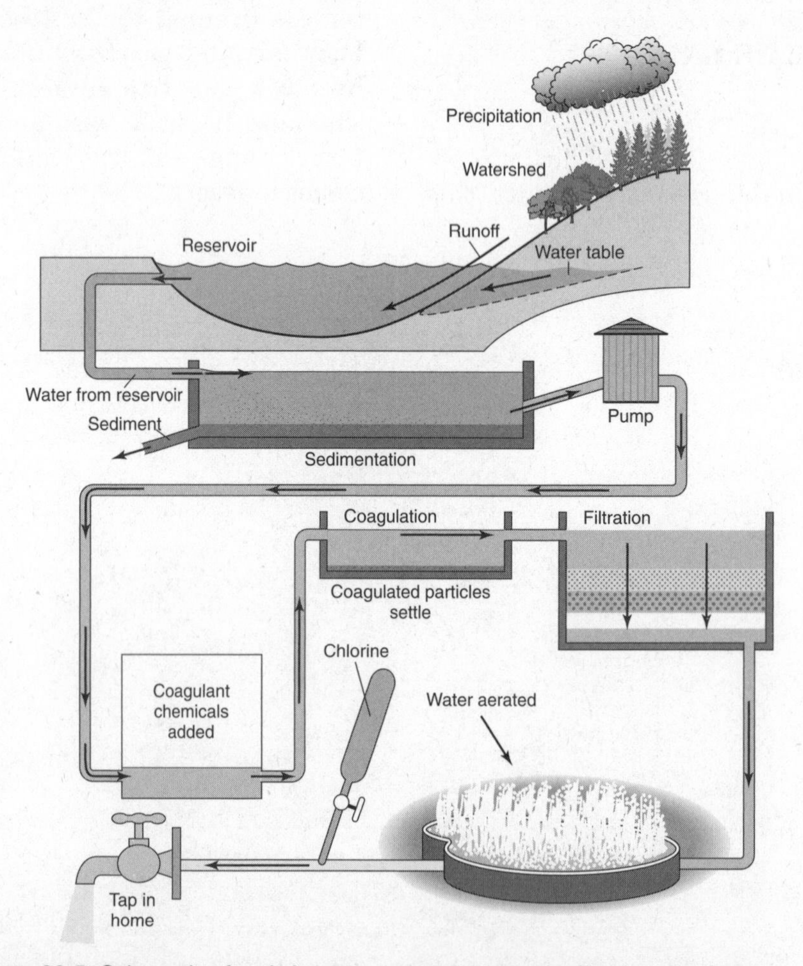


**Figure 29-5** Schematic of a city's water treatment process: Fresh water from a reservoir goes through several physical and chemical processes before it is considered clean enough to pipe into people's homes.

In addition, the types of wastes deposited have changed. Industries sometimes dump toxic chemicals into rivers. Industrial activities also create another type of water pollution, **thermal pollution**, when heat is added to the water by factories and power-generating plants. Fish that need clean, well-oxygenated, cooler water are replaced by fish that can live in warmer water with lower oxygen levels. If the levels of pollutants keep increasing, these fish will die, too. A major advance in dealing with the problem of water pollution has been the development of sewage treatment plants. In these treatment plants, human organic wastes are treated in large tanks. Wastes in the water are chemically digested by bacteria. The remaining solids, including dead bacteria, then settle to the bottom and are removed. Chlorine is added to the water to kill bacteria. Finally, the purified water is released into a river or stream.

Many cities obtain their water from underground wells. The water from these wells, called *groundwater*, accumulates over time and is stored naturally between layers of rock. There are above-ground reservoirs of fresh water, too. New York City relies on a system that directly transports clean water from upstate reservoirs through underground pipes. Other cities have their water treated first. (See Figure 29-5.)

## CHANGES TO THE AIR

No one owns the air; we all share the air, which forms a continuous blanket over Earth. If the air becomes polluted in one place, that pollution can easily spread to another place. Gases and tiny solid particles are constantly added to the air by human activities. If these substances are not normally found in the air and are harmful, they are called air *pollutants*. The burning of *fossil fuels*—coal, oil, and natural gas—to power cars, heat homes and offices, and produce electricity creates air pollution. In addition, many industries release pollutants into the air from huge smokestacks. (See Figure 29-6.)

Major improvements have been made in the efforts to reduce air pollution. Today, laws require factories to reduce or prevent the release of pollutants from smokestacks. Devices called "scrubbers" are installed, which reduce the emission of harmful compounds. Car engines, too, have built-in pollution control devices that reduce the amount of pollutants added to the air when fuel is burned.

**Figure 29-6** Factories that produce enormous quantities of manufactured goods are typical of our industrial society. Unfortunately, these factories may also release some air and water pollutants.

New technologies for producing energy also have been developed. Solar collectors and photovoltaic cells can provide us with heat or electricity without polluting the air. The Clean Air Acts of 1970 and 1977 began many of these changes. As a result, the air is now cleaner than it was just a few decades ago.

## GLOBAL AIR POLLUTION PROBLEMS

*Acid rain* is a form of air pollution that produces far-ranging effects. Sulfur dioxide and nitrogen oxides are produced when fossil fuels are burned. Winds carry these gases high into the atmosphere and over long distances. They combine with water droplets in the air, which fall back to the ground as acid rain. Many forests and lakes in North America and Europe have been severely damaged by acid rain. (See Figure 29-7 on page 238.)

Perhaps even more important are the effects of **global warming** and ozone depletion. Carbon dioxide ($CO_2$) in the **atmosphere** helps keep Earth warm by trapping heat. This is called the "greenhouse effect." But the amount of $CO_2$ has been increasing in the atmosphere due to the burning of fossil fuels and deforestation, in particular, the destruction of countless trees in rain forests, which

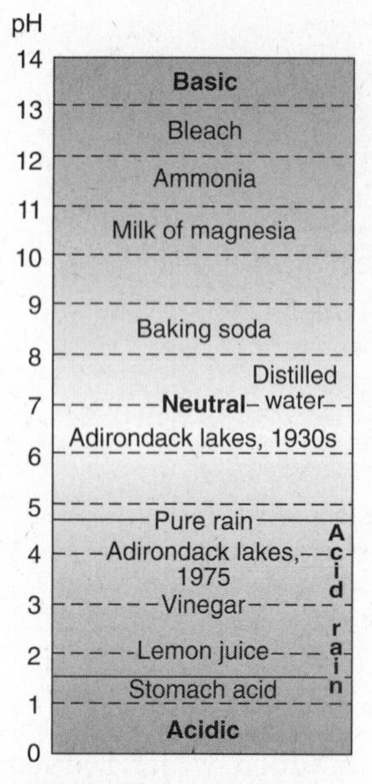


**Figure 29-7** Due to acid rain, lakes in New York's Adirondack Mountains have become more acidic, causing harm to wildlife. This pH chart shows how much the lakes' acidity increased in 40 years.

used to absorb $CO_2$. With more $CO_2$ in the atmosphere, more heat is trapped. Many people are concerned that, as a result, Earth's climate is getting warmer. Such a change in climate could have major effects on habitats and organisms everywhere. In December 2009, leaders from more than 100 nations met at the UN Climate Change Conference in Copenhagen, Denmark, to address the problem of $CO_2$ and global warming. (See Figure 29-8.)

Scientists are also concerned about the effects of certain air pollutants that damage the **ozone shield**, causing it to thin. This effect is known as ozone **depletion**. The layer of ozone gas that surrounds Earth high in the atmosphere blocks out harmful ultraviolet (UV) radiation. The UV rays are part of the energy that reaches Earth from the sun, and they can damage the DNA in our cells, causing skin cancer.

Chlorofluorocarbons (CFCs), found in air conditioners and refrigerators, are suspected of causing the most ozone depletion. In 1987, an agreement was signed by many countries to protect the ozone layer by limiting or banning the use of these chemicals. Progress has been made since the agreement was signed. Atmospheric concentrations of these chemicals have been declining. It is now expected that the atmosphere over the Antarctic, where the loss of ozone was the worst, will have completely recovered sometime after 2050.

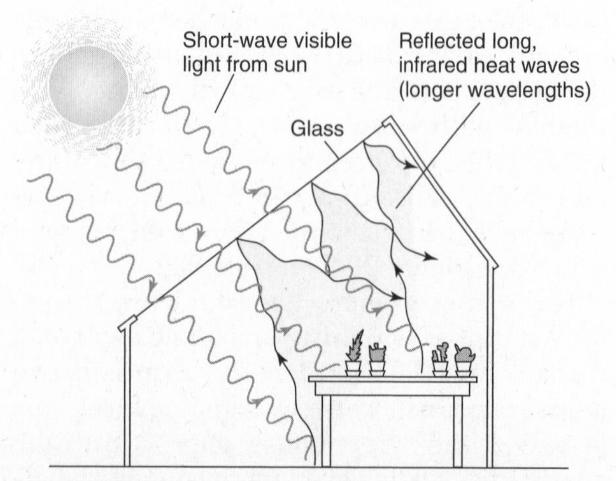


**Figure 29-8** The greenhouse effect: Carbon dioxide in the air traps infrared energy, warming the atmosphere. This is similar to the way that the glass roof of a greenhouse traps heat, keeping the plants warm.

## HUMAN POPULATION GROWTH

The most serious problem that now affects all life on Earth is the rate at which the human population is increasing. Over most of its history, the human population increased slowly. However, the rise of agriculture caused a rapid increase as people settled down with a more secure source of food. More recently, the Industrial Revolution, combined with scientific advances in farming and medicine, caused an explosion in human population size. (See Figure 29-9.)

What is Earth's carrying capacity for humans? It is now known that the growth rate for the human population peaked in 1990 and is now declining. However, the population is still growing and it is believed it will reach a peak of about 9 billion around 2050. Others think that the population is already past

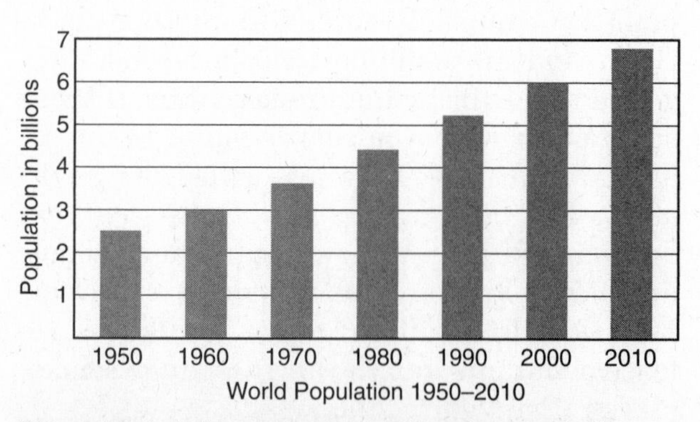


**Figure 29-9** Industrial and scientific advances in the last 60 years have caused the human population to more than double in size—an increase that may put us past Earth's carrying capacity and cause environmental problems.

Earth's carrying capacity and that serious environmental problems have already begun.

An exploding human population may lead to the extinction of vast numbers of species. Many organisms are already endangered due to the loss of habitat and other human factors. Industrialization (which causes more air, water, and land pollution), acid rain, global warming, and ozone shield depletion are worldwide concerns. Earth—as one large, complex, intact ecosystem—is threatened by an ever-increasing human population. (See Figure 29-10.)

In 1994, participants from 160 countries at an international conference agreed that Earth's population cannot continue to grow at its current rate; yet they disagreed about how to lower the growth rate. What is clear is that our population, like that of any other organism, cannot increase forever. Either we will find a way to control human population size or nature will do it for us. Polluted soil, air, and water; lack of food and space; and widespread disease may ultimately limit human population size. However, individual choices and government planning could also limit it.

**Figure 29-10** The exploding human population leads to overcrowding, pressure on limited resources, loss of wildlife habitat, increased pollution, and other problems that may threaten Earth's health as well as our own.

## Chapter 29 Review

### *Part A—Multiple Choice*

**1.** Over time, human populations have

1 produced fewer and fewer wastes
2 increased the amount of wastes produced
3 prevented harmful chemicals from being produced
4 decreased the amount of waste in landfills

**2.** Which animal has had the greatest negative effect on Earth's ecosystems?

1 gypsy moth  3 zebra mussel
2 human  4 shark

**3.** Which phrase would be appropriate for area A in the chart below?

| Technological Device | Positive Impact | Negative Impact |
|---|---|---|
| Nuclear power plant | Provides efficient, inexpensive energy | A |

1 produces radioactive wastes
2 provides light from radioactive substances
3 results in greater biodiversity
4 reduces dependence on fossil fuels

**4.** The most dangerous of all toxic wastes are

1 solid wastes
2 nutrient-rich soils
3 radioactive substances
4 plastic garbage

**5.** When humans cut down forests in an area,

1 soil is lost through erosion
2 the soil becomes richer
3 new soil forms quickly
4 flooding is prevented

**6.** When too many wastes are dumped into a river, the wastes

1 eventually disappear through dilution
2 gradually cause harm to the river ecosystem

3 are carried away, where they can cause no harm
4 are broken down immediately

**7.** Which is *not* a cause of increased water pollution?

1 dumping sewage into streams and rivers
2 addition of chlorine to treated water
3 pouring industrial wastes into rivers
4 eroded soil washing off land into streams

**8.** Dumping raw sewage into a river will lead to a reduction in the dissolved oxygen in the water. This condition will most likely cause

1 an increase in all fish populations
2 an increase in the depth of the water
3 a decrease in most fish populations
4 a decrease in water temperature

**9.** In a sewage treatment plant, bacteria are

1 added to the water before it is released into a river
2 killed by chlorine at the beginning of the process
3 used to chemically digest wastes in the water first
4 left in the purified water because they are harmless

**10.** "Natural ecosystems provide an array of basic processes that affect humans." Which statement does *not* support this quotation?

1 Bacteria of decay help recycle materials.
2 Treated sewage is less damaging to the environment than untreated sewage.
3 Trees add to the amount of atmospheric oxygen.
4 Lichens and mosses on rocks help to break down the rocks, forming soil.

**11.** A negative result of technology is an increase in the

1 development of new products
2 availability of different foods
3 wastes released into the environment
4 societal awareness of the environment

**12.** Which type of waste will decompose most quickly?

1 foam cup
2 plastic bag
3 glass bottle
4 banana peel

**13.** Which practice will best protect the soil?

1 removing excess trees from it
2 planting vegetation on it
3 allowing cattle to feed on the land
4 adding lots of chemicals to it

**14.** An increase in the use of fossil fuels is an indication of which type of society?

1 hunter-gatherer
2 agricultural
3 industrial
4 horticultural

**15.** Acid rain forms when

1 carbon dioxide traps heat near Earth
2 ozone is depleted from the atmosphere
3 gases from fossil fuels combine with water droplets in the air
4 chlorine is added to waste water

**16.** Methods used by people to reduce the emission of pollutants from smokestacks are an attempt to

1 lessen the amount of insecticides in the environment
2 lessen the formation and harmful effects of acid rain
3 eliminate diversity in natural habitats
4 use nonchemical controls on pest species

**17.** Changes in the chemical composition of the atmosphere that may produce acid rain are most closely associated with

1 insects that excrete acids
2 factory smokestack emissions
3 runoff from acidic soils
4 flocks of migrating birds

**18.** How is the Industrial Revolution related to the greenhouse effect?

1 It caused the start of the greenhouse effect on Earth.
2 It marked the end of the greenhouse effect on Earth.
3 It caused a decrease in the amount of $CO_2$ released into the atmosphere.
4 It caused an increase in the amount of $CO_2$ released into the atmosphere.

**19.** People can have a large negative effect on ecosystems when they

1 conserve natural resources
2 modify the environment
3 restrict the use of chemicals
4 pass laws to protect habitats

**20.** By causing atmospheric changes through activities such as polluting and careless tree harvesting, humans have

1 caused the destruction of habitats
2 established equilibrium in ecosystems
3 affected global stability in a positive way
4 replaced nonrenewable resources

**21.** The effect of $CO_2$ and other greenhouse gases on the atmosphere can best be likened to that of a

1 blanket
2 balloon
3 pitcher of water
4 crowd of people

**22.** Deforestation will most directly result in an increase in

1 atmospheric carbon dioxide
2 wildlife populations
3 atmospheric ozone
4 renewable resources

**23.** Which human activity would have the most direct effect on the oxygen–carbon dioxide cycle?

1 reducing the rate of ecological succession
2 destroying large forested areas
3 decreasing the use of water
4 banning the use of leaded gasoline

**24.** Chlorofluorocarbons are harmful to the environment because they

1 kill fish in lakes
2 form acid rain
3 cause ozone depletion
4 increase the greenhouse effect

**25.** The ozone layer of Earth's atmosphere helps to filter ultraviolet radiation. As the ozone layer is depleted, more ultraviolet radiation reaches Earth's surface. This increase in ultraviolet radiation may be harmful because it can directly cause

1 photosynthesis to stop in all marine plants
2 mutations in the DNA of organisms
3 abnormal migration patterns in waterfowl
4 sterility in most species of mammals and birds

**26.** Which factor is often responsible for the other three?

1 increase in levels of toxins in fresh water
2 increased poverty and malnutrition
3 increase in human population
4 increased depletion of finite resources

### *Part B—Analysis and Open Ended*

**27.** Briefly explain how humans change the land through agriculture. Your answer should include the impact on the following:

- topsoil
- forests
- wildlife

**28.** How has industrialization changed the types of wastes produced by humans?

**29.** Why might building a landfill near an aquatic ecosystem cause harm to it?

*Base your answers to questions 30 and 31 on the graph below, which shows pollution from nitrogen-containing compounds (nitrates) in a brook flowing through a forested area and a deforested area between 1965 and 1968.*

**Hubbard Brook Nitrate Pollution Study**

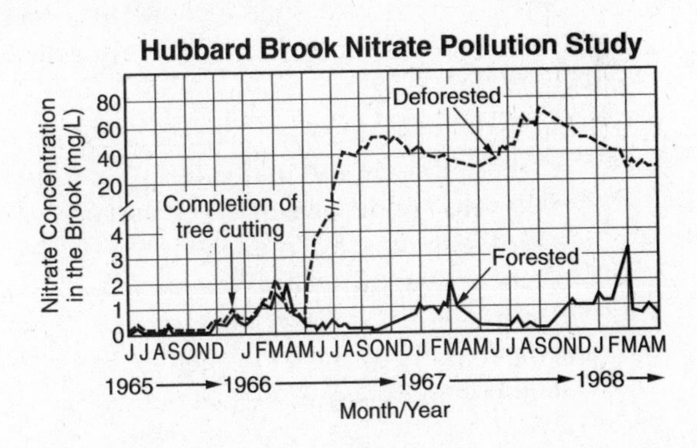


**30.** State how nitrate pollution in the brook changed after it flowed through the deforested area.

**31.** Explain how deforestation contributed to this change.

**32.** Explain why toxic waste dumps are most harmful to the environment.

**33.** Why do scientists consider good topsoil so valuable? List *three* human activities that cause loss of soil.

*Refer to the following graphs to answer questions 34 and 35.*

**Oxygen content and fish population in a lake**

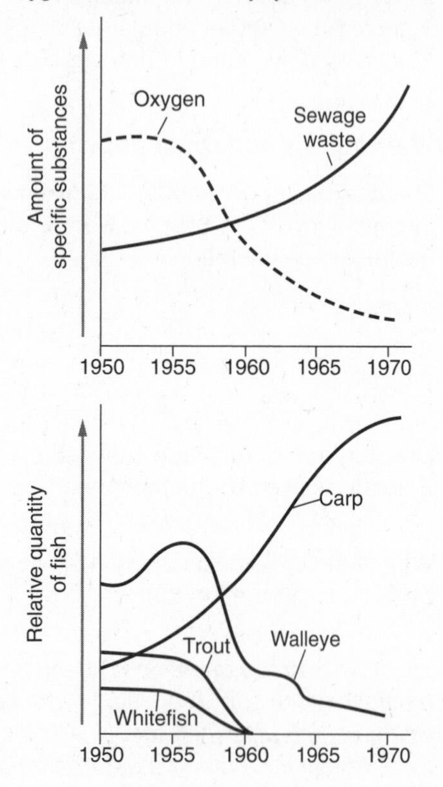


**34.** According to the graphs, an increase in sewage waste in a lake over time would be associated with

1 an increase in dissolved oxygen and an increase in most fish populations
2 a decrease in dissolved oxygen and an increase in most fish populations
3 an increase in dissolved oxygen and a decrease in most fish populations
4 a decrease in dissolved oxygen and a decrease in most fish populations

**35.** According to the graphs, the fish species that adapted most successfully to the change in oxygen content over time was the

1 trout, because it can live in highly oxygenated water
2 carp, because it can live in poorly oxygenated water
3 walleye, because it can live in highly oxygenated water
4 whitefish, because it can live in poorly oxygenated water

**36.** Explain how a new power plant built on the banks of the Rocky River could have an environmental impact on the Rocky River ecosystem downstream from the plant. Your explanation must include the effects of the power plant on:

- water temperature
- dissolved oxygen
- local fish species

**37.** Describe the importance of sewage treatment to both people and wildlife.

**38.** Both car exhaust and factory emissions add pollutants to the air. For each case, tell *how* it adds to air pollution and *what* is being done to reduce the problem.

**39.** The map below shows the movement of some air pollution across part of the United States. Which statement is a correct inference that can be drawn from this information?

**Movement of Air Pollution**

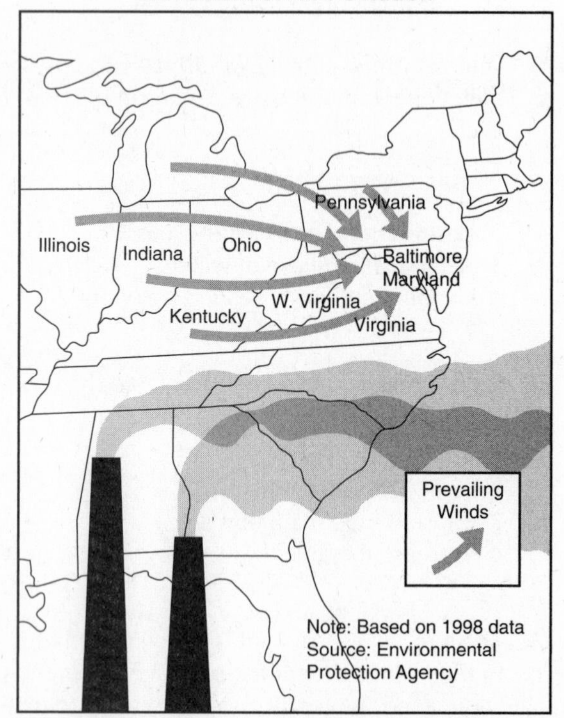


1 Illinois produces more air pollution than the other states shown.
2 The air pollution problem in Baltimore is increased by the addition of pollution from other areas.
3 There are no air pollution problems in the southern states.
4 The air pollution problems in Virginia clear up quickly as the air moves toward the sea.

*Refer to the illustration below to answer questions 40 and 41.*

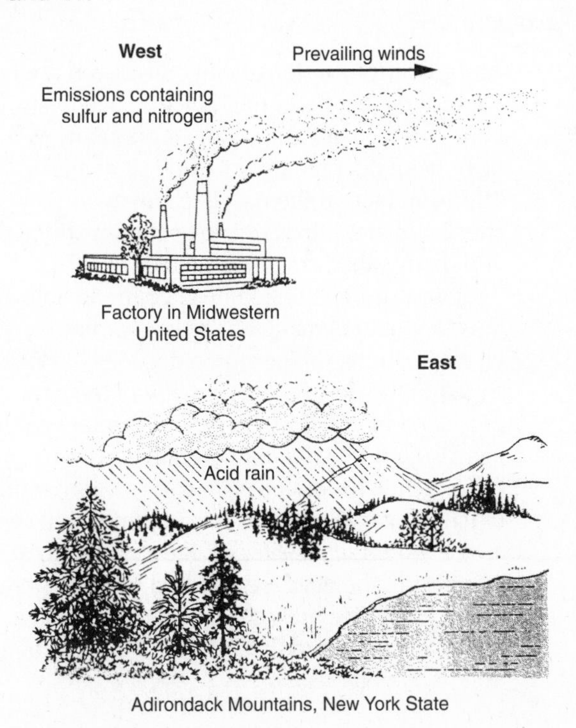


**40.** According to the illustration, acid rain is both an air pollutant and a water pollutant. Explain why this is true.

**41.** For each type of pollution (air and water), give *one* example of the kind of habitat the acid rain affects. By what means does it reach these different ecosystems?

*Base your answers to questions 42 and 43 on the information below and on the following diagram.*

Acid rain can have a pH between 1.5 and 5.0. The effect of acid rain on the environment depends on the pH of the rain and the characteristics of the environment. It appears that acid rain has a negative effect on plants. The following scale shows the pH of normal rain.

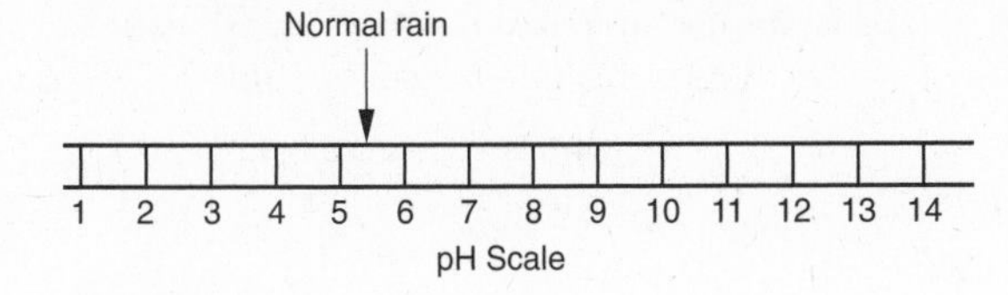


**42.** Provide the information that should be included in a research plan to test the effect of pH on the early growth of bean plants in the laboratory. In your answer be sure to:

- state a hypothesis
- identify the independent variable
- state *two* factors that should be kept constant

**43.** Construct a data table in which you could organize the research results.

*Refer to the following graph to answer questions 44 and 45.*

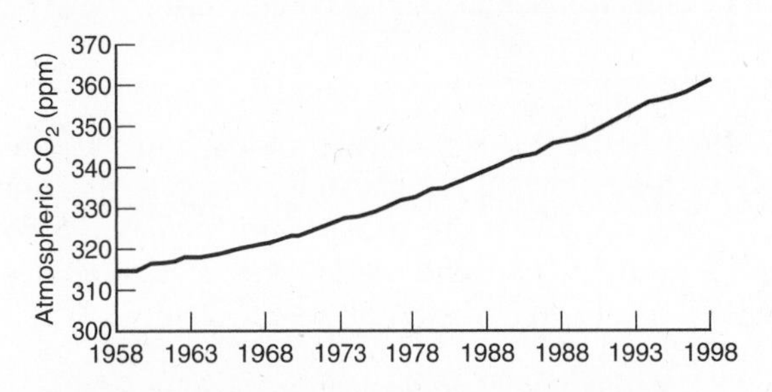


**44.** According to the graph, from the late 1950s to the late 1990s, the amount of $CO_2$ in Earth's atmosphere has been

1 steadily decreasing
2 steadily increasing
3 staying about the same
4 going up and down

**45.** Changes in the amount of $CO_2$ in Earth's atmosphere have been correlated with steadily increasing average global temperatures over the past 50 years. Based on this statement and the data in the graph, you could reason that

1 as the amount of $CO_2$ in the air increases, the average temperature decreases
2 as the amount of $CO_2$ in the air increases, the average temperature increases
3 as the amount of $CO_2$ in the air decreases, the average temperature stabilizes
4 as the amount of $CO_2$ in the air decreases, the average temperature increases

**46.** Some scientists are urging that immediate action be taken to stop activities that contribute to global warming. Discuss the effects of global warming on the environment and describe some human activities that may contribute to it. Your answer *must* include:

- an explanation of what is meant by the term *global warming*
- *one* human activity that is thought to be a major contributor to global warming
- an explanation of *how* the human activity may contribute to the problem
- *one* negative effect of global warming if it continues for many years

**47.** What substance is thought to cause the depletion of Earth's ozone shield? What has been done to solve this problem?

**48.** In the early 1980s, scientists discovered "holes" in the ozone shield that surrounds Earth. What is *one* possible health threat to humans that this environmental change could cause?

**49.** Choose *one* ecological problem to discuss from the following list: *global warming; destruction of the ozone shield; acid rain; increased nitrogen and phosphorous in lakes; loss of biodiversity*. In your answer be sure to state:

- the type of ecological problem you have chosen
- *one* human action that may have caused the problem
- *one* way in which the problem may negatively affect humans
- *one* way in which the problem may negatively affect the ecosystem
- *one* positive action that people can or did take to reduce the problem

**50.** Describe *two* specific methods that have been recently used by people to reduce the amount of chemicals being added to the environment.

**51.** All living organisms are dependent on a stable environment. Describe how humans have made the environment *less* stable for other organisms by:

- changing the chemical composition of air, soil, and water
- reducing the biodiversity of an area
- introducing advanced technologies

*Base your answers to questions 52 and 53 on the information below and on your knowledge of biology.*

Amphibians have long been considered an indicator of the health of life on Earth. Scientists are concerned because amphibian populations have been declining worldwide since the 1980s. In fact, in the past decade, twenty species of amphibians have become extinct and many others are endangered.

Scientists have linked this decline in amphibians to global climatic changes. Warmer weather during the last three decades has resulted in the destruction of many of the eggs produced by the Western toad. Warmer weather has also led to a decrease in rain and snow in the Cascade Mountain Range in Oregon, reducing the water level in lakes and ponds that serve as the reproductive sites for the Western toad. As a result, the eggs are exposed to more ultraviolet light. This makes the eggs more susceptible to a water mold that kills the embryos by the hundreds of thousands.

**52.** The term that is commonly used to describe the worldwide climatic changes mentioned in the passage is

1 global warming
2 deforestation
3 mineral depletion
4 industrialization

**53.** State *two* ways in which the decline in amphibian populations could disrupt the stability of the ecosystems they inhabit.

*Refer to Figure 29-9 on page 238 to answer the following question.*

**54.** By the year 2010, the worldwide human population had reached more than

1 three times the size it was in 1950
2 three times the size it was in 1960
3 two times the size it was in 1970
4 two times the size it was in 1960

### *Part C—Reading Comprehension*

*Base your answers to questions 55 to 57 on the information below and on your knowledge of biology.*
Source: *Science News*, Vol. 162, p. 400.

> Satellite observations of the Arctic Ocean show that the amount of sea ice there this year [2002] was the lowest it's been in more than 20 years.
>
> In September, the extent of the sea ice—defined as the area in which ice covers at least 15 percent of the ocean's surface—was 5.27 million square kilometers, says Julienne C. Stroeve, a climatologist at the National Snow and Ice Data Center in Boulder, Colo. Of that area, sea ice actually covered about 3.6 million square kilometers, a figure 17 percent lower than normal for that time of year and 9 percent below the previous minimum for a September. The earlier record low was set in 1998, during the late stages of the strongest El Niño ever seen and when the global average temperature had been much higher than normal for several months.
>
> Satellites have been monitoring Arctic sea ice since 1978. Since then, annual ice coverage has dropped about 3 percent per decade, and September ice coverage has declined 8 percent per decade.
>
> Several factors contributed to the low ice cover this year [2002], says Stroeve. From March through May, southerly winds pushed ice away from the northern shores of Eurasia and North America. Because the open water absorbed more radiation than snow-covered ice would have, the near-shore waters warmed and accelerated melting at the edges of the ice packs. From June through August, unusually warm and persistently stormy conditions blanketed the Arctic Ocean, fracturing the [sea ice] and further fostering melting.
>
> Unlike earlier years with low ice coverage, this year [2002] the sea northeast of Greenland was relatively free of ice.

**55.** State the method that is used to conduct observations of the amount of sea ice on the Arctic Ocean.

**56.** Explain what was unusual about the amount of sea ice observed in the Arctic in September of 2002.

**57.** State the *two* reasons why the sea ice coverage may have been so low during the year of this study.

# 30
# Saving the Biosphere

**Vocabulary**

biosphere
nonrenewable resources
renewable resources
sustainable development

## WHAT NEEDS TO BE SAVED?

In April 1970, the first Earth Day marked the beginning of the modern environmental movement. Many environmental organizations were founded at that time and the government has passed several environmental protection laws since then.

Can we protect the environment not only for people but for all species? The total area of land, water, and air on Earth's surface where life is found—and that needs protection—is called the **biosphere**. (See Figure 30-1.) Saving the biosphere means paying attention to local, regional, and global problems. What must we do now to protect the environment for organisms that will be alive after us?

**Figure 30-1** The biosphere is the total area of Earth's land, air, and water in which life is found. Earth is the ultimate ecosystem; although environmental problems may start out as local ones, they can become regional and, eventually, have a global impact.

## A CHANGE IN ATTITUDE

Sometimes the most important, and difficult, changes concern accepted attitudes in our society. For example, what if we thought our lifestyle should not harm the environment and other living species? Would we be willing to make the necessary changes to accomplish this?

An industrialized society mainly views Earth as a source of valuable resources for its use. In contrast, ecology teaches us that humans are just one of many interdependent species that also need resources to live. For our species to survive, we must make sure that these important relationships within ecosystems also survive.

## THINK GLOBALLY, ACT LOCALLY

The future health of the environment will depend on people's attitudes and behaviors. It has been suggested that people should learn to appreciate the "hidden costs" of many consumer goods. In other words, the environment pays a price for the products used by people in an industrial society.

Environmentalists have encouraged people to live by the "3 R's": reduce, reuse, recycle. To *reduce* consumption, you would use less of a product or resource; for example, fewer paper towels can be used to clean up a spill. You can also *reuse* a product; for example, paper or plastic grocery bags that you bring food home in can be taken back the next week and used again. Finally, many used products can be made into other products; for example, in many cities, you now have to *recycle* plastic, glass, metal, and paper. (See Figure 30-2.) These materials are used again in other products, such as benches made up of a "wooden" building material that is actually a form of recycled plastic. Such recycling helps to conserve natural resources. (See Figure 30-3.)

**Figure 30-2** Recycling of plastic, glass, and metal containers is required in some cities. As shown here, students can help in the recycling effort by sorting and recycling metal soda cans.

**Figure 30-3** The teenagers in this photograph are helping to conserve natural resources by recycling newspapers.

## NON-NATIVE SPECIES

Great care must be taken to avoid the environmental damage that can occur when a species from a distant region is introduced into a new environment. Without any predators or natural controls, such a non-native, or *introduced*, species (such as rabbits in Australia) can reproduce without limit, upsetting the stability of the ecosystem.

Much is also being learned about how the harmful effects from the chemical control of pests in the environment can be avoided by the use of natural, or *biological*, control. For example, native predators of insect pests can be used instead of pesticides as a means of biological control.

## RENEWABLE VERSUS NONRENEWABLE RESOURCES

Air, water, and sunlight are some of the important resources that are basic to life on Earth. Modern industrialized society requires other resources, too, such as coal, oil, and metal ores.

Resources can be considered renewable or nonrenewable. A **renewable resource** can be replaced within a generation. Enough of the resource is being made (by natural processes) to replace what is being used. For

Figure 30-4 The solar panels on this school's roof capture solar energy—a renewable resource—which is used to provide 70 percent of the building's heat and hot water.

example, the wood used to build houses can be replaced if enough trees are replanted. A variety of *biofuels* such as ethanol, which are produced from plants, can be used in place of gasoline. The sun's *solar energy* and the wind can be considered renewable resources, since they cannot be used up within a human lifetime. (See Figure 30-4.)

A **nonrenewable resource** cannot be replaced within our lifetime; it exists in limited amounts and takes a very long time to form. This includes such energy sources as coal, oil, and natural gas. In addition, metals such as gold, silver, iron, copper, and aluminum, and nonmetals such as sand, gravel, and limestone are nonrenewable resources.

Figure 30-5 Windfarms, such as this one in California, harness wind power—another renewable resource—to generate electricity.

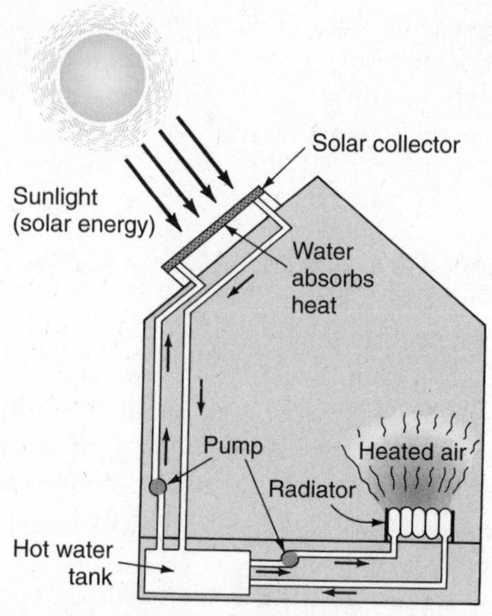


Figure 30-6 The diagram illustrates how energy from sunlight can be collected by a solar panel and used to heat a building.

One way to protect the biosphere is to use renewable energy sources. For example, electricity can be made in a dam from the power of falling water, rather than from the burning of coal. Wind power can turn the blades on giant windmills to generate electricity. (See Figure 30-5.) Sunlight can be used to heat water and buildings, and to produce electricity. (See Figure 30-6.) Finally, hot water from deep beneath Earth's surface supplies *geothermal energy*, which can be used to heat buildings and to make electricity.

## SUSTAINABLE DEVELOPMENT

If the economy is not growing, people think that something is wrong. Yet, unlimited growth is not possible. We cannot use more and more of Earth's resources indefinitely; and we cannot add more and more pollution to our environment. We need to find a way to live that is sustainable—a way that does not ruin Earth's ability to support life in the future. Improving the way we live without harming the environment is called **sustainable development**. By making these changes now, we may secure a healthier future for our planet.

## SUSTAINABLE DEVELOPMENT OF FORESTS

Some of the greatest damage to the environment is done by cutting down trees. The most economical way to harvest trees is to cut down an entire forest, a method called *clear-cutting*. (See Figure 30-7.) Such deforestation causes animals to lose habitat; and the rain—no longer absorbed by tree roots—flows right into streams, carrying topsoil with it. Due to soil erosion, the land cannot support plant growth; and the freshwater habitat for fish is disturbed, too.

Sustainable development in forestry means replacing every tree that is cut with a seedling and making sure that the seedling survives. A forest—with trees of all sizes and different ages, continually regrowing—would provide a healthy habitat for other woodland species and would prevent soil erosion.

Another good method for protecting forests involves using different plants as a replacement for tree-derived products. For example, the kenaf plant, which looks like a tall bamboo and grows very quickly, makes excellent pulp for paper. By growing kenaf on farms and using the plants as a source of paper pulp, we would be able to cut fewer forests. This would help us live in a more sustainable manner with the forests.

**Figure 30-7** The wood from forests has many uses: lumber, paper pulp, and fuel. Although cutting all the trees in one area is economical for loggers, the environment pays a high price in terms of loss of habitats.

## ENVIRONMENTAL PROTECTION IN A DEVELOPED COUNTRY

Like other industrialized, developed countries, the United States uses up more than its share of energy and resources. Also, since the environment has been seriously affected, environmental awareness has increased since 1970.

One important result of this increased awareness was the formation of the Environmental Protection Agency (EPA), which is responsible for safeguarding the environment for future generations.

Another response to environmental issues was passage of the Endangered Species Act, which regulates a wide range of activities that affect threatened or endangered plant and animal species. (See Figure 30-8.)

**Figure 30-8** The spotted owl of the Pacific Northwest is an endangered species. By protecting the owl, the Endangered Species Act also protects the old-growth forests it lives in.

## ENVIRONMENTAL PROTECTION IN A DEVELOPING COUNTRY

People's lives in developed countries are very different from those in developing countries. In industrialized nations, the average standard of living is high and most people can expect to live for 70 or more years. Some populations in rapidly developing countries such as India and China also now enjoy these higher standards of living. But in many developing countries, most people are poor and have a much shorter life span. So, saving the biosphere means different things in rich na-

**Figure 30-9** In Africa, near Kenya's Ewasu River, local guides take tourists on camel safaris. This is one way in which "parks for people" programs can help local people earn a living from their natural environment without doing it harm.

tions and poor nations. Environmental leaders are now learning about these differences.

Conservationists realize that it is very difficult to set aside parks for endangered animals if doing so stops people from getting enough food and housing. One solution that has worked—called "parks for people"—directly involves local people in protecting their environment. For example, in Kenya, villagers work as guides for tourists who come to see the wildlife. (See Figure 30-9.)

## SAVING THE BIOSPHERE: A WORLDWIDE EFFORT

In 1992, representatives from 178 countries attended the largest environmental meeting ever held, known as the Earth Summit. The main theme of the meeting was sustainable development. Work has continued since then, but differences between countries have interfered with more progress. After several years, it was recognized that progress in implementing sustainable development had been extremely disappointing since the 1992 Earth Summit. The United Nations convened a World Summit on Sustainable Development in South Africa in 2002. The leaders of 104 countries were present; however, the United States was absent from this summit. Ultimately, the protection of Earth's biosphere for the future requires the efforts of people from all around the world.

# Chapter 30 Review

### *Part A—Multiple Choice*

**1.** The biosphere is the total area where life exists on or in Earth's

1 land only
2 water only
3 land and water
4 land, water, and air

**2.** A major reason that humans have negatively affected the environment in the past is that they

1 often lacked an understanding of how their activities affect the environment
2 attempted to control their population growth
3 passed laws to protect certain wetlands
4 discontinued the use of certain chemicals used to control insects

**3.** Recycling of materials such as glass, metal, and plastic helps to

1 keep the cost of groceries low
2 conserve our natural resources
3 prevent natural resources
4 build more wooden houses

**4.** Which human activity would be *least* likely to disrupt the stability of an ecosystem?

1 disposing of wastes in the ocean
2 increasing the human population
3 using more fossil fuels
4 recycling bottles and cans

**5.** An industrialized society views Earth mainly as a

1 home for wildlife
2 hazardous place to live
3 source of natural resources
4 barren landscape

6. Coal and wood are found in nature. They are both examples of

   1 enzymes
   2 resources
   3 metals
   4 proteins

7. An example of a renewable resource is

   1 natural gas
   2 silver
   3 coal
   4 wood

8. The definition of a renewable resource is that it

   1 can be replaced by nature within a generation
   2 cannot be replaced by nature within a generation
   3 is manufactured by humans
   4 is not too expensive to use

9. A nonrenewable resource is one that

   1 is replaced by nature as fast as it is used
   2 exists in a limited supply that can run out
   3 is recycled naturally by Earth's systems
   4 does not pollute ecosystems when it is used

10. Suppose that the average life span of most people is about 70 years and that there is only a 70-year supply of fossil fuel left on Earth. Then imagine that you are a member of a government panel that is deciding on how to handle the fuel situation. The best possible decision you could suggest would be to

    1 use up all the fuel in the present generation and not worry about the future
    2 find some alternative energy sources so that the fossil fuel lasts longer
    3 destroy the remaining fossil fuel so that no nations will fight over it
    4 have people return to a farming society so that they do not need the fuel

11. Which practice would most likely deplete a nonrenewable natural resource?

    1 harvesting pine trees on a tree farm
    2 restricting water usage during a period of water shortage
    3 burning coal to generate electricity in a power plant
    4 building a dam and a power plant to use water to generate electricity

12. All of the following are nonrenewable energy sources *except*

    1 coal
    2 gas
    3 falling water
    4 natural gas

13. Which statement is true about endangered species and nonrenewable resources?

    1 They are both living things that need to be protected.
    2 They are both nonliving things that need to be protected.
    3 They both need to be protected so they do not disappear.
    4 They both can be renewed quickly if they do disappear.

14. The goal of sustainable development is to

    1 achieve unlimited economic growth at any cost
    2 make sure the economy expands at a steady rate
    3 improve the way we live without harming the environment
    4 expand our lifestyle even if we run out of natural resources

15. Which method would you *not* use to solve environmental problems?

    1 promote global awareness
    2 cooperation among people
    3 "parks for people" programs
    4 increasing the population

16. Which action would best illustrate people's concern for the biosphere?

    1 passing game laws that limit the number of animals that may be hunted
    2 increasing the use of pesticides that may drain off farms into river systems
    3 allowing air to be polluted by only those factories that use new technology
    4 removing resources from nature at a faster rate than they are being replaced

17. One way to help provide a suitable environment for the future is to urge individuals to

    1 apply ecological principles when making decisions that have an impact
    2 agree that population controls have no effect on environmental matters
    3 control all aspects of natural environments
    4 work toward increasing global warming

### *Part B—Analysis and Open Ended*

18. What is meant by "think globally, act locally" in terms of environmental protection?

**19.** Define the "3 R's" proposed by environmentalists. Give *one* example of each.

**20.** Why is it important for industries to be involved in recycling programs?

**21.** Compare and contrast renewable and nonrenewable resources. Your answer should include:

- a definition of *renewable* and of *nonrenewable*
- *one* example of a renewable resource
- *two* examples of a renewable *energy* source
- *two* examples of a nonrenewable resource

**22.** Recycling can extend the use of nonrenewable resources but *cannot* restore them. Humans can restore renewable resources to reduce some negative effects of increased consumption. Identify *one* resource that is renewable, and describe *one* specific way people can restore this resource if it is being depleted.

**23.** Use the following terms to complete the flowchart below, which lists a variety of natural resources: *flowing water; copper; wind power; wood (charcoal); oil; gold; trees (lumber); limestone; geothermal energy; gravel; natural gas; coal; sunlight; sand; silver.*

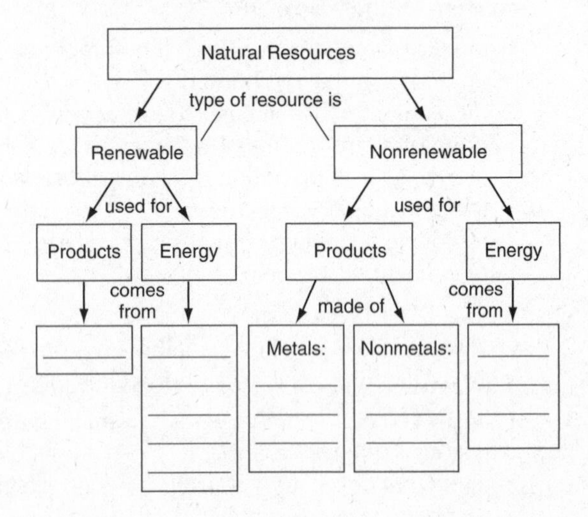


**24.** Explain what is meant by the term "sustainable development." Give *one* example of sustainable use of a natural resource.

**25.** Refer to Figure 30-3 on page 247 to answer this question. How might the activity these young people are involved in actually help to conserve forests and wildlife?

**26.** In what way does the protection of habitats help enforce the Endangered Species Act?

**27.** Refer to Figure 30-7 on page 249 to answer this question. How does the method of "clear-cutting" shown here damage habitats both on land and in the water?

**28.** Why are people in developing nations sometimes more directly affected by conservation programs than are people in developed (industrialized) nations?

**29.** How do "parks for people" programs help local people and their environment at the same time?

**30.** Give at least one reason why *all* nations—developed and developing—should be involved in global efforts to save the biosphere.

**31.** List *five* changes that could be made in your own home that could help protect the environment.

**32.** List *five* changes that could be made in your community that could help protect the environment.

### *Part C—Reading Comprehension*

*Base your answers to questions 33 to 35 on the information below and on your knowledge of biology.*
Source: *Science News*, Vol. 162, p. 237.

The great white shark, river dolphins, several types of whales, and an unusual two-humped camel are among animals that were designated on Sept. 24 [2002] to receive new or heightened protection under the Convention on Migratory Species, a UN treaty.

Meeting in Bonn, Germany, delegates placed the great white shark in the treaty's Appendix I. Such a listing bans the 80 nations that have ratified the treaty from catching or harming the species inside their boundaries. It also requires the nations to protect the species' habitat.

Three whale species—fin, sperm, and sei—also made it into Appendix I, along with the nearly extinct Ganges and Indus River dolphins (*Platanista gangetica gangetica*) in India, Pakistan, and Bangladesh. These virtually blind freshwater dolphins, which navigate by sonar, are threatened by pollution, hunting, and entanglements in fishnets.

Perhaps the most unusual animals added to Appendix I are the 300 hairy-kneed camels, found in three isolated pockets of China and Mongolia, including a former nuclear test range. They survive on salt water, and scientists suspect they're a new species.

Even nations that haven't ratified the Convention on Migratory Species, such as the United States, Japan, Russia, and China, sometimes adopt its policies.

**33.** State *two* ways in which animals get protection when they are listed in Appendix I of the UN treaty.

**34.** Explain why the river dolphins of India, Pakistan, and Bangladesh need special Appendix I levels of protection.

**35.** State *two* facts about the hairy-kneed camels that support the decision to protect them under the UN treaty.

# APPENDIX A

# MST Learning Standards 4 and 1

The Regents Living Environment Examination is based on the New York State Learning Standards for Mathematics, Science, and Technology (MST). In particular, the examination will test student performance on the commencement (high school) level of **MST Learning Standard 4: The Living Environment and MST Learning Standard 1: Scientific Inquiry.**

## MST LEARNING STANDARD 4: THE LIVING ENVIRONMENT

There are seven key ideas for MST Learning Standard 4: The Living Environment. The material in the six themes of this review book covers all the required knowledge of these seven key ideas, which are as follows:

1. Living things are both similar to and different from each other and nonliving things.
2. Organisms inherit genetic information in a variety of ways that result in continuity of structure and function between parents and offspring.
3. Individual organisms and species change over time.
4. The continuity of life is sustained through reproduction and development.
5. Organisms maintain a dynamic equilibrium that sustains life.
6. Plants and animals depend on each other and their physical environment.
7. Human decisions and activities have had a profound impact on the physical and living environment.

## MST LEARNING STANDARD 1: SCIENTIFIC INQUIRY

There are three key ideas for MST Learning Standard 1: Scientific Inquiry, which are as follows:

1. The central purpose of scientific inquiry is to develop explanations of natural phenomena in a continuing, creative process.
2. Beyond the use of reasoning and consensus, scientific inquiry involves the testing of proposed explanations involving the use of conventional techniques and procedures and usually requiring considerable ingenuity.
3. The observations made while testing proposed explanations, when analyzed using conventional and invented methods, provide new insights into phenomena.

What students should know and be able to do in relation to each of these key ideas is described below. This content may be tested on the Regents Living Environment Examination.

### Introduction to Scientific Inquiry

Science is a body of knowledge about the world. It is also a thinking process that has been designed by people in order to learn about how the world works. Based on observations and evidence collected from experimentation, and using what is already known, scientific explanations are developed about the world. These explanations are always subject to change as new observations and evidence are presented. Scientific methods exist to constantly test and re-test what we know in relation to existing explanations. In this way, scientific knowledge advances toward a more complete understanding of the world around us.

**Key Idea 1:** *The central purpose of scientific inquiry is to develop explanations of natural phenomena in a continuing, creative process.*

Natural phenomena, events, and occurrences are understood by people on the basis of existing explanations. To think about these explanations, it is necessary to be able to visualize, that is, to create mental pictures and to develop mathematical models. To develop scientific explanations, one uses evidence that can be observed as well as what people already know about the world. It is also important to learn about the history of science and the particular individuals who have contributed to scientific understanding. While science provides knowledge about the world, it also challenges people to develop the values to use this knowledge ethically and effectively.

To develop scientific ideas, one needs to think, do library research, and discuss one's ideas with others, including experts. Scientific inquiry involves asking questions and locating information from a variety of sources. It also involves making wise judgments about how reliable and relevant the information is.

At times there may be more than one explanation for the same phenomenon. A science student needs to work to resolve the differences. This is done through the use of evidence from experiments and direct observation. Also, an explanation that results in predictions that turn out to be accurate is more likely to be the correct explanation. All scientific explanations may be changed if new evidence suggests the change. This leads to a continually better understanding of how things work in the world.

It is necessary to develop explanations about small things and large things, that is, at different levels of scale. Also, one must consider phenomena from different points of view and degrees of complexity. In order to do this, experts from different subject areas often need to be consulted.

**Key Idea 2:** *Beyond the use of reasoning and consensus, scientific inquiry involves the testing of proposed explanations involving the use of conventional techniques and procedures and usually requiring considerable ingenuity.*

To test an explanation that has been put forward, a scientist must design ways to make observations related to the explanation. The design of the research must make use of library investigation in order to review scientific literature and also through discussions with other scientists. The research plan requires that one understand the big concepts being investigated. The plan should also include a variety of techniques, proper equipment, and safety procedures.

In creating a research plan, scientists use hypotheses to test the proposed explanations. A hypothesis is a prediction of what should be observed under specific conditions if the explanation is true. Hypotheses are very useful in science for helping one to determine what data should be collected and how the data should be interpreted. The research plan to test a hypothesis must include procedures to make sure that there is a fair interpretation of the data. This is called avoiding bias. These procedures include repeated trials, large sample size, and objective data-collection techniques.

Once a research plan is created, it must be carried out. This means doing the actual experiment, obtaining and putting together the necessary equipment, and recording one's observations.

**Key Idea 3:** *The observations made while testing proposed explanations, when analyzed using conventional and invented methods, provide new insights into phenomena.*

Observations made during scientific research usually need to be analyzed to see what they mean. The data may be organized and represented in a variety of ways in order to do this; for example, diagrams, tables, charts, graphs, equations, and matrices. When the data are interpreted, the result may be the statement of a new hypothesis. Another result may be the conclusion that a general understanding or explanation of a natural phenomenon is, in fact, correct.

The mathematical processes of statistical

analysis are used to determine if the results obtained might have been simply due to chance. Statistics also allows one to conclude the degree to which predicted results based upon the hypothesis match the actual results. From this matching comes the conclusion as to whether the proposed explanation is, or is not, supported by the data.

The analysis of the data, followed by public discussion, can lead to a revision of the explanation, the development of new hypotheses, and the design of new research plans.

When claims are made based on the collected evidence, the claims should be questioned if the design of the experiment was at fault; for example, if there were small sample sizes, incomplete or misleading use of data, or the lack of controlled conditions. Also, great care should be taken to not confuse fact with opinion.

When all research and data analysis is concluded, a written report is prepared for the public to study. The report includes a literature review, the research, the results and suggestions for further research. One purpose of making the results public is to allow the research to be repeated. Science assumes that through the collection of similar evidence, different individuals will come to the same explanations of nature. Peer review—the study of research reports by fellow scientists—is important as a check on the quality of the research. It also results, at times, in the suggestion of alternative explanations for the same observations.

# APPENDIX B

# Required Laboratory Activities for the Regents Exam, Part D

The *Regents Examination in Living Environment* includes multiple-choice and open-ended questions based on a series of required laboratory activities completed during the school year. Part D of the Living Environment examination (given in January, June, and August each year) will test at least three of the four laboratory activities that are required for that year. There are now four different laboratory activities scheduled for implementation and testing through 2010. (*Note:* Over time, new labs may be introduced to replace the current lab activities. Lab #4: Adaptations for Reproductive Success in Flowering Plants may be implemented in the near future to replace Lab #3: The Beaks of Finches.) The first four labs are described below.

Lab #1: Relationships and Biodiversity

Lab #2: Making Connections

Lab #3: The Beaks of Finches

Lab #5: Diffusion Through a Membrane

While completing the laboratory activities, you will record your results and answers to questions in the Student Laboratory Packets. You are to keep these sheets for review before taking the Regents examination. You will also transfer your answers to separate Student Answer Packets, which will be used and kept by the school as evidence of your completion of the laboratory requirement for the Living Environment Regents exam. All directions to the teachers and printed materials for the students have been prepared and distributed by the New York State Education Department.

### Required Laboratory Activity #1: Relationships and Biodiversity

This activity is a simulation that consists of six tests done in the lab as well as a seventh task, a reading assignment. Your goal is to collect and analyze data on several different plant species in order to determine which of the species is most closely related to a valuable but endangered species. The evidence, both structural and molecular, is used to develop a hypothesis about the evolutionary relationships between the plant species. The final task of the activity, the reading passage, focuses on the importance of preserving biodiversity.

This laboratory activity is most closely correlated with topics covered in Theme V—Genetics and Molecular Biology, Theme VI—Evolution: Change Over Time, and Theme VII—Interaction and Interdependence. The activity could be used during the teaching of these themes or as a performance assessment of laboratory skills. The lab requires that you have an understanding of DNA and protein synthesis.

### Required Laboratory Activity #2: Making Connections

The purpose of this activity is to help you learn how to design and use a controlled experiment in order to draw a conclusion. In particular, you are to determine which of two conflicting claims is supported by your experimentation. The laboratory activity consists of two parts. In Part A, you will practice two simple techniques—taking a person's pulse and measuring muscle fatigue by squeezing a clothespin. Part B is the main portion of the activity. You will design your own investigation after reviewing guidelines for conducting a controlled experiment. You will use your experiment to determine which claim is supported, namely that a person can squeeze a clothespin more times by exercising first or more times by *not* exercising first. Results are put in writing and some students make oral presentations of their reports to the class for peer review.

This laboratory activity is most closely correlated with topics of human physiology covered in "Chapter 6: Getting Food to Cells: Nutrition" and "Chapter 7: Gas Exchange and Transport" in Theme II—Energy, Matter, and Organization. However, a minimum of content knowledge is required for the activity. The main focus of the activity is to understand the concepts involved in experimental design, and this activity could be used to introduce this topic.

**Required Laboratory Activity # 3: The Beaks of Finches**

The purpose of this activity is to use a simulation to study how structural differences affect the survival rate of members within a species. The lab is based on the observations of the many finch species on the Galápagos Islands that Charles Darwin used in support of the process of natural selection. You will work in pairs to represent a finch. Each pair is randomly assigned a grasping tool—such as forceps, tongs, pliers, or tweezers—that represents a type of beak to be used to pick up seeds of different sizes. The efficiency of the tool-beaks at picking up small seeds such as lentils determines whether the "finches" survive and stay in the same "environment" with these seeds or "migrate" in search of food to a different environment with larger seeds such as lima beans. The survivors in the activity now compete with others to continue to explore the efficiency of their "beaks."

This laboratory activity is most closely correlated with topics covered in "Chapter 21: The Process of Evolution" within Theme VI—Evolution: Change Over Time. You will need to be familiar with the concepts of adaptation, variation, and natural selection.

**Required Laboratory Activity #5: Diffusion Through a Membrane**

In this laboratory activity, you will study the process of diffusion by using a model "cell" to test selective permeability of the cell membrane. The "cell" is made of dialysis tubing or a plastic bag that contains a glucose and starch solution, which is immersed in water for a period of time. The water in the beaker is then tested for the presence of glucose and starch. In the second part of the laboratory activity, you will use a microscope to observe the effects of salt water and distilled water on red onion cells. You will see the effects of the diffusion (osmosis) of water out of the cells when they are surrounded by salt water.

This laboratory activity is most closely correlated with sections on cell processes found in "Chapter 4: Chemical Activity in the Cell" within Theme II—Energy, Matter, and Organization, as well as topics in "Chapter 8: The Need for Homeostasis" and "Chapter 10: Excretion and Water Balance" within Theme III—Maintaining a Dynamic Equilibrium.

# APPENDIX C

# Living Environment Part D— Sample Lab Questions

Beginning with the June 2004 administration, the Regents Examination in Living Environment has included a new section, Part D. The questions on Part D consist of a combination of multiple-choice and open-ended questions related to at least three of the four required living environment laboratory activities and comprise approximately 15% of the examination.

These sample questions are provided to help teachers and students become familiar with the format of questions for this part of the examination. They provide examples of ways the required laboratory experiences may be assessed. A rating guide is also included.

---

**Sample Items Related to Lab Activity #1: *Relationships and Biodiversity***

1 In the *Relationships and Biodiversity* laboratory activity, students were instructed to use a clean dropper to place each of four different samples of plant extracts on the chromatography paper. A student used the same dropper for each sample without cleaning it between each use. State one way this student's final chromatogram would be different from a chromatogram that resulted from using the correct procedure. [1]

2 State one reason that safety goggles were required during the indicator test for enzyme *M*. [1]

Base your answers to questions 3 through 6 on the information and data table below and on your knowledge of biology.

A student was told that three different plant species are very closely related. She was provided with a short segment of the same portion of the DNA molecule that coded for enzyme *X* from each of the three species.

**Information Regarding Enzyme X**

| | | | |
|---|---|---|---|
| DNA sequence from plant species *A* | CAC | GTG | GAC |
| Amino acid sequence for enzyme *X* coded for by that DNA | Val | His | Leu |
| DNA sequence from plant species *B* | CAT | GTG | CAA |
| Sequence of bases in mRNA produced by that DNA | ____ | ____ | ____ |
| Amino acid sequence for enzyme *X* coded for by the DNA | Val | His | Val |
| DNA sequence from plant species *C* | CAG | GTA | CAG |
| Sequence of bases in mRNA produced by that DNA | GUC | CAU | GUC |
| Amino acid sequence for enzyme *X* coded for by the DNA | ____ | ____ | ____ |

3 The correct sequence of mRNA bases for plant species *B* is

(1) GUA CAC GUU
(2) GTA CAC GTT
(3) CAU GUG CAA
(4) TCG TGT ACC

4 Use the mRNA Codon Chart on the next page to determine the amino acid sequence for enzyme *X* in plant species *C* and record the sequence in the appropriate place in the data table. [1]

5 Is it possible to determine whether species *B* or species *C* is more closely related to species *A* by comparing the amino acid sequences that would result from the three given DNA sequences? Support your answer. [1]

6 Determine whether species *B* or species *C* appears more closely related to species *A*. Support your answer with data from the data table. (*Base your answer only on the DNA sequences provided* for enzyme *X* in these three plant species.) [1]

## Universal Genetic Code Chart

Messenger RNA codons and the amino acids they code for.

| FIRST BASE | SECOND BASE U | | SECOND BASE C | | SECOND BASE A | | SECOND BASE G | | THIRD BASE |
|---|---|---|---|---|---|---|---|---|---|
| U | UUU | PHE | UCU | SER | UAU | TYR | UGU | CYS | U |
| | UUC | | UCC | | UAC | | UGC | | C |
| | UUA | LEU | UCA | | UAA | STOP | UGA | STOP | A |
| | UUG | | UCG | | UAG | | UGG | TRP | G |
| C | CUU | LEU | CCU | PRO | CAU | HIS | CGU | ARG | U |
| | CUC | | CCC | | CAC | | CGC | | C |
| | CUA | | CCA | | CAA | GLN | CHA | | A |
| | CUG | | CCG | | CAG | | CGG | | G |
| A | AUU | ILE | ACU | THR | AAU | ASN | AGU | SER | U |
| | AUC | | ACC | | AAC | | AGC | | C |
| | AUA | | ACA | | AAA | LYS | AGA | ARG | A |
| | AUG | MET or START | ACG | | AAG | | AGG | | G |
| G | GUU | VAL | GCU | ALA | GAU | ASP | GGU | GLY | U |
| | GUC | | GCC | | GAC | | GGC | | C |
| | GUA | | GCA | | GAA | GLU | GGA | | A |
| | GUG | | GCG | | GAG | | GGG | | G |

## Sample Items Related to Lab Activity #2: *Making Connections*

Base your answers to questions 7 through 9 on the information and data table below and on your knowledge of biology.

In the Making Connections laboratory activity, a group of students obtained the following data:

| Student Tested | Pulse Rate at Rest | Pulse Rate After Exercising |
|---|---|---|
| 1 | 70 | 97 |
| 2 | 75 | 106 |
| 3 | 84 | 120 |
| 4 | 60 | 91 |
| 5 | 78 | 122 |

7 Explain how this change in pulse rate is associated with homeostasis in muscle cells. [1]

8 Identify the system of the human body whose functioning is represented by this data. [1]

9 Identify *one* other system of the human body whose functioning would be expected to be altered as a direct result of the exercise. Describe how this system would most likely be altered. [1]

Base your answers to questions 10 and 11 on the information below and on your knowledge of biology.

A biology class performed an investigation to determine the influence of exercise on pulse rate. During the investigation, one group of twelve students, Group *A*, counted how many times they could squeeze a clothespin in a 1-minute period, then exercised for 4 minutes, and repeated the clothespin squeeze for an additional 1 minute. Another group of twelve students, Group *B*, also counted how many times they could squeeze a clothespin in a 1-minute period, but then they rested for 4 minutes, and repeated the clothespin squeeze for an additional 1 minute. The data table below shows the average results obtained by the students.

**Effect of Exercise on Number of Clothespin Squeezes**

| Groups of Student | Average Number of Clothespin Squeezes During First Minute | Average Number of Clothespin Squeezes During Second Minute |
|---|---|---|
| Group A (exercise) | 75 | 79 |
| Group B (rest) | 74 | 68 |

10 State *two* specific examples from the description of the investigation and the data table that support this investigation being a well-designed experiment. [2]

11 The chart below shows relative blood flow through various organs during exercise and at rest.

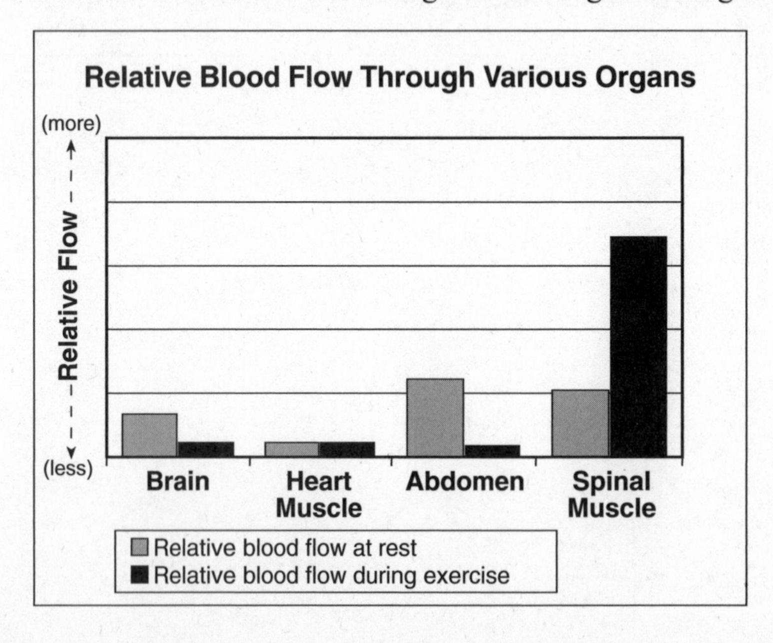


Using information from both the data table and the chart, explain how muscle fatigue and blood circulation could account for the results the students obtained. [2]

## Sample Items Related to Lab Activity #5: *Diffusion Through a Membrane*

Base your answers to questions 12 and 13 on the diagrams below and on your knowledge of biology.

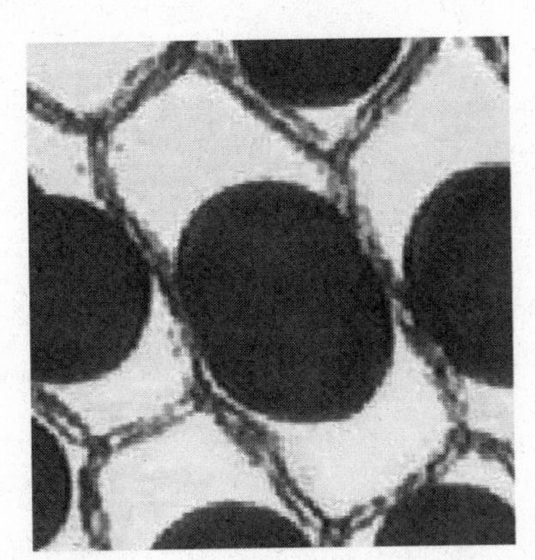

Diagram 1: red onion cells Diagram 2: red onion cells

12 Describe how to prepare a wet-mount slide of red onion cells with the cell membrane shrinking away from the cell wall, as shown in diagram 1. The following materials are available: microscope slide, pipettes, cover slips, paper towels, water, salt solution, and red onion sections. [3]

13 List the laboratory procedures to follow that would cause the cells in diagram 1 to resemble the cells in diagram 2. [2]

---

14 A student places an artificial cell, similar to the one used in the laboratory activity *Diffusion Through a Membrane*, in a beaker containing water. The artificial cell contains starch and sugar. A starch indicator is added to the water in the beaker. Explain how the student will know if the starch is able to diffuse out of the artificial cell. [1]

# GLOSSARY

**abiotic** describes the nonliving parts of an organism's environment

**absorption** the process that occurs when food molecules are tiny enough to pass through the wall of the small intestine and into the blood vessels

**acidity** describes a low pH level due to dissolved acids, such as in acid rain

**active transport** movement of substances across a membrane from an area of lower concentration to an area of higher concentration; requires energy

**adaptations** special characteristics that make an organism well suited for a particular environment

**adaptive radiation** the process by which several populations evolve from an original parent population, each adapted to different niches

**AIDS (acquired immunodeficiency syndrome)** an immunodeficiency disease, caused by HIV in humans

**algae** plantlike organisms, often single-celled, that carry out photosynthesis

**alleles** the two different versions of a gene for a particular trait

**allergic reactions** conditions caused by an overreaction of the immune system

**alveoli** microscopic air sacs in the lungs; their lining acts as the respiratory surface

**amino acids** organic compounds that are the building blocks of proteins

**amnion** the membrane that surrounds, protects, and holds in a fluid that cushions the embryo

**antibiotic** a chemical that kills specific microorganisms; frequently used to combat infectious diseases

**antibodies** molecules that individuals produce as a defense against foreign objects in the body; antibodies bind to specific antigens

**antigens** proteins on a foreign object that stimulate the immune system to produce antibodies

**arteries** vessels that have thick, muscular walls and which carry the oxygen-rich blood through the body

**artificial selection** see *selective breeding*

**asexual reproduction** form of reproduction that requires only one parent to pass on genetic information, e.g., budding

**atmosphere** the blanket of gases that covers Earth; usually called "air"

**atoms** the smallest units of an element that can combine with other elements

**ATP (adenosine triphosphate)** the substance used by cells as an immediate source of chemical energy for the cell

**autoimmune diseases** illnesses caused when an overactive immune system begins to attack its own normal body tissues

**autosomes** in humans, the 22 perfectly matched pairs of chromosomes; i.e., not the X and Y chromosomes

**autotrophic** describes a self-feeding organism that obtains its energy from inorganic sources, e.g., plants (producers)

**bacteria** single-celled organisms that have no nuclear membrane to surround and contain their DNA molecule

**behavior** every action that an animal does, either learned or instinctive; usually to aid survival

**bile** substance produced by the liver and secreted into the small intestine, it breaks down lipids into smaller droplets of fat

**binary fission** form of asexual reproduction used by single-celled organisms, such as the ameba

**biochemistry** the chemistry of living things, found to be similar in all organisms

**biodiversity** the variety of different species in an ecosystem or in the world

**biology** the study of all living things, including humans

**biome** a very large area characterized by a certain climate and types of plants and animals

**biosphere** the total area of land, water, and air on Earth's surface where life is found

**biotechnology** describes new procedures that utilize discoveries in biology, particularly recombinant DNA technology; see *genetic engineering*

**biotic** describes the living parts of an organism's environment

**bipedalism** the ability to walk upright on two feet; an important trait of early hominids

**blastula** the hollow ball of cells that a zygote becomes after its first series of mitotic cell divisions

**budding** a form of asexual reproduction in which the offspring grows out of the side of the parent

**cancer** a disease that results from uncontrolled cell division, which damages normal tissues

**capillaries** the smallest blood vessels; where the exchange of nutrients and gases between the blood and the cells takes place

**carbon** one of the six most important chemical elements for living things; carbon atoms form the backbone of nearly all organic compounds

**carbon dioxide** the inorganic molecule from which plants get carbon for photosynthesis; waste product of cellular respiration; a greenhouse gas

**carnivores** animals that obtain their energy by eating other animals; see also *consumers* and *heterotrophic*

**carrier** a female who has a recessive defective allele on one X chromosome but not on the other; she may carry the genetic disorder, but not actively have the condition
**carrying capacity** the size of a population that an ecosystem can support
**catalysts** substances that increase the rate of a chemical reaction, but are not changed during the reaction
**cell division** method by which a single cell reproduces; it divides into two new cells
**cell membrane** a selectively permeable plasma membrane that separates and regulates substances that pass between the inside and outside of a cell
**cells** the smallest living units of an organism; all organisms are made up of at least one cell
**cell theory** that cells are the basic units of structure and function of all living things
**cellular respiration** the process that uses oxygen to create ATP for energy use
**cell wall** the rigid layer that surrounds an entire plant cell
**centrifuge** machine in which a liquid is spun around; used to separate materials of different densities from one another
**chloroplast** the organelle in a plant that contains the pigment chlorophyll and carries out photosynthesis
**chromosomes** structures composed of DNA that contain the genetic material
**circulation** the movement of blood throughout the body of an animal
**cloning** the production of identical individuals from the cell of another individual
**codon** a genetic "word" made up by each different combination of three nucleotide bases; each represents a specific amino acid
**community** populations of different species that interact within a particular area
**competition** the struggle between organisms for limited resources such as food and space
**compound light microscope** a tool that uses two lens types to magnify the image of a specimen
**compounds** molecules formed by the combination of different elements; a small number are necessary for living things
**consumers** organisms that obtain their energy by feeding on other organisms; heterotrophic life-forms
**coordination** the means by which body systems work together to maintain homeostasis; a property of living things
**covalent bond** chemical bond in which atoms share electrons with each other, thus making each atom more stable
**cytokinesis** the process that occurs after mitosis, when the cytoplasm and contents of the cell divide in half
**cytoplasm** the watery fluid that fills a cell, surrounding its organelles

**decomposers** heterotrophic organisms that obtain their energy by feeding on dead and decaying organisms
**deforestation** the cutting down and clearing away of forests; clear-cutting
**dependent variable** in an experiment, the change that occurs because of the independent variable
**deplete** to use up natural resources that cannot be replaced within our lifetimes
**depletion (ozone)** a reduction in the amount of ozone in Earth's ozone layer
**development** the changes in an organism that occur from fertilization until death
**deviations** changes in the body's normal functions that are detected by control mechanisms, which maintain a balanced internal environment
**diaphragm** a large flat muscle that lies across the bottom of the chest cavity; it expands and contracts during breathing
**differentiation** the development of specialized cells from less specialized parent cells through controlled gene expression
**diffusion** the movement of molecules from an area of higher concentration to an area of lower concentration
**digestion** the process of breaking down food particles into molecules small enough to be absorbed by cells
**diversity** the variety of different traits in a species or different species in an ecosystem
**DNA (deoxyribonucleic acid)** the hereditary material of all organisms, which contains the instructions for all cellular activities
**dynamic equilibrium** in the body, a state of homeostasis in which conditions fluctuate yet always stay within certain limits

**ecology** the study of the interactions of living things with their environment
**ecosystem** an area that contains all the living and nonliving parts that interact
**egg cell** the female gamete that supplies half the genetic information to the zygote
**embryo** an organism in an early stage of development before it is hatched, born, or germinated
**endocrine system** a system of glands and hormones that function to maintain homeostasis
**energy** found in different forms, all living things require a continuous input to stay alive
**energy pyramid** describes the flow of energy through an ecosystem; most energy is at the base (producers), and decreases at each higher level (consumers)
**environment** the physical surroundings of an organism, with which it interacts
**enzymes** proteins that act as catalysts for a biological reaction
**equilibrium** in ecosystems, an overall stability in spite of cyclic changes
**estrogen** in females, along with progesterone, a major sex hormone that affects secondary sex characteristics and reproduction
**evolution** the change in organisms over time due to natural selection acting on genetic variations that enable them to adapt to changing environments
**excretion** the removal of metabolic wastes from the body
**experiment** an investigation, conducted by using the scientific method of inquiry
**expression** see *gene expression*
**extinct** describes the death of all living members of a species; i.e., no longer alive anywhere on Earth

**extinction** the complete disappearance of a species from Earth; occurs when a species no longer produces any more offspring

**feedback mechanism** a system that reverses an original response that was triggered by a stimulus
**fertilization** in sexual reproduction, process by which an egg cell and a sperm cell unite to form a zygote
**fetus** a developing embryo after the first three months of development
**fixed action patterns** inherited behaviors (instincts) that do not change, even if repeated many times
**food chain** the direct transfer of energy from one organism to the next
**food web** the complex, interconnecting food chains in a community
**fossils** the traces or remains of a long-dead organism, preserved by natural processes
**fungi (*singular*, fungus)** heterotrophic organisms that obtain their energy by feeding on decaying organisms, e.g., yeast and mushrooms

**gametes** the male and female sex cells that combine to form a zygote during fertilization
**gastrula** the three-layered structure that a zygote becomes (after blastula phase); each layer gives rise to particular body parts and systems of the developing embryo
**gene** the segment of DNA that contains the genetic information for a given trait or protein
**gene expression** the use of genetic information in a gene to produce a particular trait, which can be modified by interactions with the environment
**genetic code** the triplet code, in which each different combination of three bases makes up a codon (for an amino acid)
**genetic cross** the type of experiment in which two organisms with different traits are bred
**genetic engineering** recombinant DNA technology; i.e., the insertion of genes from one organism into the genetic material of another; see also *biotechnology*
**genetic variations** the differences among offspring in their genetic makeup
**geologic time** Earth's history divided into vast units of time by which scientists mark important changes in Earth's climate, surface, and life-forms
**global warming** an increase in the average atmospheric temperature of Earth due to more heat-trapping $CO_2$ in the air; causes the "greenhouse effect"
**glucose** a simple sugar that has six carbon atoms bonded together; a subunit of complex carbohydrates
**Golgi complex** cell structures that package and distribute many materials for the cell

**habitat** the place in which an organism lives; a specific environment that has an interacting community of organisms
**herbivores** animals that obtain their energy by eating plants; see also *consumers* and *heterotrophic*
**hereditary** describes the genetic information that is passed from parents to offspring
**heterotrophic** describes an organism that obtains its energy by feeding on other living things, e.g., animals (consumers)
**hibernation** a type of behavior in which an animal retreats to a secluded place within its habitat, such as a den or a cave, to sleep during the winter months.
**homeostasis** in the body, the maintenance of a constant internal environment
**hominids** classification that includes humans, human ancestors, and African great apes
**homologous structures** similar structures, or characteristics, that different (species of) organisms both inherited from a common ancestor
**hormones** chemical messengers that bind with receptor proteins to affect gene activity, resulting in long-lasting changes in the body
**host** the organism that a parasite uses for food and shelter by living in or on it
**hydrogen** one of the six most important chemical elements for living things
**hypothesis** a possible answer to a specific scientific question, tested by doing an experiment

**immune system** recognizes and attacks specific invaders, such as bacteria, to protect the body against infection and disease
**immunity** the ability to resist or prevent infection by a particular microbe
**immunodeficiency diseases** illnesses that occur when the body's immune system is underactive because it is weakened, e.g., by HIV
**independent variable** in an experiment, the difference that might explain the observation
**inheritance** the process by which traits are passed from one generation to the next
**inorganic** in cells, substances that allow chemical reactions to take place; in ecosystems, substances that are cycled between living things and the environment
**insulin** substance secreted by the pancreas that maintains normal blood sugar levels
**internal development** occurs when the embryo develops within the female's body
**internal fertilization** occurs when the sperm fertilizes the egg cell within the female's body

**karyotype** a photograph of the chromosomes in a human cell, paired up and numbered from the largest to smallest
**kidneys** in land vertebrates, the organ primarily responsible for regulating the chemical composition of blood (and water balance)
**kingdom** the major (i.e., largest) grouping into which scientists categorize living things

**level of organization** a scale for looking at the structure of a system, e.g., from atoms to cells to tissues to organs to organisms to populations to ecosystems
**limiting factors** different environmental conditions that can limit where an organism lives, e.g., sunlight and water
**linear sequence** the order of the (nucleotide) subunits that make up each long strand of a DNA molecule

**linkage** occurs when the genes for one type of trait are inherited along with the genes for another particular type of trait, because they are located on the same chromosome

**lipids** the group of organic compounds that includes fats and oils

**liver** the largest organ in the body, it assists most important body systems, e.g., excretion and digestion; removes ammonia, toxins, and breaks down old red blood cells

**lysosomes** organelles scattered throughout the cell that contain digestive enzymes and are involved in breaking down food

**malfunction** occurs when an organ or body system stops functioning properly, which may lead to disease or death

**meiosis** the division of one parent cell into four daughter cells; reduces the number of chromosomes to one-half the normal number

**membrane** see *cell membrane*

**metabolism** describes the chemical activities (building-up and breaking-down reactions) that take place in an organism

**microbes** microscopic organisms that may cause disease when they invade another organism's body; microorganisms, e.g., bacteria and viruses

**microscope** tool used to magnify the image of tiny objects in order to study them

**migration** a behavior pattern in which groups of animals travel with the seasons, e.g., northward and then southward each year, following resources such as rain and food

**mitochondria** the organelles at which the cell's energy is released

**mitosis** the division of one cell's nucleus into two identical daughter cell nuclei

**molecules** the smallest unit of a compound, made up of atoms

**movement** the flow of materials between the cell and its environment; a property of living things, i.e., locomotion

**multicellular** describes organisms that are made up of more than one cell

**mutation** an error in the linear sequence (gene) of a DNA molecule

**natural selection** the process by which organisms having the most adaptive traits for an environment are more likely to survive and reproduce

**nerve cells** in animals, the cells that transmit nerve impulses to other nerve cells and to other types of cells

**nervous system** a system that enables the detection of, and response to, a stimulus; in vertebrates, a complex organization of cells and organs

**niche** an organism's role in, or interaction with, its ecosystem; includes all the things an organism does to survive

**nitrogen** one of the six most important chemical elements for living things

**nitrogenous wastes** metabolic wastes produced when amino acids are broken down

**nonrenewable resources** describes resources that cannot be replaced within a lifetime, such as coal and natural gas

**nucleotides** the building blocks, or subunits, of DNA; they include four types of nitrogen bases, which occur in two pairs

**nucleus** the dense region of a (eukaryotic) cell that contains the genetic material

**nutrients** important molecules in food, such as lipids, proteins, and vitamins

**nutrition** the life process by which organisms take in and utilize nutrients

**observation** made when researchers intentionally watch and study things around them

**organ** describes a level of organization in living things, i.e., a structure made up of similar tissues that work together to perform the same task, e.g., the liver

**organelles** structures within a cell that perform a particular task, e.g., the vacuole

**organic** refers to substances found in living things; see *organic compounds*

**organic compounds** describes those substances that contain carbon and hydrogen (in living things)

**organisms** living things; life-forms

**organ system** a group of organs that works together to perform a major task, e.g., the digestive system

**osmosis** the diffusion of water molecules across a cell membrane; a type of passive transport

**ovaries** the female reproductive organs that produce the mature egg cells

**ovulation** the release of one mature egg cell from the ovaries every month

**oxygen** one of the six most important chemical elements for living things; released as a result of photosynthesis; essential to cellular (aerobic) respiration

**ozone depletion** see *depletion (ozone)*

**ozone shield** the layer of ozone gas that surrounds Earth high in the atmosphere and blocks out harmful ultraviolet (UV) radiation

**pancreas** gland that secretes pancreatic juice (containing enzymes that aid digestion), and insulin (maintains normal blood sugar levels)

**parasite** the organism that lives in or on another organism (a host), causing it harm

**passive transport** movement of substances across a membrane; requires no use of energy

**pathogens** microscopic organisms that cause diseases, such as certain bacteria and viruses; see also *microbes*

**pesticides** chemicals used to kill agricultural pests, mainly insects, some of which have evolved resistance to the chemicals

**pH** a measurement (on a scale of 0 to 14) of how acidic or basic a solution is

**photosynthesis** the process that, in the presence of light energy, produces chemical energy (glucose) and water

**placenta** the organ that forms in the uterus of mammals to nourish a developing embryo and remove its waste products

**plasma** liquid in blood that is 90 percent water, plus many proteins, salts, vitamins, hormones, gases, sugars, and other nutrients

**platelets** fragments of cells in blood that plug leaks when an injury occurs; they begin the complex chemical process that results in formation of a blood clot

**polygenic inheritance** occurs when a trait, such as height, is determined by several genes

**population** all the individuals of the same species that live in the same area

**predator** an organism that feeds on another living organism (the prey); a consumer

**predator-prey** an interaction in which the prey is usually killed right away

**pregnancy** in animals, the condition of having a developing embryo within the body

**prey** an organism that is eaten by another organism (the predator)

**producers** organisms on the first trophic level, which obtain their energy from inorganic sources, e.g., by photosynthesis; autotrophic life-forms

**progesterone** in females, along with estrogen, a major sex hormone; see *estrogen*

**proteins** a group of organic compounds that are made up of chains of amino acids

**puberty** sexual maturation; i.e., when sperm and egg cells are produced and secondary sex characteristics develop

**radiation** a form of energy that can cause genetic mutations in sex cells and body cells

**receptor molecules** proteins that play an important role in the interactions between cells, e.g., molecules that bind with hormones

**recombinant DNA** pieces of DNA that can be removed and joined with other pieces of DNA; the genes can then be moved from one cell (or organism) into another

**recombination** the formation of new combinations of genetic material due to crossing-over during meiosis or due to genetic engineering

**recombining** during meiosis, the process that causes an increase in genetic variability due to the exchange of material between chromosomes

**reflex** a behavior that occurs automatically, such as pulling your hand away from a hot stove; aids survival

**renewable resources** describes resources that can be replaced within a lifetime, such as wood and biofuels

**replication** the process by which DNA makes a copy of itself during cell division and protein synthesis

**reproduction** the production of offspring (i.e., passing on of hereditary information), either by sexual or asexual means

**residue** the remains of dead organisms, which are recycled in ecosystems by the activities of bacteria and fungi

**response** an organism's reaction to a stimulus; can be inborn or learned

**respiration** in the lungs, the process of exchanging gases; in cells, the process that releases the chemical energy stored in food; see also *cellular respiration*

**restriction enzyme** a molecule that recognizes a small sequence of base pairs within a DNA strand, which it cuts (to make recombinant DNA)

**ribosomes** the organelles at which protein synthesis occurs, and which contain RNA

**scavenge** to gather the remains of a kill, rather than to hunt living animals

**science** a body of knowledge about our natural world

**scientific inquiry** the process of understanding the nature of scientific thinking.

**scientific method** an organized approach to problem solving

**selective breeding** the process by which humans encourage the development of specific traits by breeding the plants or animals that have those traits; see *artificial selection*

**sex cells** the male and female gametes; they have one-half the normal chromosome number as a result of meiosis

**sex chromosomes** in a human karyotype, the last two chromosomes, which determine the person's sex; i.e., the X and the Y chromosomes

**sex-linked** in traits, occurs when a particular allele with a defective portion of DNA is present on the X chromosome and the Y chromosome lacks alleles for these traits

**sexual reproduction** describes reproduction that requires two parents to pass on genetic information

**societies** organized social groups in which many animals (from insects to mammals) live; usually aids survival and reproduction

**simple sugars** single sugars that have six carbon atoms, e.g., glucose

**solar energy** radiant energy from the sun that is a renewable resource

**speciation** occurs when a population changes until its members are no longer able to reproduce with members of any other population; i.e., it has become a new species

**species** a group of related organisms that can breed and produce fertile offspring

**sperm cell** the male gamete that supplies half the genetic information to the zygote

**stability** the ability of an ecosystem to continue and to remain healthy; usually, the greater the species diversity, the more stable the ecosystem

**starches** complex carbohydrates made up of many glucose molecules; used for energy storage in plants

**statistical analysis** the mathematical processes used to determine if the experimental results obtained are valid or if they might have been simply due to chance

**stimulus (*plural*, stimuli)** any event, change, or condition in the environment that causes an organism to make a response (i.e., to react)

**stomata (*singular*, stoma)** openings in the surface of a leaf that are adapted to control the loss of water; each stoma is surrounded by two guard cells

**subunits** the four types of nucleotide bases that make up the DNA molecule

**succession** the gradual replacement of one ecological community by another until it reaches a point of stability

**sustainable development** refers to improving the way we live without harming the environment

**sweat glands** excretory glands that contain some nitrogenous wastes; also produce perspiration to help regulate temperature by cooling the body through evaporation

**symbiosis** a close relationship between two or more different organisms that live together, which is often but not always beneficial

**synapse** a gap that separates adjoining nerve cells and through which nerve impulses are transmitted
**synthesis** the building of compounds that are essential to life, e.g., protein synthesis
**system** describes a level of organization in living things, e.g., a group of organs that work together to perform the same task; see also *organ system*

**taxonomy** is the science of naming and classifying organisms according to their evolutionary relationships and shared characteristics
**technology** the process of using scientific knowledge and other resources to develop new products and processes
**template** in DNA replication, the original molecule that is used to make a copy
**territory** the area in which an animal lives, and which it usually defends
**testes** the pair of male reproductive organs that produces the sperm cells
**testosterone** in males, the main sex hormone that influences secondary sex characteristics and reproduction
**theory** a general statement that is supported by many observations and experiments; the most logical explanation of the evidence; generally accepted as fact
**theory of evolution** see *evolution*
**thermal pollution** a type of water pollution that occurs when heat is added to the water by factories and power-generating plants; can have effect on aquatic wildlife
**tissues** describes a level of organization in living things, i.e., groups of similar cells that work together to perform the same function
**toxins** chemicals that can harm a developing fetus if taken in by the mother during pregnancy; also, chemicals that may get passed from one trophic level to the next (and increase in each organism) as they move up the food chain
**trophic levels** each of the feeding levels on a food chain or in a food web

**unicellular** describes organisms that are made up of just one cell; single-celled
**uterus** in female mammals, the reproductive organ that holds the developing embryo

**vaccinations** injections that prepare the immune system to better fight a specific disease in the future
**vacuoles** the organelles that store materials, including wastes, for the cell
**variability** see *genetic variation*
**vectors** usually a small circular piece of bacterial DNA (plasmid) or a virus that can move pieces of DNA from one organism to another
**veins** the thin-walled blood vessels that get larger as they get closer to the heart, and which carry the oxygen-depleted blood
**vertebrates** animals with backbones, all of which follow a common plan in their early stages of development due to a shared ancestry
**viruses** particles of genetic material that can replicate only within a host cell, where they usually cause harm

**water balance** the regulation of the body's salt levels and water levels, done by the kidneys
**white blood cells** several types of cells that work to protect the body from disease-causing microbes and foreign substances

**zygote** the fertilized egg cell that is formed when the nuclei of two gametes (a male and a female) fuse

# INDEX

# PHOTO CREDITS

*Photographs are provided courtesy of the following:*

**National Aeronautics and Space Administration (NASA):** 246

**N.Y. Public Library/Science, Industry, and Business Library:** 153

**N.Y.S. Department of Environmental Conservation:** 237

**Photo Researchers, Inc.:** 38, Van Bucher; 46, Michael Austin; 89 (top), Meckes/Ottawa; 89 (bottom), Biophoto Associates; 92, Francis Leroy, Biocosmos/Science Photo Library (artwork based on SEM); 99, Omikron; 106, Biophoto Associates; 124 (four images), Dr. Yorgos Nikas; 125, Biophoto Associates; 131, A. Barrington Brown; 148, Mary Eleanor Browning; 155, Biophoto Associates; 156, Doug Martin; 160, Laguna Design (computer artwork); 165, Biophoto Associates; 167, Leonard Lee Rue III; 170, Mary Eleanor Browning; 176, Tom McHugh; 192, A. Cosmos Blank; 193, Stephen Dalton; 197, Robert Hermes; 198, Tim Davis; 199, John Reader; 201, Tim Davis; Michael Austin; 203, Tom McHugh; 204 (bottom), Ylla; 204 (top), Fletcher & Baylis; 206, Richard Parker; 212, NASA; 213 (left), Eric Husking; 213 (right), Tom McHugh; 214, M. P. Kahl; 225, Carl Frank; 226, Karl Weidmann; 235, Michael Hayman; 236, Michael P. Gadomski; 239, Rafael Macia; 247 (bottom), Jeff Isaac Greenberg; 247 (top), Christa Armstrong Rapho; 248 (top), Tom McHugh; 248 (bottom), Georg Gerster; 249 (left), Simon Fraser/Science Photo Library; 249 (right), Doug Plummer; 250, William & Marcia Levy.

**Visuals Unlimited, Inc., © SIU:** 162

# Sample Examinations

# The Living Environment/REVIEWING BIOLOGY

# August 2017

## Part A

**Answer all questions in this part.** [30]

*Directions* (1–30): For *each* statement or question, record on the separate answer sheet the *number* of the word or expression that, of those given, best completes the statement or answers the question.

1 A fruit fly is classified as a consumer rather than as a producer because it is unable to

(1) reproduce asexually
(2) synthesize its own food
(3) release energy stored in organic molecules
(4) remove wastes from its body

2 Which change is an example of maintaining dynamic equilibrium?

(1) A plant wilts when more water is lost from the leaves than is lost by the roots.
(2) A plant turns yellow when light levels are very low.
(3) Insulin is released when glucose levels in the blood are high.
(4) A person sweats when the environmental temperature is low.

3 Organisms contain compounds such as proteins, starches, and fats. The chemical bonds in these compounds can be a source of

(1) amino acids
(2) simple sugars
(3) energy
(4) enzymes

4 Phosphorus is necessary for the growth of healthy plants. Scientists are developing plants that can grow in phosphorus-poor soil. Some of these new varieties, produced in a lab, make extra copies of a protein that helps them obtain more phosphorus from the soil. The process being used to develop these new varieties is most likely

(1) paper chromatography
(2) natural selection
(3) direct harvesting
(4) genetic engineering

5 Which life function is *not* necessary for an individual organism to stay alive?

(1) nutrition
(2) reproduction
(3) regulation
(4) excretion

6 Lobsters prey on sea hares, which are marine animals. The lobsters find their prey through a sense of smell. The sea hares defend themselves by squirting ink at the lobster, as shown in the photo below. The ink sticks to the lobster, interfering with its sense of smell.

The most likely reason the sea hare can escape is because the sea hare ink

(1) pushes the sea hare away rapidly as the ink is expelled
(2) blocks a receptor on certain cells in the lobster
(3) causes the lobster to change its prey
(4) prevents movement of the lobster

7 Which statement is an accurate description of genes?

(1) Proteins are made of genes and code for DNA.
(2) Genes are made of proteins that code for nitrogen bases.
(3) DNA is made of carbohydrates that code for genes.
(4) Genes are made of DNA and code for proteins.

8 The bobolink is a small blackbird that nests in fields of tall grass. It breeds in the summer across much of southern Canada and the northern United States. It migrates long distances, wintering in southern South America. The numbers of these birds are declining due to disruption of the areas where they live.

In order to save these birds from extinction, the best course of action would be to

(1) prevent the birds from migrating to South America
(2) encourage farmers to let their hay fields undergo succession
(3) work to protect bobolink habitats in South and North America
(4) capture all the bobolinks and keep them safe in zoos

9 A child with cystic fibrosis has an altered protein in his cells that stops chloride ions from leaving the cells. This protein most likely affects the functioning of

(1) cell membranes
(2) nuclei
(3) mitochondria
(4) ribosomes

10 Which row in the chart below shows a direct relationship that can exist between two living organisms?

| Row | Relationship |
|---|---|
| (1) | producer – carnivore |
| (2) | predator – prey |
| (3) | parasite – prey |
| (4) | carnivore – host |

11 Scientists have studied the return of plant life on Mount St. Helens ever since the volcano erupted in 1980. Wildflowers began colonizing the area, followed by shrubs and small trees. Scientists predict that it will likely take hundreds of years before the area returns to a forest dominated by fir and hemlock trees. These changes are an example of

(1) humans degrading an ecosystem by removing wildflowers
(2) the loss of genetic variation in a plant species
(3) the growth of a forest through ecological succession
(4) the biological evolution of wildflowers, shrubs, and trees

12 The most likely result of completely removing carbon dioxide from the environment of a plant is that sugar production will

(1) continue at the same rate
(2) increase and oxygen production will also increase
(3) increase and oxygen production will stay the same
(4) decrease and eventually stop

13 Before a new shopping center can be built on previously undeveloped land, the builders must submit a proposal to the local government for approval. Which statement identifies an environmental concern associated with the development of the shopping center?

(1) Building the center would decrease resources needed by local organisms.
(2) The new shopping center would increase competition with already existing businesses.
(3) Building the center would decrease the amount of pollution in the area.
(4) The new shopping center would increase the biodiversity of the area.

14 Homeowners have been encouraged to learn how to identify invasive plants and to remove them if they find them. The most likely reason for removing invasive plants is to

(1) allow only one type of native plant to grow
(2) preserve biodiversity
(3) eliminate unfamiliar food sources
(4) increase the rate of ecological succession

15 Which row in the chart below correctly pairs a human activity with its impact on the environment?

| Row | Human Activity | Impact |
|---|---|---|
| (1) | decrease in the use of pesticides | erosion of rock in the soil |
| (2) | increase in housing developments | improvement in air quality |
| (3) | increase in human population | reduction in water usage |
| (4) | decrease in recycling | reduction in amount of available resources |

16 The diagram below represents some steps in a procedure used in the field of biotechnology.

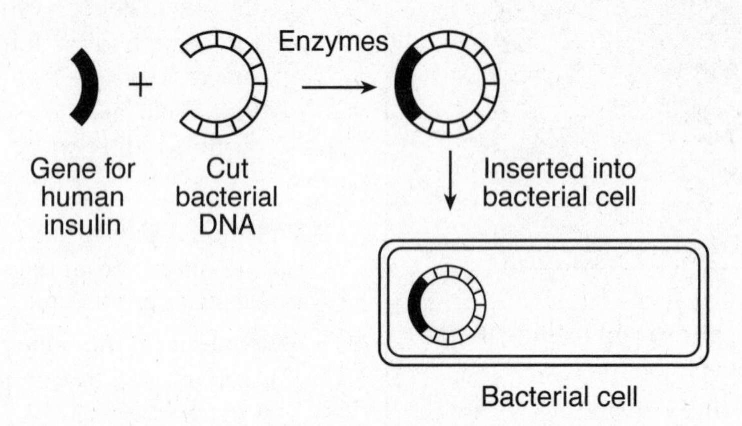


This bacterial cell can now be used to produce

(1) the bacterial gene for insulin that can be inserted into humans
(2) human genes for enzymes that can be inserted into humans
(3) insulin that can be used by humans
(4) enzymes necessary to treat human diseases

17 The graph below represents the number of brown and green beetles collected in a particular ecosystem.

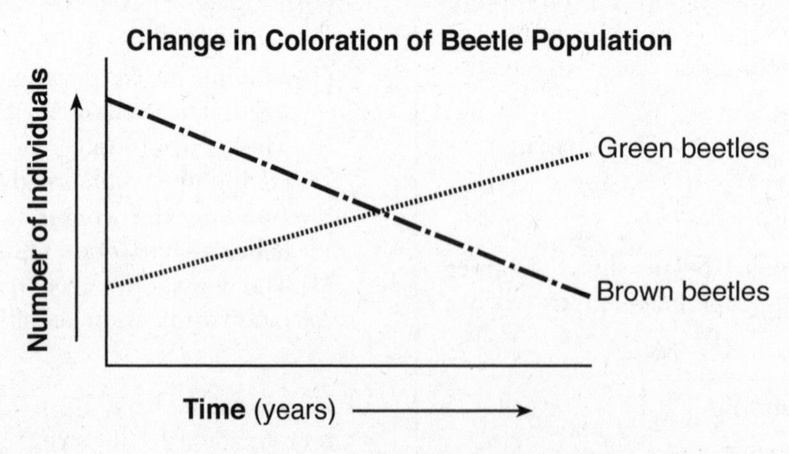


The change observed in the number of green and brown beetles in the population is most likely due to

(1) natural selection
(2) selective breeding
(3) gene manipulation
(4) a common ancestor

18 A reproductive system is represented in the diagram below.

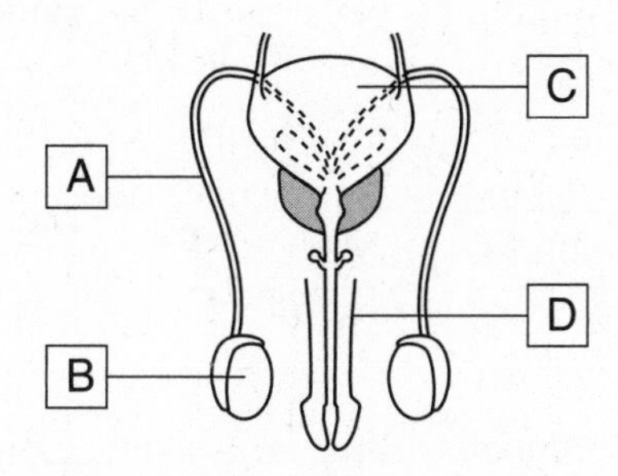


Which structure is correctly paired with its reproductive function?

(1) *A* – pathway of gametes
(2) *B* – synthesis of progesterone
(3) *C* – production of sperm
(4) *D* – regulation of homeostasis

19 For centuries, humans have used resources from coastal areas and open ocean waters. An example of an activity that would promote the conservation of coastal areas and ocean resources is

(1) harvesting large numbers of different fish species
(2) allowing all-terrain vehicles access to beach areas
(3) creating protected zones of natural grasses and shrubs in beach areas
(4) encouraging the construction of factories along the ocean shoreline

20 Which activity would eventually result in a stable ecosystem?

(1) deforestation in an area to increase space for the species living there
(2) mowing a large field so it can be used for recreation
(3) allowing native plants to grow undisturbed in an abandoned field
(4) spraying pesticides on a field at the end of each growing season

21 Some states require shoppers to pay a deposit on certain beverage containers made of plastic and glass. When shoppers return the containers, their deposits are returned to them. How is this system intended to help the environment?

(1) It encourages people to buy products that do not have a deposit.
(2) It reduces the amount of money shoppers actually spend.
(3) It reduces the amount of plastics and glass put into landfills.
(4) It forces manufacturers to reduce air pollution when they are making the containers.

22 The diagram below represents a food web.

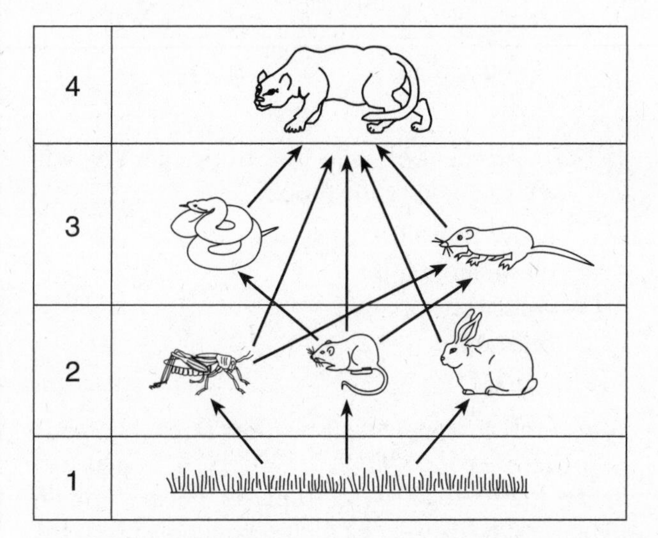


Which level contains organisms that carry out autotrophic nutrition?

(1) 1
(2) 2
(3) 3
(4) 4

23 Mad cow disease is a fatal disease that destroys brain tissue. Researchers have found that a prion protein, which is an abnormally constructed molecule, is responsible. Which statement best describes the characteristics a protein must have to function correctly?

(1) A protein is a long chain of amino acids folded into a specific shape.
(2) A protein is a long chain of simple sugars folded into a specific shape.
(3) A protein is made of amino acids synthesized into a short, circular chain.
(4) A protein is made of simple sugars synthesized into a short, circular chain.

24 The diagram below represents the results of the net movement of a specific kind of molecule across a living cell membrane.

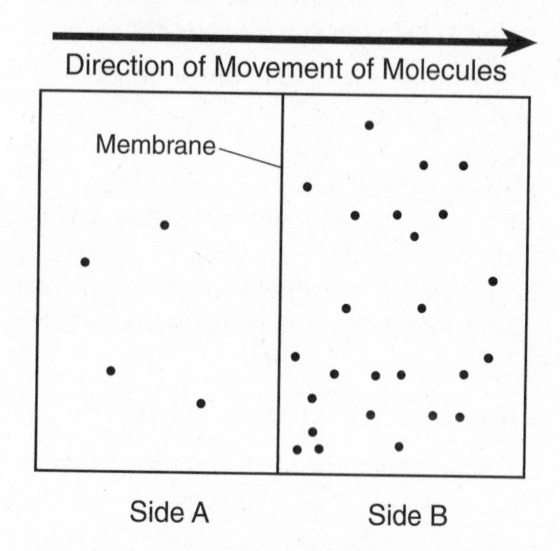


The movement of molecules from side *A* to side *B* is an example of the process of

(1) active transport
(2) chromatography
(3) cellular respiration
(4) diffusion

25 Several companies now offer DNA "banking services," where DNA is extracted from a pet and is stored so that a "replacement pet" might be produced using cloning techniques when the original pet dies. Which statement best explains why the replacement pets that are produced in this way might *not* look or act like the original?

(1) The new animal must get the DNA from two different parents, not just one cell.
(2) Mutations could occur that change the cloned animal into a completely different species.
(3) Recombination of the cells as they are cloned will make the resulting pet act differently.
(4) The environment could influence how genes are expressed, changing how the animal looks and acts.

26 It is recommended that people avoid excessive use of tanning beds. Exposure to the radiation emitted by tanning beds can cause skin cancer. This cancer is the direct result of a

(1) change in a starch molecule
(2) mutation in the genetic material
(3) mutation in a protein
(4) change in a fat molecule

27 The diagram below represents a developing fetus in a human.

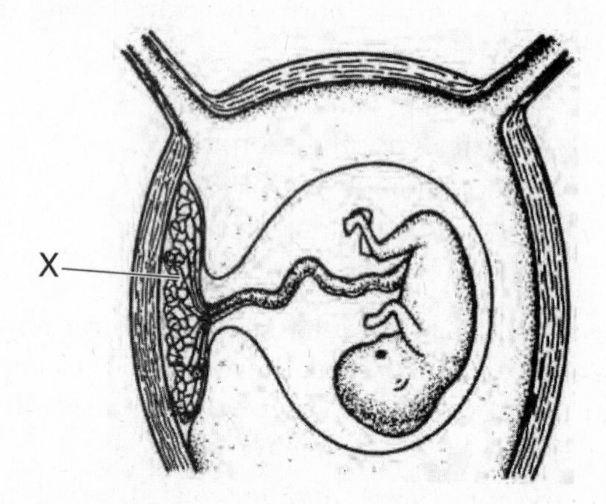


What would most likely happen if structure *X* were damaged in the early stages of pregnancy?

(1) The genes from the mother would not be turned on in the fetus.
(2) The nutrients necessary for development would not be able to reach the fetus.
(3) The fertilized egg would not be able to travel from the ovary to the uterus.
(4) Development would take longer since the fetus would have to synthesize nutrients.

28 The reproductive structure in a female mammal that produces sex cells is the

(1) ovary
(2) testes
(3) uterus
(4) placenta

29 Fungi are decomposers that play an important role in the maintenance of an ecosystem. The role of fungi is important because they

(1) synthesize energy-rich compounds that are directly used by producers
(2) break down materials that can then be used by other organisms
(3) limit the number of plants that can perform photosynthesis in an area
(4) are competitors of other consumers such as herbivores

30 In 2011 and 2012, scientists working on the Banana River in Florida recorded a dramatic increase in the number of manatee deaths. Over the past 50 years, this area has also seen the human population increase by more than 500,000 people. It is believed that pollution from numerous sewage tanks leaked into the water, eliminating the manatees' food source, replacing it with an alga that is toxic to the manatee. This is an example of

(1) a natural cycle in an ecosystem
(2) the effect of increased biodiversity on an ecosystem
(3) direct harvesting in an ecosystem
(4) human actions altering ecosystems with serious consequences

---

GO ON TO THE NEXT PAGE ⇨

## Part B–1

**Answer all questions in this part.** [13]

*Directions* (31–43): For *each* statement or question, record on the separate answer sheet the *number* of the word or expression that, of those given, best completes the statement or answers the question.

31 A student performed an experiment to see if water temperature affects the level of activity in aquatic snails. The student set up four tanks with five snails in each tank. All four of the setups were identical in every way, except for the temperature of the water. In order to make the conclusions more valid, the student could

(1) alter the pH of the water
(2) change the size of the tank
(3) carry out the experiment for a shorter period of time
(4) use a larger number of snails

32 The following events occur during sexual reproduction:

A. mitosis
B. meiosis
C. fertilization
D. birth

Which sequence represents the correct order of these events during sexual reproduction?

(1) $A \rightarrow C \rightarrow B \rightarrow D$
(2) $B \rightarrow C \rightarrow A \rightarrow D$
(3) $C \rightarrow B \rightarrow A \rightarrow D$
(4) $B \rightarrow A \rightarrow C \rightarrow D$

33 A broad body of evidence, subject to revisions, supported by different kinds of scientific investigations and often involving the contributions of scientists from different disciplines is necessary to develop

(1) an inference
(2) a fact
(3) a theory
(4) a prediction

34 The diagrams below represent portions of two genes that code for leaf structure in the same species of clover. Gene 1 was taken from the cells of a clover plant with 3 leaves and gene 2 was taken from the cells of a clover plant with 4 leaves.

Gene 1
(3 leaves)

G C
A T
T A
T A
C G

Gene 2
(4 leaves)

G C
A T
A T
T A
T A
C G

The clover plant having gene 2 (4 leaves) was most likely the result of

(1) an insertion
(2) a deletion
(3) a substitution
(4) normal replication

35 Increased concern over the number of heat-related illnesses among football players has led to a possible change in uniform design. Shoulder pads were designed that constantly blew cool, dry air underneath the shoulder pads. Tests showed that the use of the device during rest and recovery periods resulted in a reduction of body temperature and heart rate. This new device would help the athlete to

(1) control the rate of muscle activity
(2) increase muscle strength
(3) maintain homeostasis
(4) eliminate the release of heat from the body

Base your answers to questions 36 and 37 on the information and data table below and on your knowledge of biology.

A student wanted to investigate the effect of light on the rate of ripening of tomatoes. She set up four pots of the same size with identical amounts of soil, water, and type of tomato plants. Each plant was exposed to a different intensity of light as shown in the table below.

| Plant | Light Intensity (lumens) | |
|---|---|---|
| 1 | 0 | |
| 2 | 1000 | |
| 3 | 5000 | |
| 4 | 10,000 | |

36 To report the final results, which label would be most appropriate for the third column of the data table?

(1) Height of Tomato Plants (cm)
(2) Average Ripening Time (days)
(3) Average Weight of Tomatoes per Plant (grams)
(4) Acidity of Tomatoes (pH)

37 The independent variable in this experiment is the

(1) type of tomato plant
(2) amount of soil provided
(3) color of tomatoes
(4) light intensity

---

GO ON TO THE NEXT PAGE ⇨

Base your answers to questions 38 and 39 on the diagram below and on your knowledge of biology. The diagram illustrates activities taking place in the body of a human.

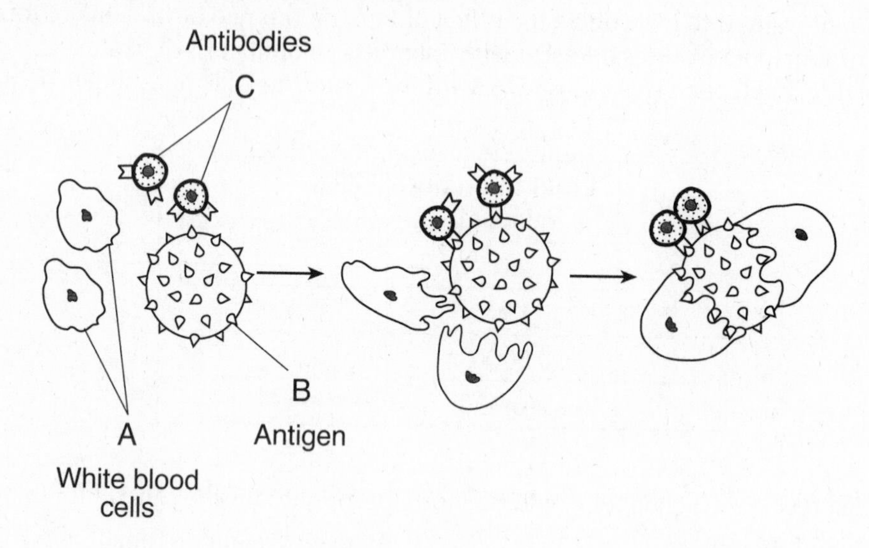


38 Vaccinations usually stimulate the body to produce more of

(1) structure *A*, only
(2) structure *B*, only
(3) structures *A* and *C*, only
(4) structures *A*, *B*, and *C*

39 Which structure normally stimulates an allergic response?

(1) *A*, only
(2) *B*, only
(3) *C*, only
(4) *A*, *B*, and *C*

---

40 Which population in the chart below has the best chance for survival in a rapidly changing environment?

| Population | Type of Reproduction | Average Life Span of Individuals | Total Number of Offspring Produced |
|---|---|---|---|
| (1) | sexual | 13 days | 100 |
| (2) | asexual | 13 days | 100 |
| (3) | sexual | 12 weeks | 25 |
| (4) | asexual | 12 weeks | 25 |

41 The table below represents a segment of a DNA molecule found in a stomach cell, both before and after undergoing replication.

**DNA Segment Before and After Replication**

| **Before replication** | TGT | ATG | AAA | CAC | AAT | TAT |
|---|---|---|---|---|---|---|
| **After replication** | TGT | ATT | AAA | CAC | AAT | TTT |

Which statement best describes a change that would most likely be observed in the cells formed as a result of this mitotic division?

(1) An enzyme the cell produces might no longer function.
(2) The cells would begin to form gametes to be released.
(3) Many new hormones would be synthesized by the cells.
(4) Chloroplasts would be produced by the ribosomes.

Base your answers to questions 42 and 43 on the information and diagram below and on your knowledge of biology.

The setup below shows four test tubes. Tube 1 contains water only. Tube 2 contains a live snail. Tube 3 contains a live green water plant. Tube 4 contains both a live green water plant and a live snail.

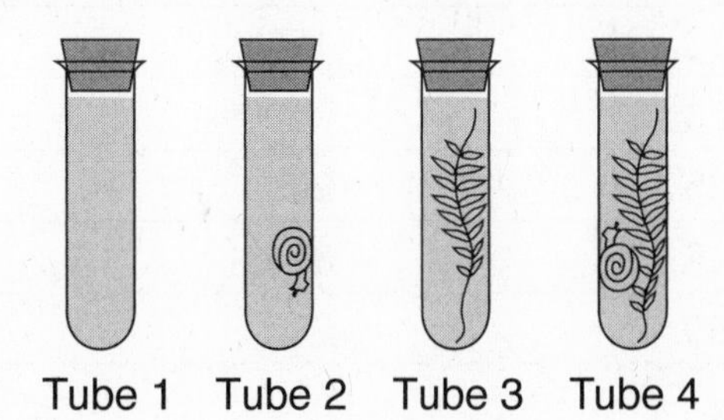


42 In this setup, which tubes contain at least one organism carrying on cellular respiration?

(1) tubes 1 and 2, only
(2) tubes 2 and 4, only
(3) tubes 3 and 4, only
(4) tubes 2, 3, and 4, only

43 Which compound that directly provides energy in living cells is being produced in every tube where cellular respiration is occurring?

(1) oxygen
(2) glucose
(3) DNA
(4) ATP

## Part B–2

**Answer all questions in this part.** [12]

*Directions* (44–55): For those questions that are multiple choice, record on the separate answer sheet the *number* of the choice that, of those given, best completes each statement or answers each question. For all other questions in this part, follow the directions given and record your answers in the spaces provided in this examination booklet.

Base your answers to questions 44 through 47 on the information and data table below and on your knowledge of biology.

The concentration of a specific antibody in the blood of an individual was measured at various times over a period of 50 days. The results obtained are shown in the data table below.

**Antibody Concentration in an Individual**

| Day | Antibody Concentration in Arbitrary Units (arb. units) |
|---|---|
| 5 | 0 |
| 10 | 110 |
| 16 | 120 |
| 25 | 10 |
| 35 | 200 |
| 45 | 390 |
| 50 | 200 |

*Directions* (44–45): Using the information in the data table, construct a line graph on the grid, following the directions below.

44 Mark an appropriate scale, without any breaks in the data, on each labeled axis. [1]

45 Plot the data on the grid. Connect the points and surround each point with a small circle. [1]

Example:

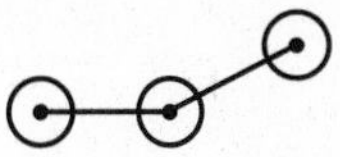

**Antibody Concentration in an Individual**

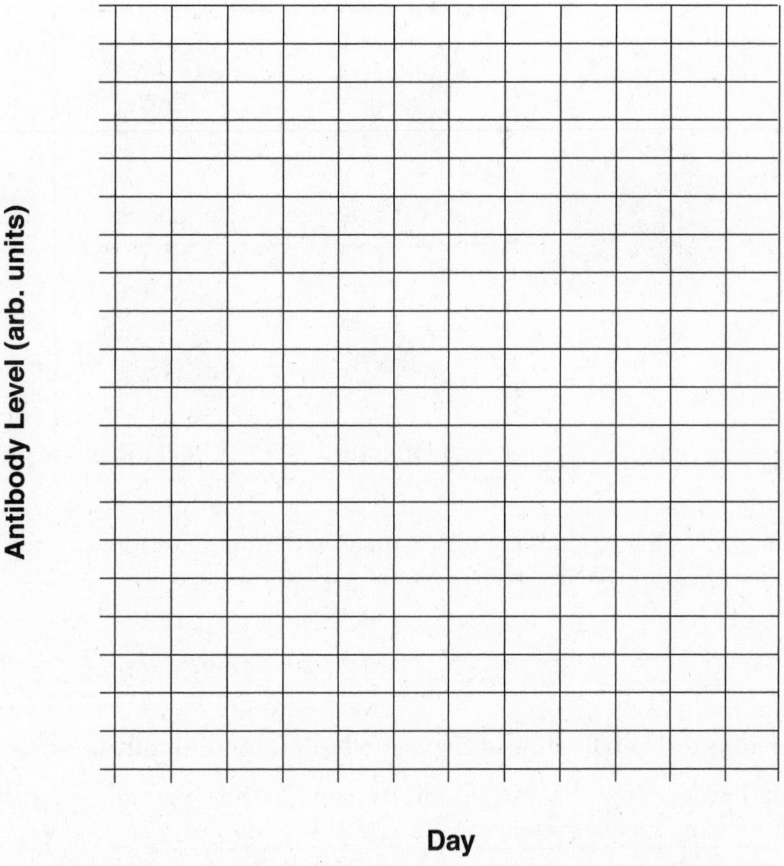

Antibody Level (arb. units)

Day

46 State *one* reason for the change in antibody production during the first 10 days. [1]

______________________________________________

______________________________________________

**Note: The answer to question 47 should be recorded on your separate answer sheet.**

47 The antibody level (in arb. units) of the individual on day 30 is closest to

(1) 30
(2) 70
(3) 110
(4) 160

Base your answers to questions 48 and 49 on the information and diagram below and on your knowledge of biology.

If a Chihuahua with short hair has a hidden gene for long hair, it can produce both long-haired and short-haired puppies when bred to a Chihuahua with long hair.

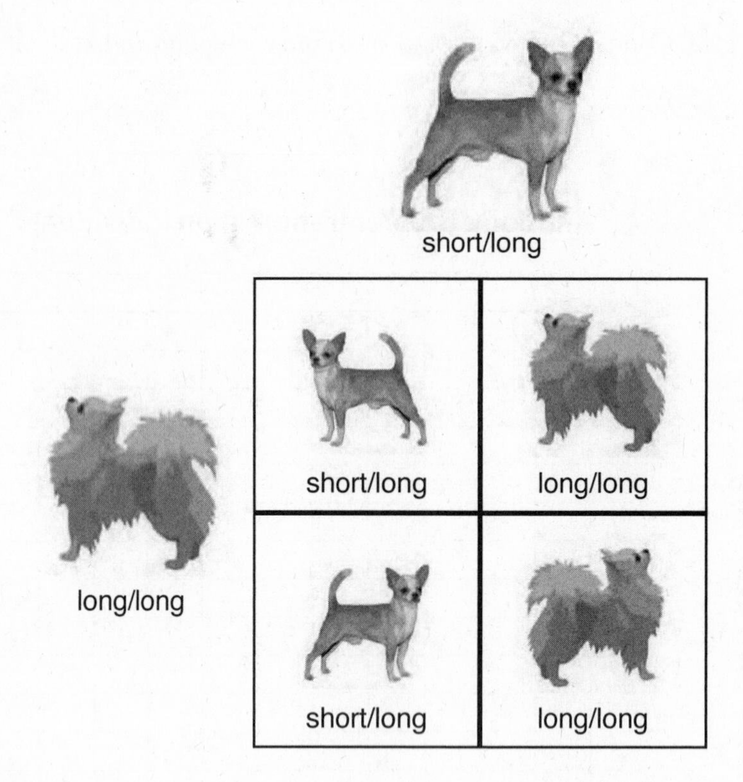


48 A family decides that they want to produce Chihuahuas with long hair. Identify a procedure that could be used to make sure that the puppies all have long hair. [1]

______________________________________________________________________________

**Note: The answer to question 49 should be recorded on your separate answer sheet.**

49 A Chihuahua is born having a trait that is different from either of its parents. A possible explanation for the difference is that the Chihuahua puppy

(1) was produced as a result of the recombination of genes during sexual reproduction
(2) was produced as a result of the process of asexual reproduction
(3) inherited a gene from one of its grandparents and not its parents
(4) had a mutation that occurred after it was born

Base your answers to questions 50 and 51 on the diagram below and on your knowledge of biology. The diagram represents a technique used by scientists today to maintain the genetic makeup of an organism.

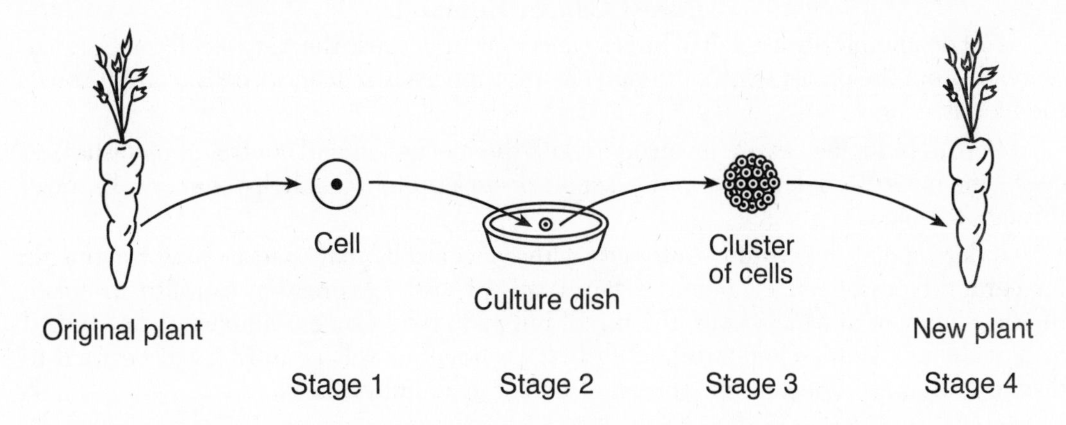


**Note: The answer to question 50 should be recorded on your separate answer sheet.**

50 Which graph below best represents the DNA content found in each cell in each of the stages in the diagram above?

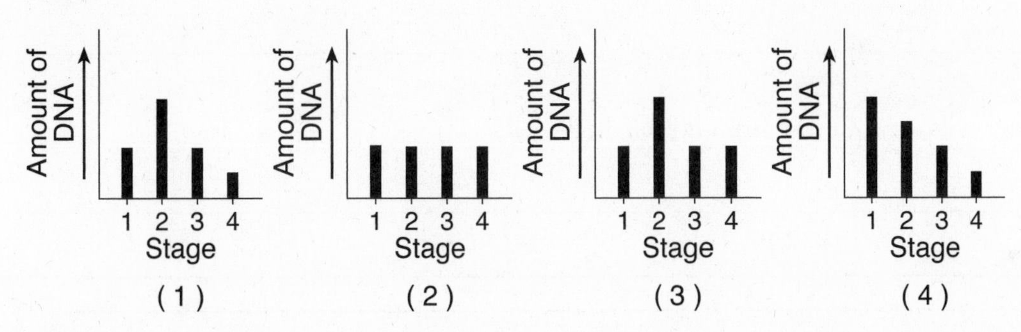


51 Describe *one* specific reason why scientists would want to maintain the genetic makeup of a particular plant. [1]

______________________________________________

______________________________________________

______________________________________________

Base your answers to questions 52 and 53 on the information below and on your knowledge of biology.

**Breast Cancer Research**

Most deaths that are a result of breast cancer occur because the cancer cells metastasize (spread) from the breast to other organs. As they metastasize, cancer cells travel through the bloodstream.

MicroRNA molecules are involved in both the movement and control of metastasized cells. One microRNA, known as miR-7, shuts down a protein that helps cancer cells travel through the blood.

Understanding how miR-7 interacts with cancer cells may lead to new treatments for certain types of cancer. Since certain levels of miR-7 expression can also stimulate the development of cancer cells, the use of miR-7 to treat cancer will have to be studied in more detail. Researchers are hoping that eventually levels of miR-7 will be used to diagnose, treat, and prevent the spread of cancer in an individual.

52 State *one negative* effect of using miR-7 as the only treatment for breast cancer. [1]

______________________________________________________________________

______________________________________________________________________

53 State *one* way cancer cells are different from normal body cells. [1]

______________________________________________________________________

______________________________________________________________________

Base your answers to questions 54 and 55 on the information and diagram below and on your knowledge of biology.

Each body cell contains the same genetic information, but can differ in appearance and size. The diagram below shows three different types of cells found in the human body.

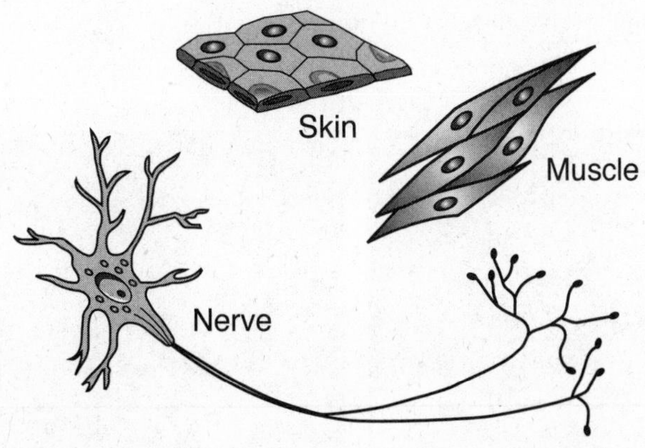


54 Identify *one* similarity, other than the genetic information, that these body cells have. [1]

________________________________________

55 Explain why differences in these human body cells are a biological advantage. [1]

________________________________________

________________________________________

## Part C

**Answer all questions in this part.** [17]

*Directions* (56–72): Record your answers in the spaces provided in this examination booklet.

Base your answers to questions 56 through 58 on the information below and on your knowledge of biology.

**Owl vs. Owl**

Barred owl Spotted owl

Federal wildlife officials plan to dispatch armed bird specialists into forests of the Pacific Northwest starting this fall to shoot one species of owl to protect another that is threatened with extinction. …

…"If we don't manage barred owls, the probability of recovering the spotted owls goes down significantly," said Paul Henson, Oregon state supervisor for Fish and Wildlife. The agency's preferred course of action calls for killing 3,603 barred owls in four study areas in Oregon, Washington and northern California over the next four years. …

…Mr. Henson said unless barred owls are brought under control, the spotted owl in coming decades might disappear from Washington's northern Cascade Range and Oregon's Coast Range, where the barred owl incursion [takeover] has been greatest.

The northern spotted owl was listed as a threatened species in 1990. Barred owls are bigger, more aggressive and less picky about food. Barred owls now cover the spotted owl's range, in some places outnumbering them as much as 5-to-1.

Source: Associated Press, 7/26/13

56 Describe how the barred owl population is having a *negative* effect on the spotted owl population. [1]

57 Explain why it is important to protect the spotted owl from extinction. [1]

58 Certain groups oppose the plan to kill barred owls, in part because they feel it will not solve the problem. They recommend that the focus should be on protecting the habitat of the spotted owl. Describe the role that the habitat plays in the survival of an animal species such as the spotted owl. [1]

Base your answer to question 59 on the information below and on your knowledge of biology.

The 1990 Federal Clean Air Act requires New York State to conduct an emissions test on most gasoline-powered automobiles in order to help reduce harmful emissions. Vehicles that fail this test must be repaired and pass inspection before they can be driven on the road. Some people did not support this legislation.

59 State *one* advantage and *one* disadvantage of automobile emission testing. [1]

Advantage: ____________________

Disadvantage: ____________________

Base your answers to questions 60 through 63 on the information below and on your knowledge of biology.

**Enzyme Investigation**

An enzyme was isolated from digestive juices taken from the small intestine. An experiment was set up to test the ability of the enzyme to break down protein. Two test tubes, labeled *A* and *B*, were placed in a hot water bath at 37°C, human body temperature.

Test tube *A* contained only protein and test tube *B* contained protein and the enzyme. The chart below shows the set-up.

| Test Tube | Contents |
|---|---|
| A | protein |
| B | protein, enzyme |

After two hours, the contents of both test tubes were analyzed. Test tube *A* showed only the presence of protein. Test tube *B* showed the presence of the end products of protein digestion, indicating the enzyme had successfully broken down the protein.

60 Identify the end products of protein digestion that made up the contents of test tube *B* after the two hours. [1]

______________________________

61 Explain the importance of temperature in the functioning of enzymes. [1]

______________________________

______________________________

62 State what the result would be if the same enzyme that was added to test tube *B* was added to a test tube containing starch. Support your answer. [1]

______________________________

______________________________

63 In the digestive system many large molecules, such as proteins, are broken down into much smaller molecules. State what happens to these smaller molecules following digestion. [1]

______________________________

______________________________

Base your answers to question 64–66 on the information below and on your knowledge of biology.

**Secondhand Smoke and Estrogen**

A fertility researcher conducted a study of pregnant women. The researcher's hypothesis was that the estrogen levels of pregnant women who were exposed to daily secondhand cigarette smoke would be higher than estrogen levels of pregnant women not exposed to daily secondhand smoke.

The researcher measured the estrogen levels of eight pregnant women each week throughout their pregnancy. Four of the women lived in houses with heavy smokers, the other four did not. The women's ages varied from 19 to 42 years old. Six of the women were pregnant with girls, one was pregnant with a boy, and one was pregnant with twin boys. The research was submitted for peer review.

64–66 Analyze this experiment. In your answer, be sure to:

- identify *one* error in the researcher's experimental design [1]
- identify *one* way, other than affecting estrogen levels, that secondhand smoke could affect a developing embryo [1]
- explain why the process of peer review is an important step in this research [1]

Base your answers to questions 67 through 69 on the information and passage below and on your knowledge of biology.

**Snowy Owls Move to the South**

Snowy owls are large white birds that normally inhabit the cold northern regions of Canada. Recently, scientists and birdwatchers have sighted the snowy owls much farther south than usual.

When snowy owls are in northern areas, they feed on lemmings (small rodents). When lemmings are not available, as in the areas further south, the owls will seek out mice or rabbits as their food source.

Several snowy owls migrated into an area represented by the food web below.

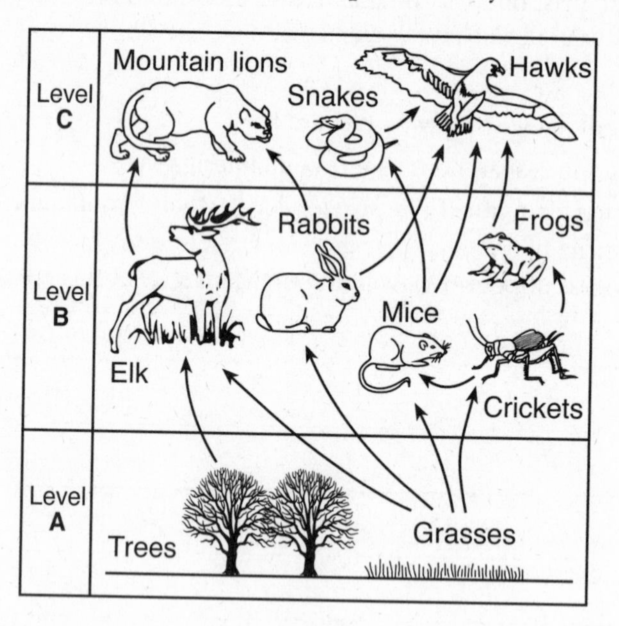


67 Identify *one* population of organisms shown in the food web, other than rabbits or mice, that would likely be affected by the introduction of the snowy owls and explain why their population would be affected. [1]

Population affected: ____________________

____________________

____________________

68 Identify *one* condition that might cause snowy owls to leave their usual habitat and move to another area. [1]

____________________

69 State which level, *A*, *B*, or *C*, contains the *least* total available energy. Support your answer. [1]

Level: ______

____________________

____________________

Base your answers to questions 70 through 72 on the information below and on your knowledge of biology.

**Pocket Mice**

Pocket mice are small rodents that feed mainly at night and are preyed upon by owls, hawks, and snakes. Scientists studied pocket mice living on dark volcanic rock in both New Mexico and fifty miles away in Arizona. They recorded their data in the chart below.

| Number of Mice on Dark Volcanic Rock | | | | |
|---|---|---|---|---|
| Year | New Mexico | | Arizona | |
| | Light Fur | Dark Fur | Light Fur | Dark Fur |
| 2000 | 120 | 122 | 16 | 125 |
| 2001 | 140 | 136 | 8 | 140 |
| 2002 | 134 | 130 | 6 | 135 |
| 2003 | 115 | 120 | 12 | 115 |
| 2004 | 122 | 126 | 8 | 129 |

70 State *one* possible hypothesis that would explain the differences in the observed data between the two locations. [1]

________________________________________

________________________________________

71 Dark fur color in pocket mice is the result of a mutation. Scientists analyzed the sequence of bases in the gene known to play a role in fur color and discovered that the mutation was identical in both the New Mexico and Arizona mouse populations. Explain how it is possible for these two different populations to have identical gene sequences for dark fur color. [1]

________________________________________

________________________________________

72 Explain what is meant by the statement: "While mutations are random, natural selection is not." [1]

________________________________________

________________________________________

________________________________________

## Part D

**Answer all questions in this part.** [13]

*Directions* (73–85): For those questions that are multiple choice, record on the separate answer sheet the *number* of the choice that, of those given, best completes each statement or answers each question. For all other questions in this part, follow the directions given and record your answers in the spaces provided in this examination booklet.

**Note: The answer to question 73 should be recorded on your separate answer sheet.**

73 The diagram below represents evolutionary pathways of seven groups of organisms alive today.

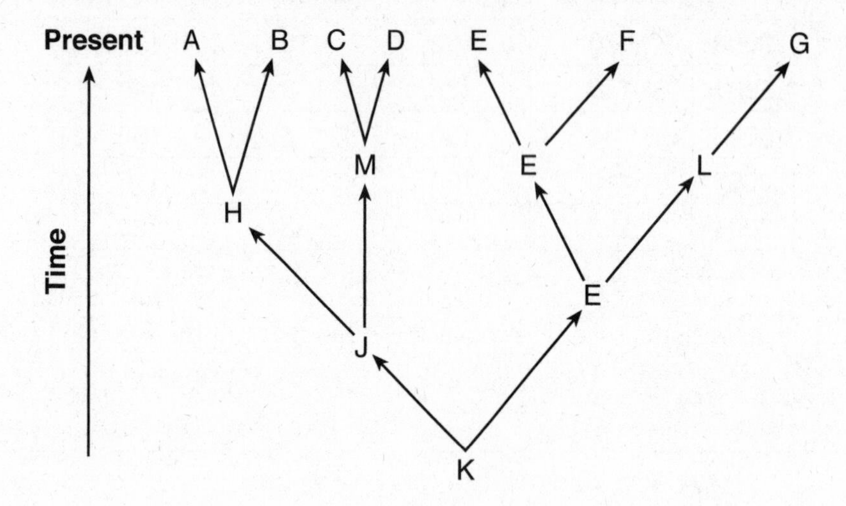


Which two living species would be expected to have the most similar proteins?

(1) *A* and *C*
(2) *B* and *C*
(3) *E* and *F*
(4) *H* and *M*

**Note: The answer to question 74 should be recorded on your separate answer sheet.**

74 Scientists recently discovered that three different types of squid, a marine animal, previously thought to be three different species, were actually all members of one species. Their earlier ideas were based on using squid carcasses (dead bodies). The new, more accepted classification is most probably based on an analysis of

(1) a greater number of squid carcasses
(2) the feeding habits of the three different species
(3) a number of newly found squid fossils
(4) the DNA present in the cells of squid

**Note: The answer to question 75 should be recorded on your separate answer sheet.**

75 The diagram below represents a laboratory experiment involving sucrose and water molecules in a cellophane bag which functions in the same way as dialysis tubing.

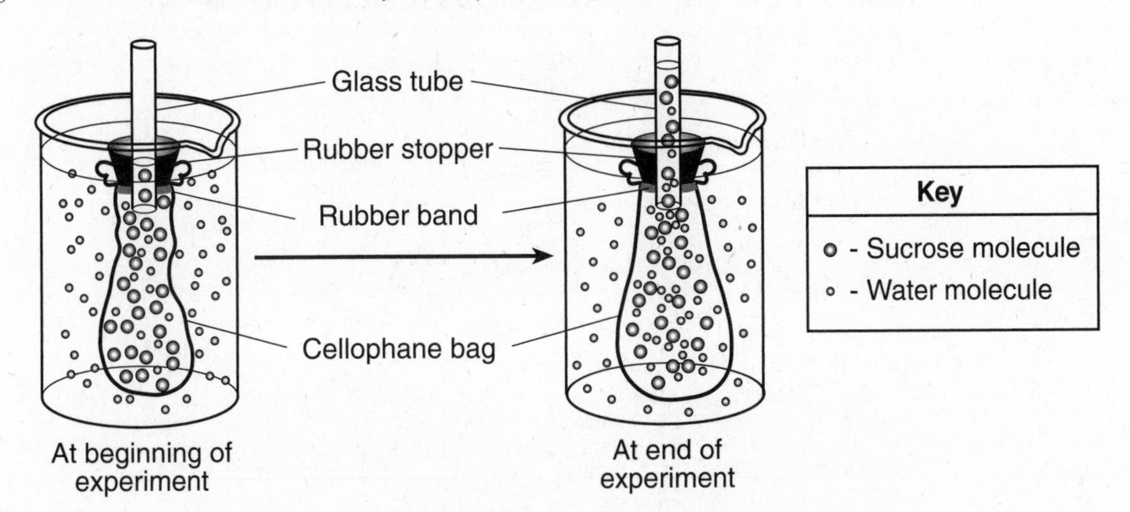


Which statement correctly explains the rise of liquid in the tube at the end of the experiment?

(1) The concentration of sucrose molecules increased as water molecules entered the bag. This concentration increase pushed the liquid up the tube.

(2) Water entered the bag due to the lower concentration of water inside. The extra water pushed the liquid up the tube as the bag filled.

(3) Sucrose indicator entered the bag and reacted with the sucrose molecules. The reaction made the bag increase in size and pushed the liquid up the tube.

(4) Sucrose molecules moved out of the bag and up the tube while water moved out, causing the rise of liquid in the tube.

Base your answers to questions 76 through 78 on the diagram below that shows variations in the beaks of finches in the Galapagos Islands and on your knowledge of biology.

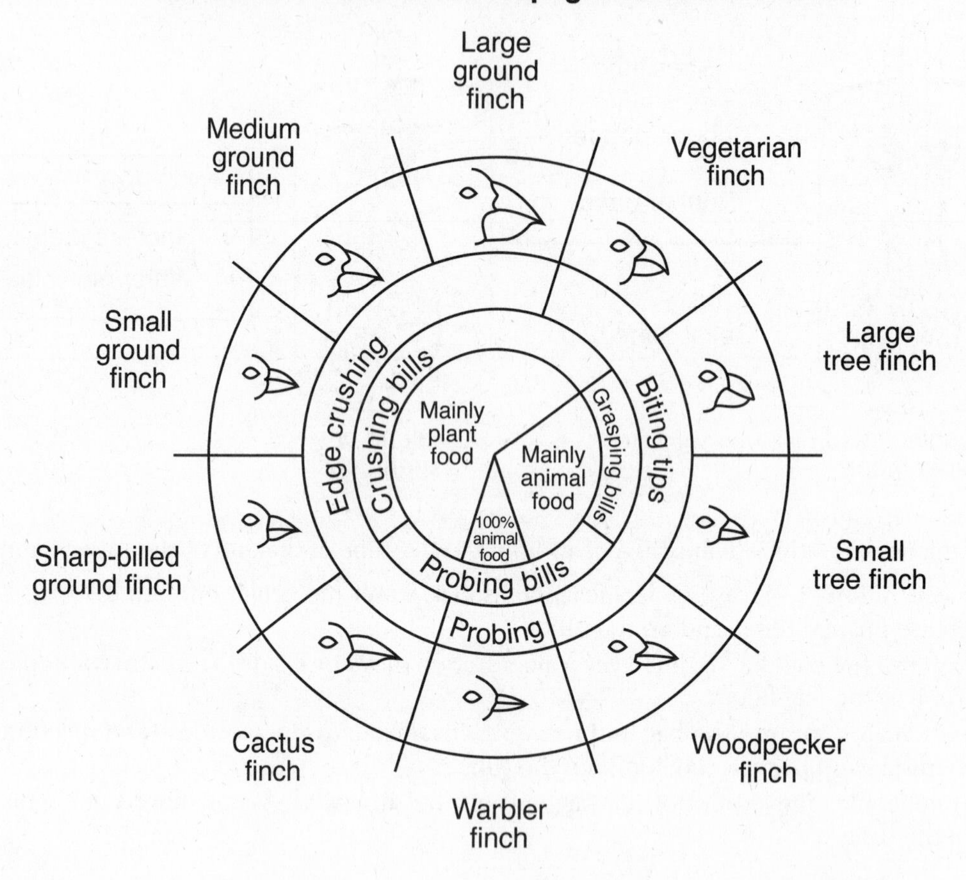


from: *Galapagos: A Natural History Guide*

**Note: The answer to question 76 should be recorded on your separate answer sheet.**

76 Which row correctly pairs a finch species with its primary nutritional role and bill type?

| Row | Finch | Bill Type | Nutritional Role |
|---|---|---|---|
| (1) | cactus finch | probing bill | carnivore |
| (2) | medium ground finch | grasping bill | herbivore |
| (3) | large tree finch | crushing bill | herbivore |
| (4) | warbler finch | probing bill | carnivore |

77 In certain years, the Galapagos plants produce many tube-shaped flowers rich in nectar. Identify the finch that is best adapted to feed on the nectar within those flowers. Support your answer. [1]

_______________________________________________

_______________________________________________

78 The number of small tree finches is increasing on an island inhabited by a large population of small ground finches. State *one* reason why the population of small ground finches has *not* been affected by the increasing number of small tree finches. [1]

___

___

___

79 Explain why glucose molecules can cross a cell membrane and starch molecules can *not*. [1]

___

___

___

Base your answers to questions 80 through 82 on the information below and on your knowledge of biology.

**Progressive Resistance Exercise**

Progressive resistance exercise (PRE) is a method of increasing the ability of muscles to generate force. The principles of PRE for increasing force production in muscles have remained unchanged for almost 60 years. These principles are (1) to perform a small number of repetitions until fatigued, (2) to allow sufficient rest between exercises for recovery, and (3) to increase the resistance as the ability to generate force increases. Traditionally, PRE has been used by young, healthy adults to improve athletic performance.

A student decided to incorporate PRE into his exercise program. He did not know how to determine when he had allowed sufficient rest between exercises for recovery. He hypothesized that waiting for his pulse to return to normal would probably be a good indication.

80 Explain why allowing his pulse rate to return to normal might be a good indication that he had waited long enough for recovery. [1]

______________________________________________

______________________________________________

**Note: The answer to question 81 should be recorded on your separate answer sheet.**

81 Students wanted to try PRE to increase their ability to rapidly squeeze a clothespin. They thought if they could do this, they could challenge another class to a clothespin squeezing competition and win. Which steps should the students take to follow the principles of PRE?

(1) Measure their pulse rate after squeezing the clothespin until fatigued. Then increase the resistance of the clothespin for the next trial.
(2) Squeeze a clothespin until fatigued, rest, and repeat. Over time, they should gradually increase the resistance of the clothespins they are squeezing.
(3) Measure their pulse rate, squeeze a clothespin for one minute, rest, and measure their pulse rate.
(4) Squeeze a clothespin for as long as they can, measure their pulse rate, rest, eat some candy. Increase the resistance of the clothespin for the next trial.

**Note: The answer to question 82 should be recorded on your separate answer sheet.**

82 Students following the principles of PRE monitored their ability to lift weights. Which observation would indicate that their exercise program was successful?

(1) They could eventually lift heavier weights than when they started.
(2) Their pulse rate increased more rapidly as they kept lifting weights.
(3) The number of weights their group could lift during competition decreased.
(4) Males and females could lift the same weight an equal number of times during competition.

83 Using the axes on the graph below, sketch a line graph showing the changes in heart rate of a person who is walking slowly, then begins running, and then sits down to rest for a few minutes. [1]

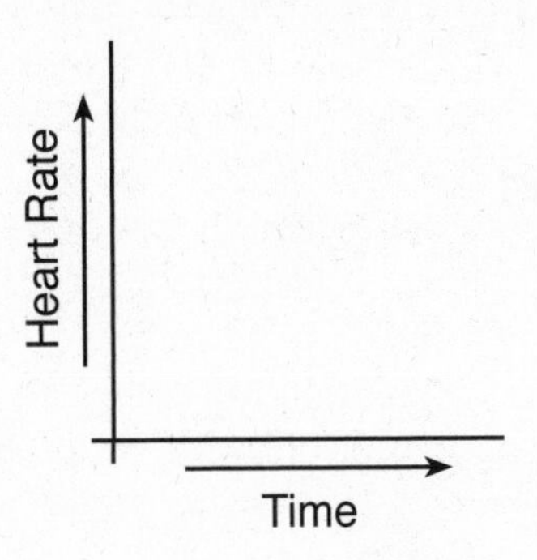


84 Identify *one* waste product that is released during exercise. Explain how this waste product leaves the body. [1]

Waste product: ____________________

______________________________________________

______________________________________________

85 State *one* way scientists could use the banding patterns produced by gel electrophoresis. [1]

______________________________________________

______________________________________________

# THE LIVING ENVIRONMENT
# AUGUST 2017

| Part | Maximum Score | Student's Score |
|---|---|---|
| A | 30 | |
| B-1 | 13 | |
| B-2 | 12 | |
| C | 17 | |
| D | 13 | |
| Total Written Test Score (Maximum Raw Score: 85) | | |
| Final Score (from conversion chart) | | |

Rater's Initials:

Rater 1 .......... Rater 2 ..........

ANSWER SHEET

Student ............................................................

Teacher ............................................................

School .................................... Grade ..............

**Multiple Choice for Parts A, B–1, B–2, and D**
**Allow 1 credit for each correct response.**

**Part A**

| | | | |
|---|---|---|---|
| 1 .......... | 9 .......... | 17 .......... | 25 .......... |
| 2 .......... | 10 .......... | 18 .......... | 26 .......... |
| 3 .......... | 11 .......... | 19 .......... | 27 .......... |
| 4 .......... | 12 .......... | 20 .......... | 28 .......... |
| 5 .......... | 13 .......... | 21 .......... | 29 .......... |
| 6 .......... | 14 .......... | 22 .......... | 30 .......... |
| 7 .......... | 15 .......... | 23 .......... | |
| 8 .......... | 16 .......... | 24 .......... | |

**Part B–1**

| | | | |
|---|---|---|---|
| 31 .......... | 35 .......... | 39 .......... | 43 .......... |
| 32 .......... | 36 .......... | 40 .......... | |
| 33 .......... | 37 .......... | 41 .......... | |
| 34 .......... | 38 .......... | 42 .......... | |

**Part B–2**

| | | |
|---|---|---|
| 47 .......... | 49 .......... | 50 .......... |

**Part D**

| | | |
|---|---|---|
| 73 .......... | 75 .......... | 81 .......... |
| 74 .......... | 76 .......... | 82 .......... |

# The Living Environment/REVIEWING BIOLOGY
# January 2018

### Part A

**Answer all questions in this part.** [30]

*Directions* (1–30): For *each* statement or question, record on the separate answer sheet the *number* of the word or expression that, of those given, best completes the statement or answers the question.

1 Which organisms and set of characteristics are correctly paired?

(1) fungi—carry out photosynthesis and heterotrophic nutrition
(2) plants—carry out respiration and autotrophic nutrition
(3) decomposers—carry out photosynthesis and autotrophic nutrition
(4) animals—carry out autotrophic nutrition and heterotrophic nutrition

2 Humans have an effect on ecosystems when they use native grasslands or forested areas for farming or urban use. One *negative* effect of these changes on the ecosystem is that there will be

(1) less biodiversity
(2) more homes
(3) successful economic growth
(4) increased food production

3 The diagram below represents structures found in the female reproductive system.

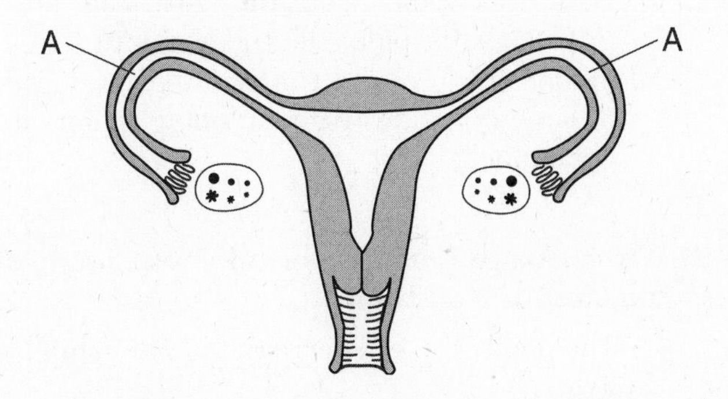


If the areas labeled *A* were completely blocked on both sides, the most likely result would be that

(1) egg and estrogen production would stop
(2) sperm and insulin production would stop
(3) fertilization would not occur
(4) an embryo would develop

4 Scientists have studied oceanic plastic garbage "patches" around the world. These are areas that accumulate plastic garbage from coastal regions. Their environmental effect ranges from killing sea life to blocking sunlight from reaching photosynthetic organisms. Without a change in human plastic usage, new garbage patches will continue to form. Which human activity would most directly *reduce* the amount of plastic garbage that enters the ocean?

(1) Ban the production and usage of all bags made from recycled plastic.
(2) Clean up plastic trash from shorelines, rivers, and other waterways that flow into the oceans.
(3) Manufacture fewer reusable water bottles, so that people will be more likely to use disposable ones.
(4) Implement a glass bottle deposit system to discourage people from recycling plastic bottles.

5 Monarch butterflies migrate from the U.S. and Canada to Mexico every winter. Over the past 10 years, there has been a drastic decrease in the number of monarch butterflies. Scientists have estimated that the population may have decreased from about 1 billion to 35 million. Which action would *not* be considered a reason for the decline in monarch butterfly populations?

(1) illegal deforestation
(2) extreme temperature changes
(3) decreasing food supplies
(4) habitat preservation

6 Finches on the Galapagos Islands express a variety of traits. Variability in the offspring of these finches is a result of

(1) mutation and cloning
(2) meiosis and mutation
(3) mitosis and asexual reproduction
(4) mitosis and genetic recombination

7 Exposure to certain environmental toxins, such as pesticides, may reduce fertility in males by interfering with their ability to produce gametes. These toxins are most likely having an effect on the

(1) testes and progesterone
(2) ovaries and testosterone
(3) ovaries and estrogen
(4) testes and testosterone

8 Which statement best describes an important process carried out by structure *X*?

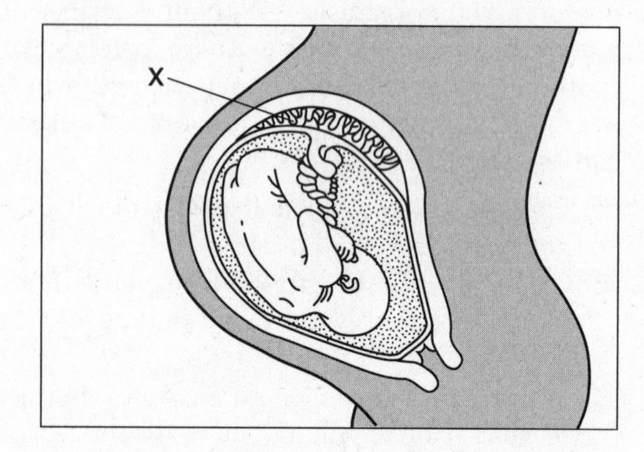


(1) Milk passes from the mother to the fetus.
(2) Materials are exchanged between fetal and maternal blood.
(3) Maternal blood is converted into fetal blood.
(4) Oxygen diffuses from fetal blood to maternal blood.

9 Traditional lightbulbs are only 10% efficient. Ninety percent of the energy they use is converted to heat. Modern lightbulbs are much more efficient, but may cost three times as much as traditional lightbulbs. Consumers who switch to modern lightbulbs are most likely

(1) spending more money for no good reason
(2) trying to stop pollution of the oceans
(3) trading a short-term cost for long-term savings
(4) helping traditional lightbulb factories employ people

10 Many oak trees are cut down and removed from an oak-hickory forest. A likely result of the direct harvesting of the oak species would be the

(1) disruption of natural cycles
(2) conservation of these natural forest resources
(3) recycling of all the nutrients in the forest
(4) prevention of the extinction of animals native to the area

11 A sequence of events is represented in the diagram below.

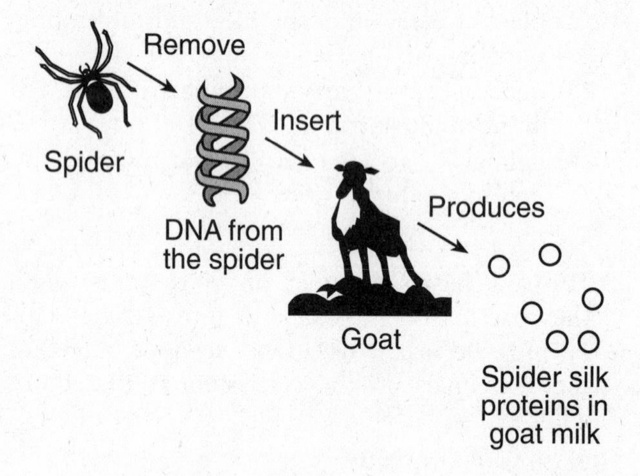


Which statement best describes a result of this process?

(1) The spider from which the DNA sample was obtained can no longer produce spider silk.
(2) The goat milk now contains DNA molecules made of spider silk proteins.
(3) Both the spider and the goat can now produce both spider silk and goat milk.
(4) Spider silk proteins can now be produced in large quantities without killing spiders to obtain them.

12 Which change is an example of a response to a stimulus?

(1) The pupil of an eye decreases in size in bright light.
(2) A leaf absorbs sunlight in the morning.
(3) The water level of a pond rises on a rainy day.
(4) A dead tree decays after many years.

13 After feeding at the surface of the ocean during the day, many ocean organisms migrate to deeper waters. While there, they release ammonia in their urine. Many bacteria use the nitrogen from the ammonia as they make amino acids, which eventually end up in food chains on both land and water. These amino acids may even be used in humans. Which statement best explains these observations?

(1) Chemical elements, including nitrogen, pass through food webs and are combined and recombined in different ways.
(2) Chemical elements, including nitrogen, are removed from food webs and eliminated from ecosystems.
(3) Nitrogen is transferred directly from bacteria to humans.
(4) All elements in the ocean remain there and are not transferred to other ecosystems.

14 Which statement describes an event that would most severely disrupt the process of ecological succession in an area?

(1) The season changes from spring into summer.
(2) Native plants are planted in an abandoned field.
(3) Plants and animals begin to colonize a newly formed volcanic island.
(4) A dam is built on a river to form a reservoir.

15 The processes of diffusion and active transport are both used to

(1) break down molecules to release energy
(2) move molecules into or out of cells of the body
(3) bring molecules into cells when they are more concentrated outside of the cell
(4) move molecules against a concentration gradient, using ATP molecules

16 Botulinum toxin is a substance that can cause paralysis in humans. The effects of the toxin are due to the blocking of a signaling molecule that is necessary for communication between nerve cells. The toxin most likely interferes with the normal functioning of a

(1) chromosome
(2) DNA molecule
(3) receptor
(4) digestive hormone

17 The bar graph below shows the number of species in four pond ecosystems.

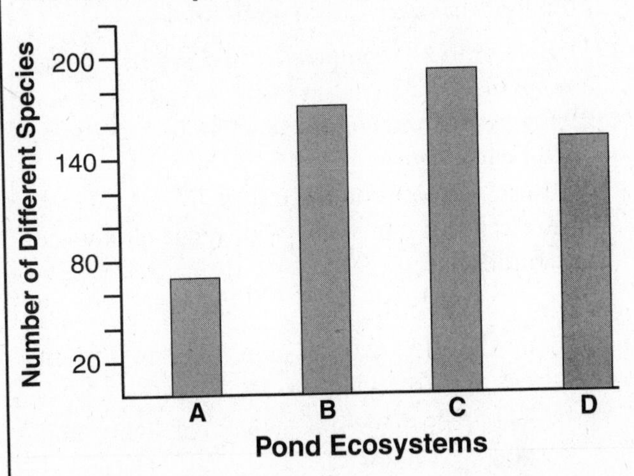


Based on this information, which ecosystem is likely to be the most stable?

(1) *A*
(2) *B*
(3) *C*
(4) *D*

18 The diagram below represents a marine food web.

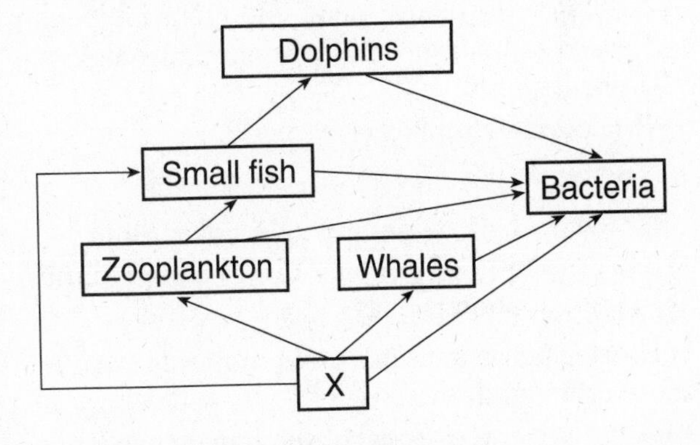


The organisms represented by *X* are

(1) decomposers
(2) producers
(3) carnivores
(4) scavengers

19 When rain forests are cut down, there is a

(1) loss of fossil fuels that could be used by industry
(2) release of excess oxygen to the atmosphere
(3) release of chemicals which cause helpful mutations
(4) loss of genetic material available for research

20 Scientists who study rock formations in caves describe some of the formations as "living rock" because, under certain conditions, they increase in size. Which statement would best dispute the claim that these rock formations are living?

(1) Rocks are not composed of cells, while living organisms are.
(2) Rocks perform complex metabolic processes, but cannot grow.
(3) Rocks cannot reproduce sexually.
(4) Rocks remain stable in a wide range of physical conditions.

21 Carbon dioxide and oxygen are important resources in ecosystems and are

(1) recycled through the activity of living and nonliving systems
(2) stored in the animals of the ecosystem
(3) lost due to the activities of decomposers
(4) released by the process of photosynthesis

22 Before they can pass from a parent cell to its offspring cells, the inherited instructions that a human cell carries must first be

(1) moved into the nucleus
(2) broken down and made into DNA molecules
(3) used to make specific protein molecules that form genes
(4) accurately replicated

23 Eye color, hair color, and skin color often vary from person to person and even within a family. One explanation is that

(1) the glucose units in a DNA molecule are often rearranged
(2) the genetic material of the female parent has the most influence on offspring
(3) the inherited traits of individuals are determined by different gene combinations
(4) some extra parts of genetic material are often gained during fertilization

24 Organic compounds are used as building blocks for

(1) water, DNA, and starches
(2) water, proteins, and oxygen
(3) proteins, DNA, and carbon dioxide
(4) proteins, starches, and fats

25 Scientists have developed the ability to manufacture hormones, such as human growth hormone, using bacteria. One benefit of this new technology is that

(1) scientists can use only one type of bacteria
(2) bacteria are relatively inexpensive and reproduce quickly
(3) patients can spend more money on their medications
(4) scientists produce drugs that cause more immune reactions

26 Although all of the cells of a plant contain the same genetic material, root cells and leaf cells are *not* identical because they

(1) use different genetic bases for the synthesis of DNA
(2) use different parts of their genetic instructions
(3) select different cells to express
(4) delete different sections of their enzymes

27 During cellular respiration, what is the direct source of the energy used in the cells of consumers in the ecosystem represented below?

(1) the Sun
(2) enzymes
(3) the atoms making up inorganic molecules
(4) the chemical bonds in organic molecules

28 Which dissolved substance do aquatic animals remove from their external environment for use in cellular respiration?

(1) carbon dioxide
(2) ATP molecules
(3) oxygen molecules
(4) nitrogen gas

29 The photographs below are of two Siamese cats.

Cat Kept Indoors

Source: Http://aboutmyrecovery.com/2008/12/13/my-very-own-siamese-pet-kitten/

Cat Kept Outdoors

Source: Http://www.superstock.com/stock-photos-images/662-220

The Siamese breed has a gene that controls fur color. The cat in the first photograph was kept indoors while the cat in the second photograph was kept outdoors. Which statement best explains the differences in fur color between these two cats?

(1) The cat kept indoors is older than the cat kept outdoors.
(2) The environment influenced the expression of fur color genes.
(3) The environment influenced the production of all the proteins in the cat kept outdoors.
(4) The cat kept outdoors has a gene mutation that prevents it from producing light-colored fur.

30 The diagram below represents a cell.

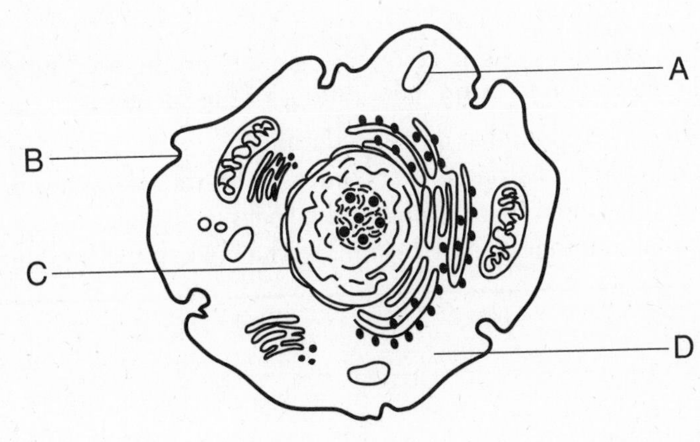


Which letter indicates the specific structure where most hereditary mutations occur?

(1) *A*
(2) *B*
(3) *C*
(4) *D*

## Part B–1

**Answer all questions in this part.** [13]

*Directions* (31–43): For *each* statement or question, record on the separate answer sheet the *number* of the word or expression that, of those given, best completes the statement or answers the question.

Base your answers to questions 31 through 33 on the information and graphs below, and on your knowledge of biology. The diagrams below show the number of fish in a lake and the average water temperature in the lake for the months of May through October.

During certain times of the year, bears feed heavily on a population of fish in a lake. At other times of the year, the bear population feeds primarily on fruits, berries, and insects.

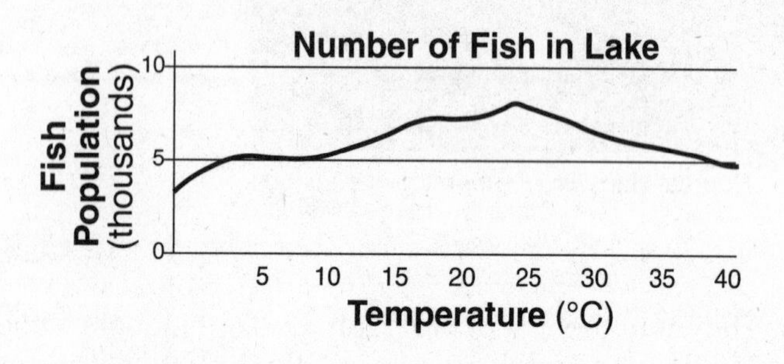


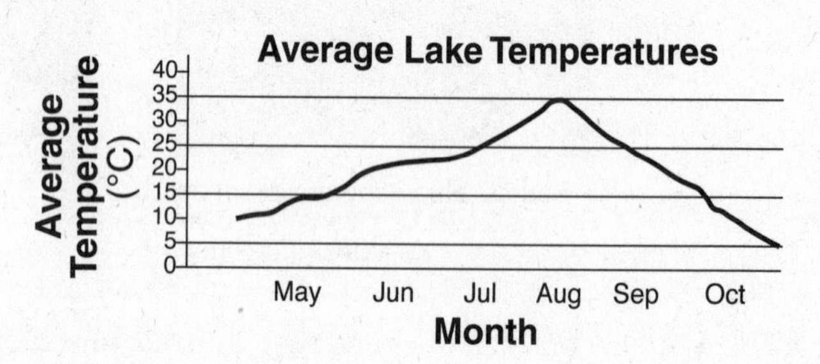


31 During which month would the bears in the area have the most fish available?

(1) May
(2) July
(3) August
(4) October

32 One of the best ways to represent the interdependence of all of the organisms in this ecosystem is

(1) an evolutionary tree
(2) a food chain
(3) an electrophoresis gel
(4) a food web

33 Within the fish population, variations exist in color, size, gamete production, and swimming speed. A variation that would most likely be passed on to future generations of the species is

(1) a swimming speed that is less than that of its predators
(2) the presence of bright, colorful markings that contrast with the lake bottom
(3) being of a size that enables them to hide among the rocks in the lake
(4) the production of a small number of gametes during the peak of the breeding season

Base your answers to questions 34 and 35 on the information below, and on your knowledge of biology.

Before conducting an experiment, two students gathered information about the effect of greenhouse gases on global warming. Student *A* found information in a newspaper article. Student *B* found information in several peer-reviewed scientific journals and on three websites.

34 Which statement most likely describes the reliability of the students' information?

(1) Information gathered by student *A* is more reliable because newspapers are always updated to reflect the most current research.
(2) Information gathered by student *B* is more reliable because some of it was gathered from peer-reviewed sources.
(3) Information gathered by student *A* is more reliable because it is from a single source without conflicting information.
(4) Information gathered by student *B* is more reliable because some of it was found on the internet.

35 After gathering the information, the students presented the information to their class. The class gave the students suggestions about how to continue with their experiment. How does this step benefit the investigation?

(1) Feedback from the class will help them design a better experiment.
(2) Feedback creates confusion, and will complicate the investigation.
(3) The students' investigation will be unaffected because the class is not carrying out the experiment.
(4) The investigation will be unchanged because students can use information only from published sources to design the experiment.

---

36 A student wondered if butterflies would show any differences in their wing color if, as caterpillars, they were grown in the dark or grown in bright white light. Which statement would be a possible hypothesis for an experiment to test this idea?

(1) Caterpillars exposed to bright white light will show more blue and green in their wings when they become butterflies than caterpillars kept in the dark.
(2) Will caterpillars kept in the dark have brighter wings when they become butterflies than caterpillars exposed to bright white light?
(3) Ten caterpillars will be kept in the dark and ten caterpillars will be exposed to bright white light and allowed to develop into butterflies.
(4) Results show that caterpillars kept in the dark and those exposed to bright white light had the same wing color when they became butterflies.

37 The chart below shows the number of differences in genetic material between individuals within the same species. Scientists can use this information to determine which populations demonstrate the greatest amount of genetic diversity.

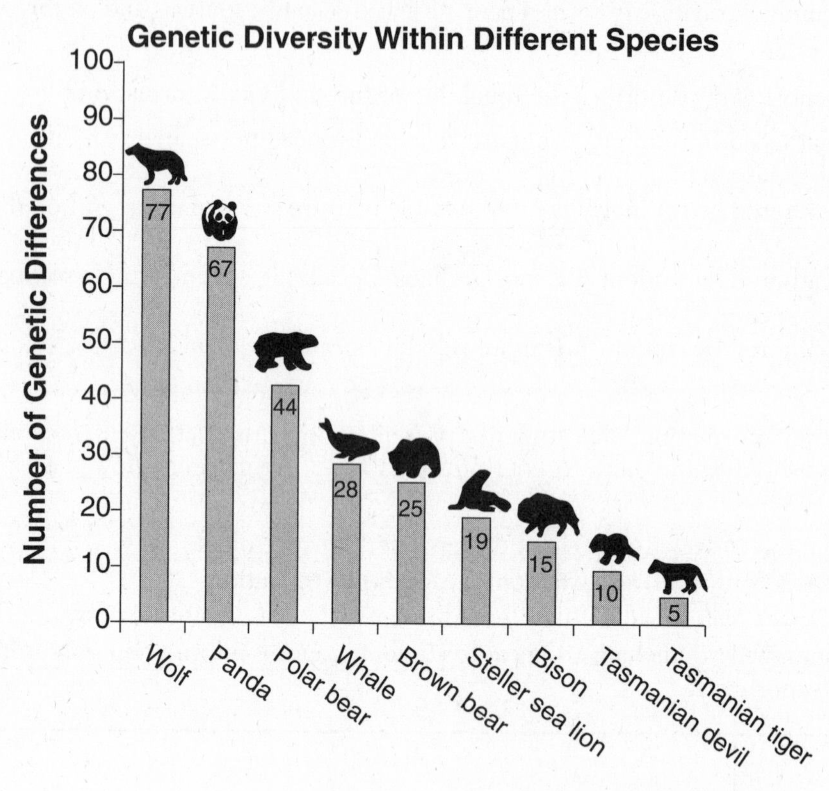


adapted from: www.pnas.org

According to the chart, which two species would be more likely to survive if their environmental conditions changed?

(1) Tasmanian tiger and Tasmanian devil
(2) brown bear and whale
(3) Tasmanian tiger and wolf
(4) panda and wolf

38 Microbeads are tiny, smooth, plastic spheres found in common household products such as facial soap. These beads, measuring from 0.0004 to 1.24 mm, roughly the size of some fish eggs, are too small to be removed by water treatment systems. Thus, they end up in rivers, lakes, and other bodies of water. The accumulation of these microbeads is an environmental concern for aquatic biologists because microbeads

(1) make the lakes and rivers cloudy and dirty, affecting their appearance
(2) may stick to some household water pipes, preventing drainage problems
(3) could be mistaken for food by some species, working their way up the food chain
(4) could clog fishing nets, affecting the ability of fishermen to catch fish

Base your answers to questions 39 and 40 on the diagrams below and on your knowledge of biology. The diagrams represent some of the systems that make up the human body.

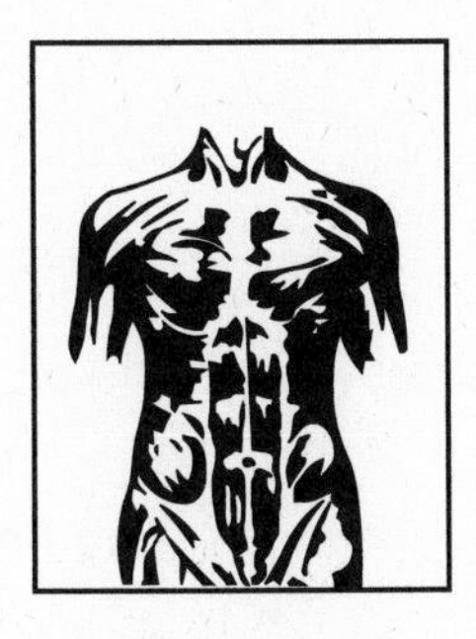

System A

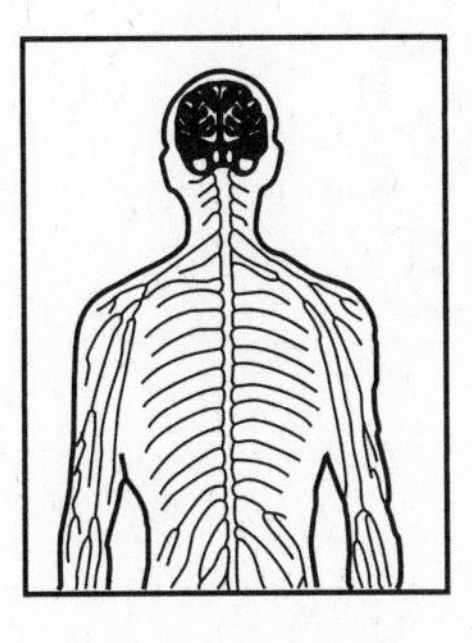

System B

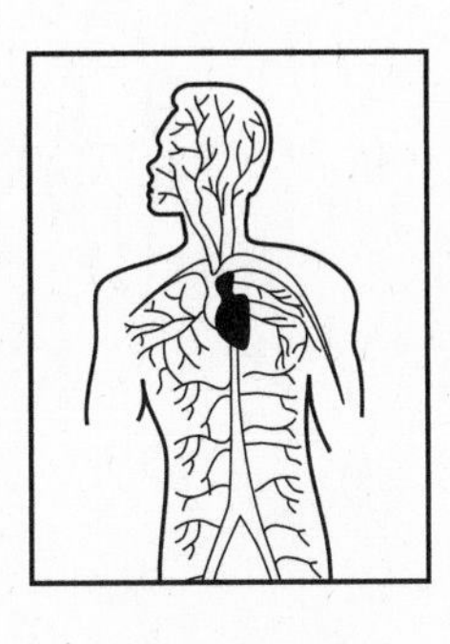

System C

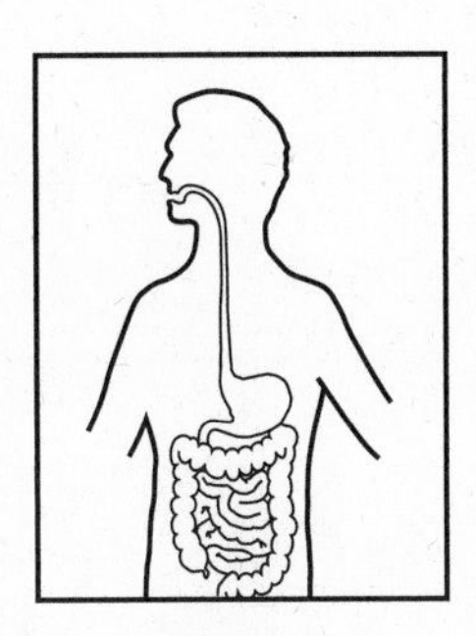

System D

39 Which row in the chart below correctly identifies the main function of these systems?

| Row | System A | System B | System C | System D |
|---|---|---|---|---|
| (1) | response | excretion | circulation | digestion |
| (2) | movement | response | circulation | digestion |
| (3) | response | circulation | excretion | digestion |
| (4) | movement | circulation | digestion | reproduction |

40 A similarity between these systems is that they all

(1) are made of cells that are identical in structure and function
(2) contain organs that work independently from other organs in that system
(3) work together to maintain a stable internal environment
(4) are separate and do not interact with other body systems

---

Base your answer to question 41 on the information below and on your knowledge of biology.

In China, farmers switched from growing conventional cotton, which required spraying with insecticides 15 times each year, to a genetically modified cotton variety called Bt cotton. The Bt cotton produces a protein toxic to the insects that destroy the cotton crop. Since the switch to Bt cotton, the use of chemical insecticides has decreased by 60%.

41 An advantage of growing the genetically modified Bt cotton instead of conventional cotton is that growing Bt cotton could

(1) result in an increase in populations of insects that are beneficial
(2) result in an increase in the size of insect populations that are resistant to the Bt protein
(3) lead to an increase in the survival rates of insects that eat cotton
(4) lead to an increase in the use of insecticides that protect cotton from insects

---

42 The diagram below represents events that occur during sexual reproduction.

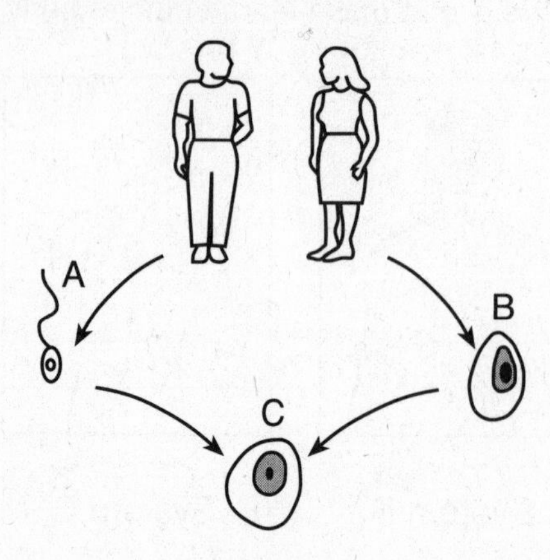


The stages labeled *A*, *B*, and *C* are necessary to ensure that the offspring will inherit

(1) half of their chromosomes from each parent
(2) double the amount of chromosomes from each parent
(3) pairs of chromosomes from each parent
(4) double the amount of chromosomes from one parent

43 A company that produces paint is planning to build a small factory in a rural community. The factory would provide many needed jobs. Before the community agrees to allow the factory to be built, the community should

(1) investigate the use of paint as a method of biological control
(2) consider just the economic advantages of building the new factory
(3) assess the risks of the new factory and compare these to the benefits
(4) insist the factory use finite resources located in the community

## Part B–2

**Answer all questions in this part.** [12]

*Directions* (44–55): For those questions that are multiple choice, record on the separate answer sheet the *number* of the choice that, of those given, best completes each statement or answers each question. For all other questions in this part, follow the directions given and record your answers in the spaces provided in this examination booklet.

Base your answers to questions 44 through 48 on the information, diagram, and data table below and on your knowledge of biology.

The laboratory setup represented below was used to investigate the effect of light on aquatic plants. Equal amounts of a green water plant were placed in beakers with gas-collecting tubes. The beakers were placed in a temperature-controlled environment. The light source was placed at different distances from the beakers. After an hour, the amount of gas collected from the plants in each tube was measured and recorded in the data table.

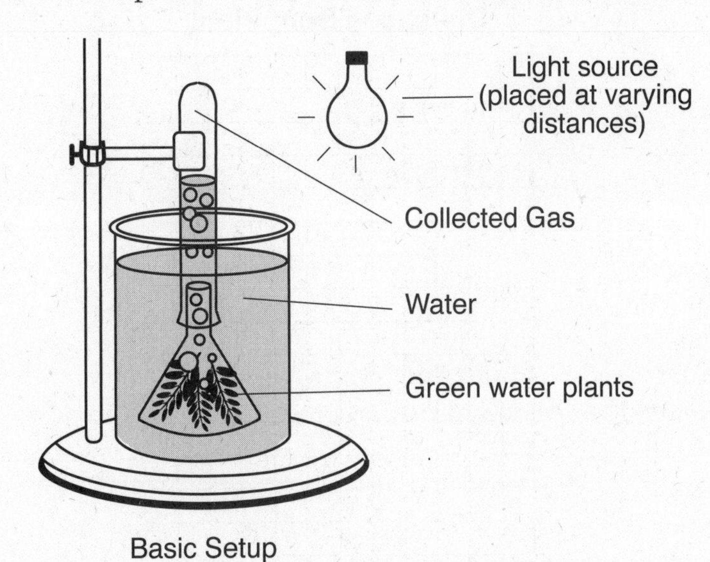


Basic Setup

**Gas Collected with Light Source at Different Distances from Plant**

| **Distance of Light Source from Plant** (cm) | **Gas Collected in Tube** (mm) |
|---|---|
| 5 | 85 |
| 10 | 37 |
| 15 | 15 |
| 20 | 8 |
| 25 | 5 |

*Directions* (44–46): Using the information given, construct a line graph on the grid following the directions below.

44 Provide an appropriate label for the $y$-axis, including units, on the line provided. [1]

45 Mark an appropriate scale, without any breaks in the data, on each labeled axis. [1]

46 Plot the data on the grid, connect the points, and surround each point with a small circle. [1]

Example:

**Gas Collected with Light Source at Different Distances from Plant**

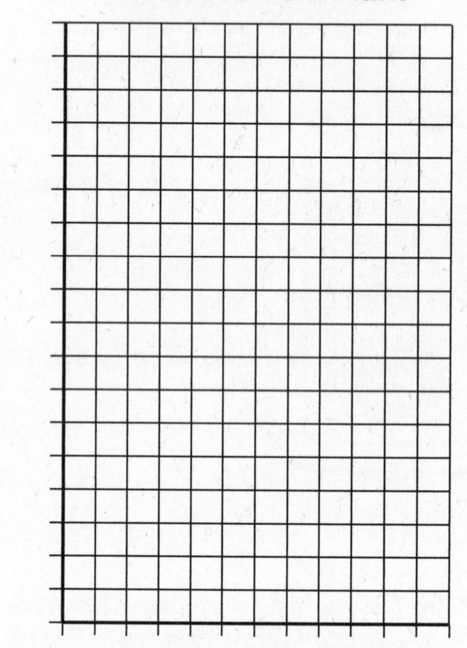

**Distance of Light Source from Plant** (cm)

**Note: The answer to question 47 should be recorded on your separate answer sheet.**

47 Which row in the chart below correctly identifies the variables in this experiment?

| Row | Independent Variable | Dependent Variable |
|---|---|---|
| (1) | amount of gas collected | distance of beaker from light source |
| (2) | number of plants in the beaker | temperature of plant |
| (3) | distance of beaker from light source | amount of gas collected |
| (4) | minutes of exposure to the light source | rate of gas collection |

48 Identify the gas being produced by the plants. [1]

______________________________

___________________________________________________________

Base your answers to questions 49 through 51 on the information below and on your knowledge of biology.

**The Bionic Pancreas**

Until recently, diabetics could rely only on regular blood sugar checks, medications, and low-carbohydrate diets in order to maintain their health.

Bioengineers at Boston University are working to create a bionic pancreas. The device includes a sensor implanted just beneath the skin that monitors blood sugar levels. It sends a wireless signal to a smartphone every five minutes. If the phone receives a signal that blood sugar is too low or too high, it then sends a different signal to a separate device also attached to the body. This device releases the appropriate hormone into the bloodstream to return blood sugar levels back to normal.

**Note: The answer to question 49 should be recorded on your separate answer sheet.**

49 According to the passage, the bionic pancreas makes corrective actions that return blood sugar levels back to normal. This artificial device helps

(1) produce more sugar
(2) break down blood cells
(3) maintain homeostasis
(4) cure their diabetes

**Note: The answer to question 50 should be recorded on your separate answer sheet.**

50 The corrective actions made by the bionic pancreas on a regular basis in response to changing blood sugar levels are similar to which natural biological process?

(1) a feedback mechanism
(2) an immune response
(3) biochemical digestion
(4) ATP production

51 The bionic pancreas sends a signal to a device to release hormones into the bloodstream to regulate blood sugar. Identify *one* hormone the device would most likely release. [1]

______________________________

___________________________________________________________

52 Many hormones are proteins used in cellular communication. Each hormone carries a specific message to specific target cells. State why each of these hormones is able to deliver a different message. [1]

______________________________________________________________________

______________________________________________________________________

53 Two different species occupy the same habitat. Identify *one* reason these two species might *not* compete. [1]

______________________________________________________________________

______________________________________________________________________

Base your answers to questions 54 and 55 on the diagram below and on your knowledge of biology. The diagram represents the energy in kilocalories (kcal) available at different feeding levels in a food chain.

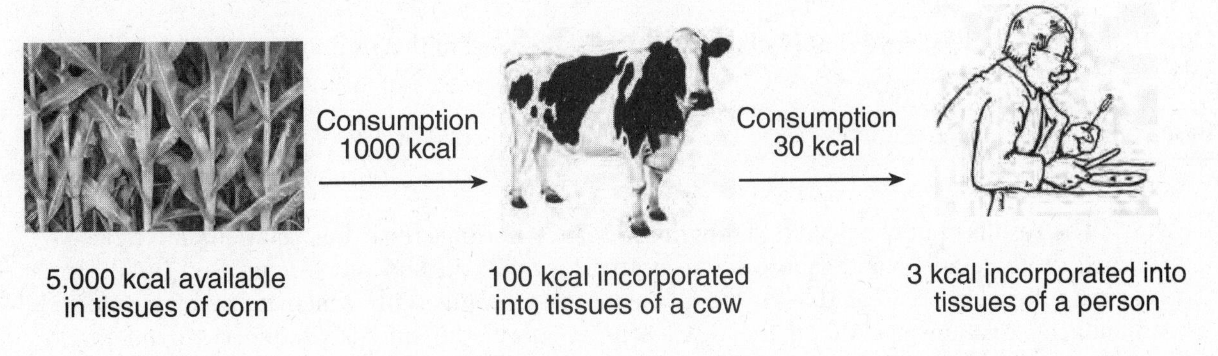


54 Complete the energy pyramid provided below by writing herbivore, plant, and carnivore in the correct locations. [1]

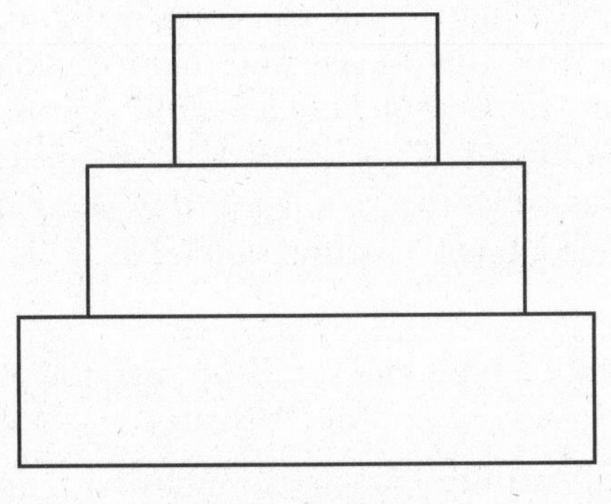

Energy Pyramid

55 Explain why there is a different amount of energy represented at each level of this energy pyramid. [1]

______________________________________________________________________

______________________________________________________________________

______________________________________________________________________

## Part C

**Answer all questions in this part. [17]**

*Directions* (56–72): Record your answers in the spaces provided in this examination booklet.

Base your answers to question 56–58 on the information below and on your knowledge of biology.

### Reindeer Drool

The results of new research highlight interesting findings regarding reindeer and moose saliva. Both reindeer and moose feed on a type of grass called red fescue. Red fescue is usually dangerous to eat due to the presence of a fungus with which it has a mutually beneficial relationship. When the red fescue is eaten, the fungus produces a toxin that decreases blood flow in the legs of the moose and reindeer. This could result in the loss of their limbs.

Since many reindeer and moose successfully feed on red fescue, scientists wondered if their saliva gave them the ability to eat the grass without suffering from circulation problems. Scientists hypothesized that moose and reindeer saliva might detoxify the grass. To conduct their experiment, the researchers smeared reindeer and moose saliva on cut red fescue that contained the fungus. They learned that the saliva slowed the growth of this fungus and detoxified the grass. The results suggest that some animal species have evolved the ability to fight back against a plant's natural defenses.

56–58 Explain the benefit of the ability moose and reindeer have to eat red fescue grass. In your answer, be sure to:

- explain why red fescue plants with the fungus normally have an advantage over red fescue plants without the fungus [1]
- explain how the moose and reindeer saliva protects them from the harmful effects of the fungus [1]
- explain how moose and reindeer (two separate, but related, mammals) could possess the same adaptation that protects them from the toxin produced by the fungus [1]

______________________________________________________________________

______________________________________________________________________

______________________________________________________________________

______________________________________________________________________

______________________________________________________________________

______________________________________________________________________

59 Cancer of the ovary is not common, but when it occurs, the cancer can cause the ovary to malfunction. Identify *one* possible result of an ovary *not* performing its intended function in the body. [1]

---

60 Several students were diagnosed with strep throat. They were all given the same antibiotic and took it for the time specified. Three weeks later, after finishing all their antibiotic, all the students except one no longer had strep throat. State *one* likely reason why the one student was still infected with strep bacteria. [1]

---

---

Base your answers to questions 61 and 62 on the information below and on your knowledge of biology.

**Project Frozen Dumbo – Saving the Elephant Population Means Using Special Breeding**

Over the last 10 years, 70 percent of Africa's wild elephant population has been killed off. The main cause is ivory poaching, in which elephants are slaughtered for their valuable tusks. At the same time, efforts to breed captive zoo elephants have not been very successful.

Now there is some good news. At zoos in Austria and England, two baby elephants were born, using sperm from South African wild elephants. For the first time, elephant sperm gathered in the wild was frozen and given to zoos. Two female zoo elephants were artificially impregnated with the sperm and went on to deliver calves. …

Source: Saving the Elephant Population Means Using Special Breeding, Pittsburgh Post-Gazette, 8/21/14

61 State *one* reason why the use of sperm from wild elephants, rather than the use of sperm from elephants in zoos in England or Austria, would be more important to the long-term survival of elephants. [1]

---

---

62 Identify *one* likely reason, other than poaching and hunting, for the decline of the elephant population. [1]

---

---

Base your answers to questions 63 through 65 on the information below and on your knowledge of biology.

**Battling Cancer with T-cell Therapy**

One reason that cancer is able to spread through tissues and organs is that cancer cells are actually the patient's own cells. The immune system of the patient does not recognize these cancer cells as foreign and, therefore, does not reject and destroy them.

Over the past eight years, immunologists have been developing a treatment for B-cell leukemia that involves using genetically engineered T cells to recognize and destroy B cells, all of which carry a protein, CD19. CD19 is found on the surface of both healthy and cancerous B cells. B cells are immune system cells that produce antibodies.

The procedure used in this treatment is outlined below:

1. T cells are removed from the patient with B-cell leukemia.
2. The T cells are genetically engineered to recognize the CD19 protein.
3. The patient is injected with the engineered T cells, which attach to cells with CD19 and destroy them.
4. The engineered T cells destroy both cancerous and healthy B cells.

This procedure has been successful in several patients. Currently, studies are continuing with more B-cell leukemia patients. It is hoped that the studies will be expanded to include other types of cancer, and that this treatment will be available to treat a variety of cancers in the future.

63 Explain why these specific T cells can be used for B-cell leukemia treatment. [1]

_______________

_______________

64 Explain why a patient needs treatments of antibodies after being injected with these modified T cells. [1]

_______________

_______________

65 Explain why the engineered T cells taken from one cancer patient will *not* work as a cancer treatment if injected into another patient with B-cell leukemia. [1]

_______________

_______________

Base your answers to questions 66 through 68 on the information below and on your knowledge of biology.

**Hydrothermal Vent Communities**

Scientists discovered a unique hydrothermal ecosystem on the sea floor at hot-water vents thousands of feet below the ocean surface. Organisms in these deep-sea regions have no access to sunlight, so they depend on the heat, methane, and high levels of sulfur-bearing minerals found in the heated fluids in which they live. Scientists were amazed to discover vent communities able to sustain vast amounts of life. The vent organisms depend on bacteria that can use the sulfur-bearing minerals to produce organic materials. These bacteria live on rock surfaces and as free-floating blobs. Some bacteria live within and provide nutrients for an unusual species of giant tubeworms that lacks a digestive system. Snails, shrimp, and clams are among the animals that feed directly on the bacteria. Crabs feed directly on other animals in the vent community.

66 Identify *one* abiotic factor that makes the hydrothermal vent ecosystem different from other ocean ecosystems. [1]

______________________________________________

67 State the relationship that exists between the crabs and the other members of the vent community. [1]

______________________________________________

______________________________________________

68 Describe *one* way the bacteria of the hydrothermal vent community differ from plants in their ability to produce organic materials. [1]

______________________________________________

______________________________________________

Base your answers to questions 69 through 72 on the information below and on your knowledge of biology.

**Transgenic Salmon**

Transgenic Atlantic salmon have been produced using DNA from other species of related fish. These genetically modified fish have an altered DNA "switch" that causes them to overproduce growth hormone. The transgenic Atlantic salmon grow to normal size, but they reach market size in half the time of conventional Atlantic salmon. As with most of the salmon consumed by people, the transgenic Atlantic salmon would be grown using aquatic farming methods. Scientists have expressed concern that transgenic fish can have undesirable effects on the natural environment. Fish growers would be expected to take steps to ensure that the transgenic salmon do not escape into the wild.

69 State *one* advantage genetic modification has over selective breeding when producing new varieties of animals or plants. [1]

70 State *one* reason the scientists altered the DNA "switch" of the Atlantic salmon to make them produce more growth hormone, rather than directly supplying the Atlantic salmon with more growth hormone. [1]

71 State *one undesirable* effect that escaped transgenic Atlantic salmon could have on the natural environment. [1]

72 State *one* benefit of raising the transgenic Atlantic salmon. [1]

## Part D

**Answer all questions in this part. [13]**

*Directions* (73–85): For those questions that are multiple choice, record on the separate answer sheet the *number* of the choice that, of those given, best completes each statement or answers each question. For all other questions in this part, follow the directions given and record your answers in the spaces provided in this examination booklet.

**Note: The answer to question 73 should be recorded on your separate answer sheet.**

73 During periods of vigorous physical activity, a person's breathing and heart rates increase. This enables the cells of the body to perform more efficiently because it helps the cells to

(1) remove waste products faster
(2) store excess glucose in muscles
(3) reduce the amount of ATP produced
(4) convert more oxygen to glucose

**Note: The answer to question 74 should be recorded on your separate answer sheet.**

74 A step in a procedure used in the *Diffusion Through a Membrane* lab is represented in the diagram below.

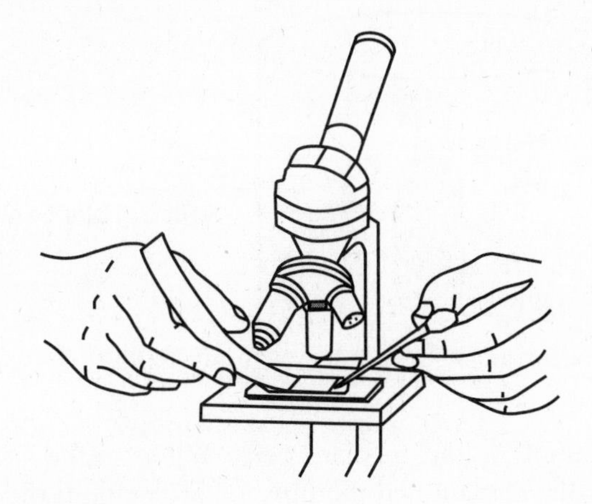

Which procedure is represented in the diagram?

(1) adding distilled water to the top of a cover glass on a slide
(2) making an artificial cell
(3) adding salt solution to a specimen under the cover glass
(4) making a thin sample to prepare a slide of red onion cells

**Note: The answer to question 75 should be recorded on your separate answer sheet.**

75 A student is opening and closing clothespins as part of a lab activity. The student begins to experience muscle fatigue, and the rate at which the student is opening and closing the clothespins slows. Which graph best represents the relationship between time and number of clothespin squeezes?

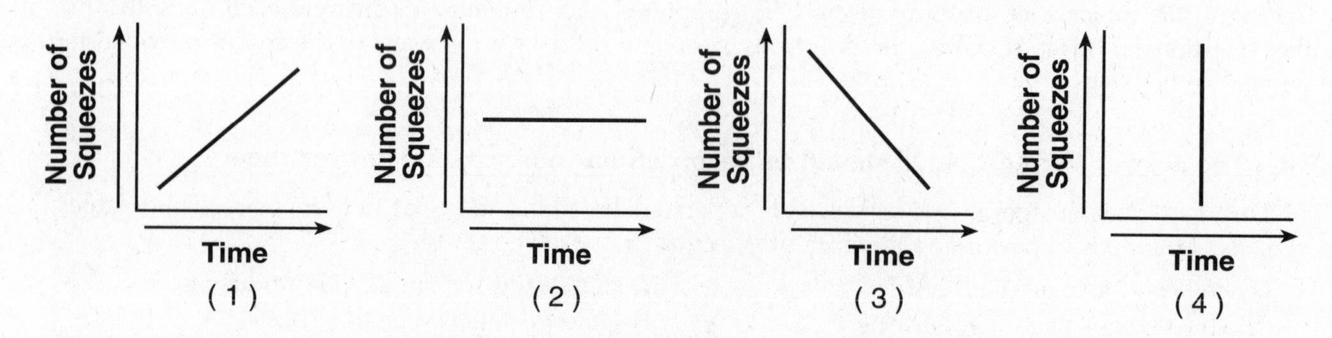


Base your answers to questions 76 and 77 on the information and diagram below and on your knowledge of biology. The diagram shows an experimental setup using an artificial plant cell.

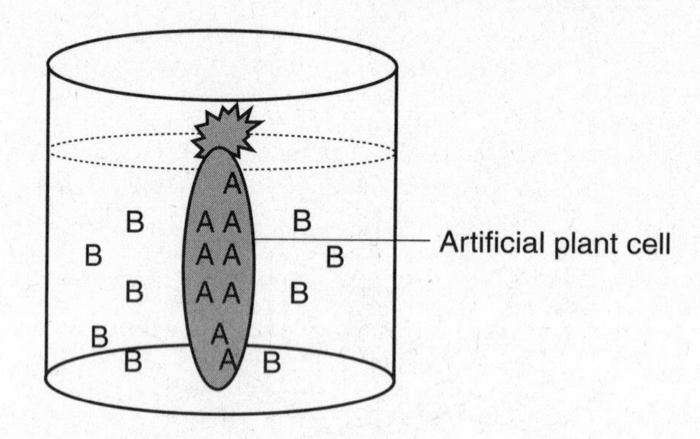


Molecules *A* and *B* are commonly found in plant cells. When tested, it was discovered that molecule *A* quickly passed through the artificial plant cell membrane. Molecule *B* did not pass through.

**Note: The answer to question 76 should be recorded on your separate answer sheet.**

76 The locations of molecules *A* and *B* at the beginning of the experiment are shown. Which statement best describes what was observed when the setup was examined 20 minutes later?

(1) Molecule *A* remained inside the artificial cell and molecule *B* remained outside.
(2) Only molecule *A* was found both inside and outside the artificial cell.
(3) Only molecule *B* was found both inside and outside the artificial cell.
(4) Both molecules *A* and *B* were found inside and outside the artificial cell.

77 State *one* way the two molecules could differ that would explain the difference in their ability to pass through the artificial plant cell membrane. [1]

______________________________________________________________________

______________________________________________________________________

______________________________________________________________________

Base your answer to question 78 on the information and diagram below and on your knowledge of biology. The diagram shows a plant called hogweed.

Hogweed is highly toxic and has become invasive in New York State. It can cause severe burns and blisters if touched.

**Hogweed Plants**

Source: http://www.washingtonpost.com

78 If you were given packaged samples of hogweed plant parts, describe *one* specific procedure you could use to determine if an unknown plant might be related to hogweed. [1]

______________________________________________________________

______________________________________________________________

Base your answers to questions 79 through 82 on the information and diagram below and on your knowledge of biology.

A human gene contains the following DNA base sequence: ACGCCCACCTTA

The gene mutated. It then contained the following DNA base sequence: ACGCGCACCTTA

**Universal Genetic Code Chart**
**Messenger RNA Codons and the Amino Acids for Which They Code**

| FIRST BASE | SECOND BASE: U | SECOND BASE: C | SECOND BASE: A | SECOND BASE: G | THIRD BASE |
|---|---|---|---|---|---|
| U | UUU, UUC — PHE<br>UUA, UUG — LEU | UCU, UCC, UCA, UCG — SER | UAU, UAC — TYR<br>UAA, UAG — STOP | UGU, UGC — CYS<br>UGA — STOP<br>UGG — TRP | U C A G |
| C | CUU, CUC, CUA, CUG — LEU | CCU, CCC, CCA, CCG — PRO | CAU, CAC — HIS<br>CAA, CAG — GLN | CGU, CGC, CGA, CGG — ARG | U C A G |
| A | AUU, AUC, AUA — ILE<br>AUG — MET or START | ACU, ACC, ACA, ACG — THR | AAU, AAC — ASN<br>AAA, AAG — LYS | AGU, AGC — SER<br>AGA, AGG — ARG | U C A G |
| G | GUU, GUC, GUA, GUG — VAL | GCU, GCC, GCA, GCG — ALA | GAU, GAC — ASP<br>GAA, GAG — GLU | GGU, GGC, GGA, GGG — GLY | U C A G |

79 In the table below, record the mRNA codons coded for by the DNA base sequence of the mutated gene ACGCGCACCTTA. [1]

80 Then, using the Universal Genetic Code Chart, record the amino acid sequence that is coded for by the mRNA codons you placed in the table. [1]

| Mutated Gene DNA Base Sequence | ACG | CGC | ACC | TTA |
|---|---|---|---|---|
| mRNA codons | ______ | ______ | ______ | ______ |
| Amino acid sequence | ______ | ______ | ______ | ______ |

**Note: The answer to question 81 should be recorded on your separate answer sheet.**

81 Which type of mutation is represented in the new gene?

(1) addition
(2) deletion
(3) inversion
(4) substitution

**Note: The answer to question 82 should be recorded on your separate answer sheet.**

82 The amino acids bond together to form which type of complex molecule?

(1) protein
(2) starch
(3) fat
(4) sugar

83 A certain small population of finches already has an "ideal" beak type for its present environment. Describe *two* specific adaptations, other than beak type, that would contribute to the ability of these finches to survive. [1]

84 In order to determine the effect of muscle fatigue on the ability of students to squeeze a clothespin, five male students did jumping jacks for three minutes and then squeezed a clothespin as many times as possible in a minute. Three other male students ran up and down the stairs for 30 seconds and then squeezed a clothespin as many times as possible for one minute. The results of the two groups were recorded. Identify *one* change that could be made to the experiment to increase the validity of the conclusion made from these results. [1]

85 There is a group of plants, known as halophytes, that has traits that enable them to survive in salty environments. Describe *one* change, other than death, that would be observed in the cells of a plant that did *not* have these traits and was planted in a salty environment. [1]

# THE LIVING ENVIRONMENT
# JANUARY 2018

| Part | Maximum Score | Student's Score |
|---|---|---|
| A | 30 | |
| B-1 | 13 | |
| B-2 | 12 | |
| C | 17 | |
| D | 13 | |
| Total Written Test Score (Maximum Raw Score: 85) | | |
| Final Score (from conversion chart) | | |

Rater's Initials:

Rater 1 ........... Rater 2 ...........

ANSWER SHEET

Student ............................................................

Teacher............................................................

School .......................................Grade ................

**Multiple Choice for Parts A, B–1, B–2, and D**
**Allow 1 credit for each correct response.**

**Part A**

1............................... 9.............................. 17.............................. 25..............................

2............................... 10.............................. 18.............................. 26..............................

3............................... 11.............................. 19.............................. 27..............................

4............................... 12.............................. 20.............................. 28..............................

5............................... 13.............................. 21.............................. 29..............................

6............................... 14.............................. 22.............................. 30..............................

7............................... 15.............................. 23..............................

8............................... 16.............................. 24..............................

**Part B–1**

31.............................. 35.............................. 39.............................. 43..............................

32.............................. 36.............................. 40..............................

33.............................. 37.............................. 41..............................

34.............................. 38.............................. 42..............................

**Part B–2**

47.............................. 49.............................. 50..............................

**Part D**

73.............................. 75.............................. 81..............................

74.............................. 76.............................. 82..............................